Ferromagnetic Materials

Structure and Properties

Ferromagnetic Materials
Structure and Properties

R. A. McCurrie
University of Bradford, UK

ACADEMIC PRESS
Harcourt Brace & Company, Publishers
London San Diego New York
Boston Sydney Tokyo Toronto

ACADEMIC PRESS LIMITED
24/28 Oval Road
LONDON NW1 7DX

United States Edition published by
ACADEMIC PRESS INC.
San Diego, CA 92101

This book is printed on acid-free paper

A catalogue record for this book is available from the British Library

ISBN: 0-12-482495-1

Typeset by P&R Typesetters, Salisbury, Wiltshire
Printed in Great Britain by The University Printing House, Cambridge

Contents

Preface

The importance of ferro- and ferrimagnetic materials in modern electrical and electronic engineering is reflected by the large and rapidly growing number of research publications on high-permeability and permanent magnet materials.

The aim of this book is to present an account of the most important magnetic materials. No attempt has been made to be comprehensive—the wealth of information on magnetic materials makes this quite impossible in a book of this size. Accordingly, the emphasis has been placed on those magnetic materials that have clearly defined engineering applications. It is intended that the scope and treatment will be suitable for postgraduate research students in physics, materials science and technology, metallurgy, electrical and electronic engineering, and for industrial researchers.

The magnetic properties of both high-permeability and permanent magnet materials are highly structure-sensitive, and depend on composition, crystallographic and metallurgical microstructure and the degree of mechanical deformation. It has not been possible to discuss the relevance of these in detail, but the relationship and dependence of the magnetic properties on microstructure has been emphasized throughout.

Although many references to the original research publications have been cited, the author would like to acknowledge the important contributions to our knowledge and understanding of magnetic materials that have been made by many researchers whose work, for reasons of space, it has not been possible to cite.

R. A. McCurrie
October 1993

Chapter 1

Introduction

1.1 FERROMAGNETISM

It is assumed that the reader has some familiarity with the basic concepts of ferromagnetism and hysteresis, so the following discussion is intended to be no more than a brief summary of the basic principles that are essential for a full appreciation of the subject matter of the book.

We begin by defining some of the basic properties of ferromagnetic materials:

1. Below the Curie temperature T_c, they possess a spontaneous magnetization $M_s(T)$, i.e. a magnetization in the absence of an applied field. This is the result of a complex quantum mechanical exchange interaction (Heisenberg exchange interaction) for which the magnetic energy is lower if the ionic magnetic moments are parallel and cooperatively aligned (Fig. 1.1). In the case of ferrimagnetics, there are two ferromagnetically aligned sublattices that are magnetized in opposite directions so that the net magnetization is the difference in the two sublattice magnetizations, i.e. $M = M_A - M_B$ (Fig. 1.2).
2. The magnitude of the spontaneous magnetization $M_s(T)$ is a function of temperature; it is a maximum at 0 K, decreases as the temperature increases, and is zero at the Curie temperature T_c (Fig. 1.3).
3. If they are heated above the Curie temperature, the spontaneous magnetization disappears and the materials become paramagnetic with effectively randomly oriented ionic magnetic moments.
4. On cooling below T_c, the spontaneous magnetization reappears with equal intensity, assuming of course that there have been no structural or chemical changes during the heating.
5. Ferromagnetic materials, though spontaneously magnetized, may show no macroscopically observable magnetization – in apparent contradiction to property 1; an example is an ordinary piece of iron or steel below T_c.
6. When ferromagnetics are in the state described by property 5, the observable magnetization may often be increased from zero to saturation by the application of a field as low as 10 A m^{-1}. However, the field required depends on the material, and, although it is very low for soft high-permeability materials, it is usually very high for hard magnetic materials, i.e. permanent magnets. For $SmCo_5$ a field of 10^6 A m^{-1} may be required to change the net magnetization significantly.

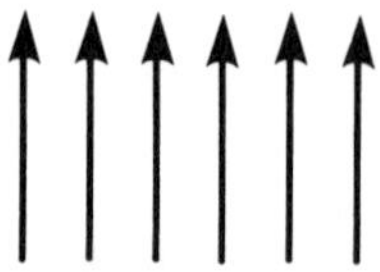

Fig. 1.1 Ferromagnetism. Alignment of ionic magnetic moments (electron spins) on a single lattice.

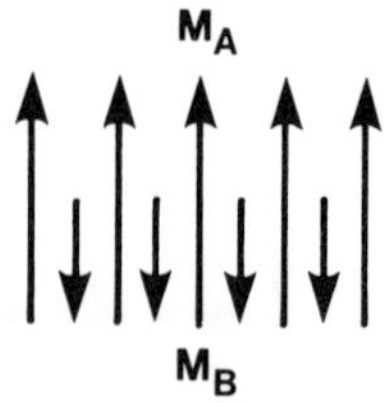

Fig. 1.2 Ferrimagnetism. Two antiferromagnetically aligned ferromagnetic sublattices.

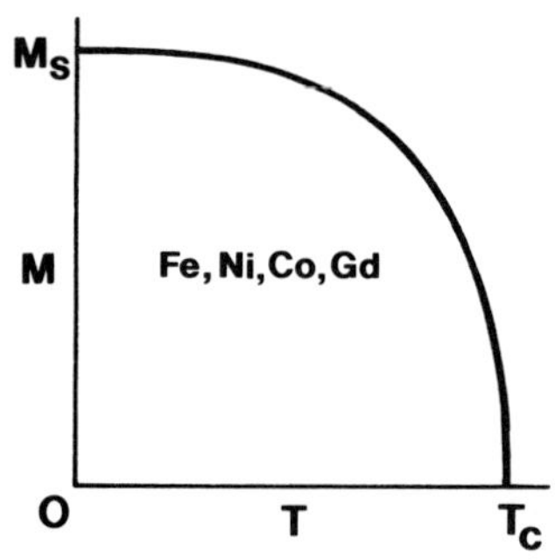

Fig. 1.3 Temperature dependence of saturation magnetization $M_s(T)$, showing a maximum at 0 K. M_s is zero at the Curie temperature T_c.

1.1.1 Ferromagnetic Domains

The existence of ferromagnetics in state 5 is due to the presence of magnetic domains (Figs 1.4b, c, d), i.e. spontaneously magnetized regions of the material separated by domain walls (Fig. 1.5). The magnetization vectors of the domains are arranged in such a way that their vector sum is zero, so that there are closed magnetic flux paths within the material and no net observable magnetization. The main domains are separated by 180° domain walls (Fig. 1.5), whereas the flux closure domains have 90° walls. The presence of domains and the behaviour of domain walls when subjected to applied fields is of fundamental importance in understanding the magnetic properties of ferromagnetic (and ferrimagnetic) materials.

Domains are spontaneously nucleated in all ferromagnetic materials in order to reduce the magnetostatic energy that would be associated with the leakage of magnetic flux into the surrounding air space. Consider a uniformly magnetized block of a ferromagnetic material with its associated magnetic flux, as shown in Fig. 1.4(a). The leakage of this flux can be considerably reduced if the material is divided into domains as shown in Figs 4(b–d), and in the final closed loop structure there is no flux leakage at all, assuming that we are considering a sufficiently large specimen. The division into domains does not continue indefinitely, but does so until the energy expended in forming a new domain wall is greater than the accompanying reduction in magnetostatic energy. Many different domain structures have been observed, even in the same material – applied fields and cooling from above the Curie temperature generate many new domain structures. Two contrasting examples (3.2% Si–Fe and $SmCo_5$) are shown in Figs 1.6(a, b).

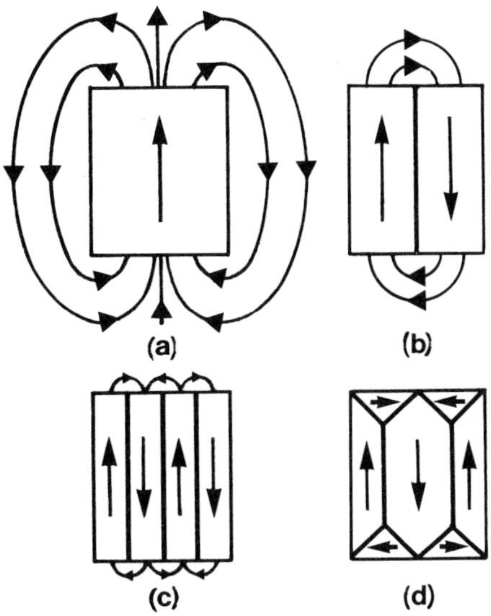

Fig. 1.4 Reduction of magnetostatic energy due to flux leakage by formation of a domain structure.

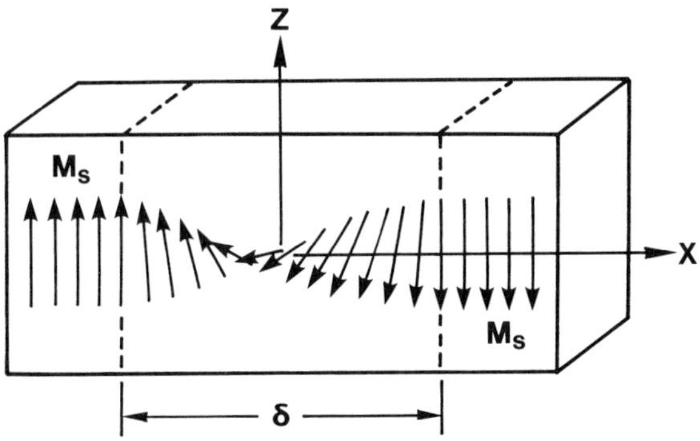

Fig. 1.5 Magnetic structure of a 180° domain wall.

1.1.2 Hysteresis Loops

An appreciation of domain structures enables us to obtain a better understanding of the mechanisms of magnetization and hysteresis in ferromagnetic materials. Consider a material with the very simple domain structure shown in Fig. 1.7. When an external field is applied, the domain whose magnetization vector is closest to the field direction grows at the expense of that which is less favourably oriented. Thus the process of magnetization is really one of domain growth and displacement of domain walls, which are ultimately 'swept out' of the material. The final stage of magnetization involves the rotation of the domain magnetization vectors away from the easy axis of magnetization (owing to magnetocrystalline anisotropy), and hence requires larger applied fields so that the rate of change of magnetization with field is considerably reduced, as we can

(a)

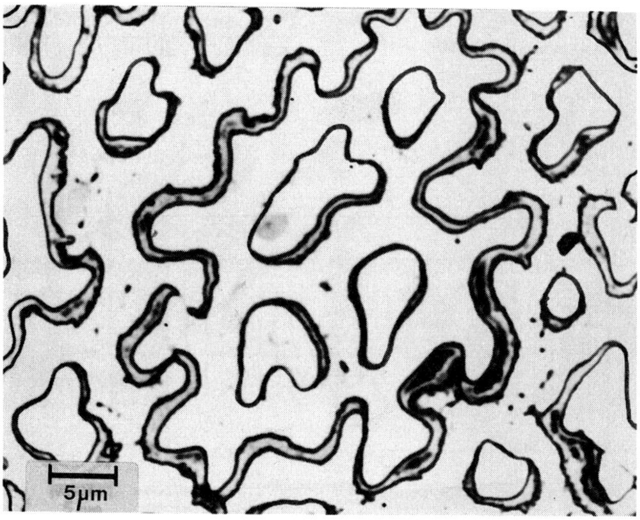

(b)

Fig. 1.6 (a) 'Fir-tree' domain structure on a (100) surface of a 3.2% silicon–iron crystal. The main domain magnetization vectors are parallel to the 'trunks' of the 'fir-trees', whereas the magnetization vectors in the leaves are perpendicular to the trunks, although both sets of vectors are approximately in the same plane (colloid technique). (b) Basal plane domain structure of a hexagonal $SmCo_5$ crystal. The regions delineated by the domain walls are alternately magnetized into and out of the plane of the photograph, i.e. parallel and antiparallel to the axis of magnetization.

see in the magnetization curve shown in Fig. 1.8. During the magnetization process, the applied field moves the domain wall through the material against various microstructural and crystallographic obstacles. The magnitude of the field required to do this determines whether the material is classified as magnetically hard or soft. If, after saturation, the field is decreased, the original magnetization curve is not retraced; instead, the magnetization remains at a higher level than expected for the particular field value and gives rise to the phenomenon of magnetic hysteresis.

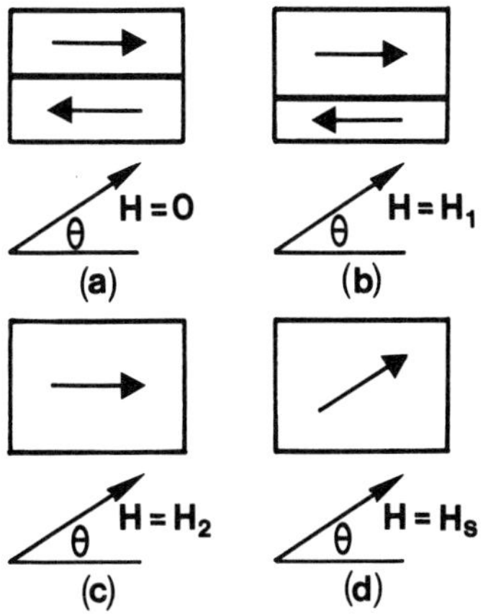

Fig. 1.7 Changes in domain structure with increasing applied field.

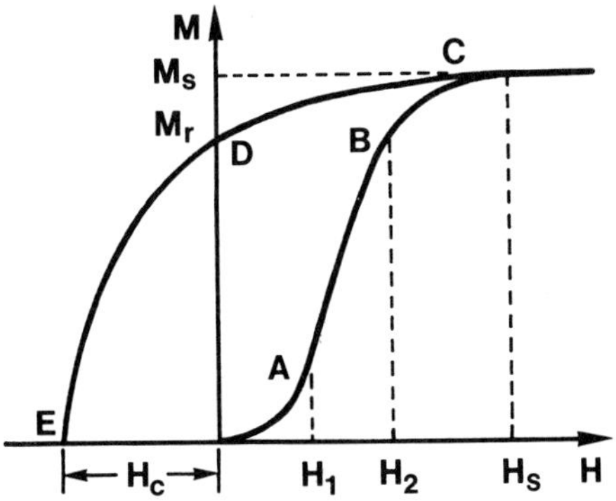

Fig. 1.8 Initial magnetization curves for a crystal with the simple domain structure shown in Fig. 1.7. Note the hysteresis as shown by the curve CDE. At E the magnetization is reduced to zero in a reverse applied field H_c known as the coercivity.

When the applied field is zero, there is a remanent magnetization, which can be decreased to zero by applying a reverse field H_c known as the coercive field or the coercivity. The coercivity is a very strongly structure-sensitive parameter, which may vary by a factor of 10^5 from a very soft high-permeability Ni–Fe alloy with $H_c = 10\ \mathrm{A\ m^{-1}}$ to a very hard material such as $SmCo_5$ with $H_c = 10^6\ \mathrm{A\ m^{-1}}$. It should be borne in mind that the above discussion is concerned with intrinsic magnetization curves, i.e. $M-H$ loops, but the same principles apply to the more useful $B-H$ loops, which include the contribution to the flux density B in the material due to the externally applied field, i.e.

$$B = \mu_0(M + H),$$

where the sign of H must be taken into account. Very significant differences in the shape of the $B-H$ loops are observed for hard permanent magnet materials, but for soft high permeability materials the $B-H$ and $M-H$ loops are very similar except that in the $B-H$ loop, B does not 'saturate', but continues to increase owing to the increasing contribution from H. This will be clear from examination of Figs 1.9 and 1.10.

The most convenient parameter for describing the ease with which a soft ferromagnetic can be magnetized by an external field H is the magnetic permeability μ, which is defined as the ratio of the flux density in the material, B, and the external field, H, that produces it, i.e.

$$\mu = B/H.$$

A typical $B-H$ hysteresis loop for soft magnetic materials is shown in Fig. 1.10. It should be appreciated that μ is not a constant, but a function of the applied field H. The typical variation of μ with H is shown in Fig. 1.11.

Permanent magnet materials (Fig. 1.12) are best described by their remanence B_r, flux coercivity $_BH_c$ and the maximum energy product $(BH)_{max}$. From Fig. 1.12, it is clear that the $B-H$ loops of permanent magnet materials are very much wider than those for

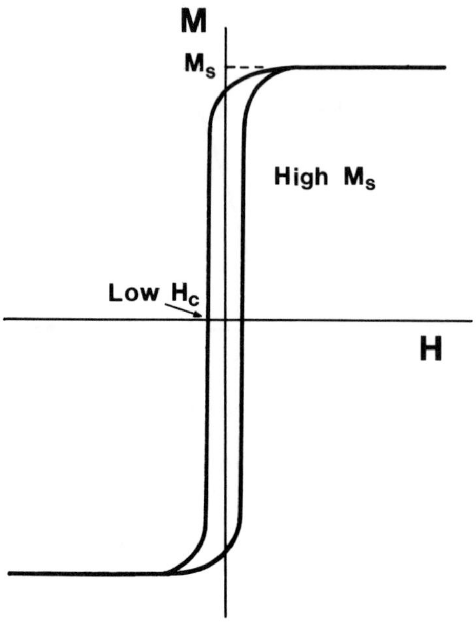

Fig. 1.9 M–H hysteresis loop.

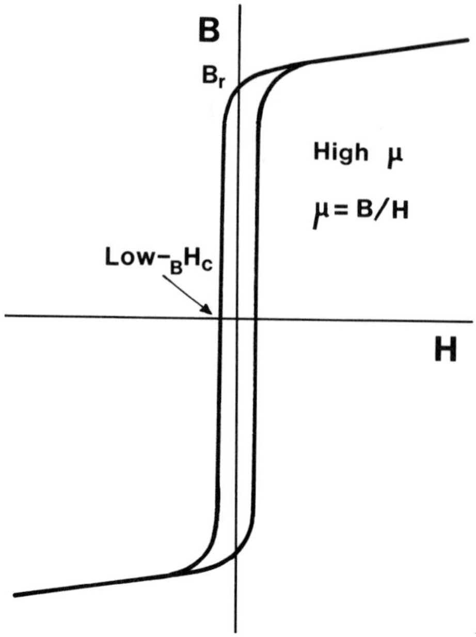

Fig. 1.10 B–H hysteresis loop. Note that this curve does not saturate, because B continues to increase with H even after the material magnetization has reached saturation, $B = \mu_0(M + H)$, so $B = \mu_0(M_s + H)$.

soft high-permeability materials, and hence enclose a significant area. The best measure of the quality of a permament magnet is its maximum energy product $(BH)_{max}$, which is given by the maximum area that can be inscribed in the demagnetization quadrant of the B–H loop. The available magnetic energy stored by the magnet is half the $(BH)_{max}$ product. The coercivity $_BH_c$ is, of course, a measure of the resistance of the permanent magnet to demagnetization; clearly it is desirable for $_BH_c$ to be as large as possible.

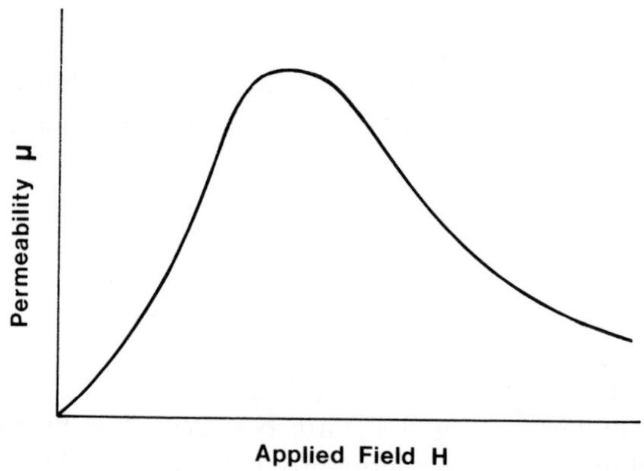

Fig. 1.11 Dependence of permeability on applied field.

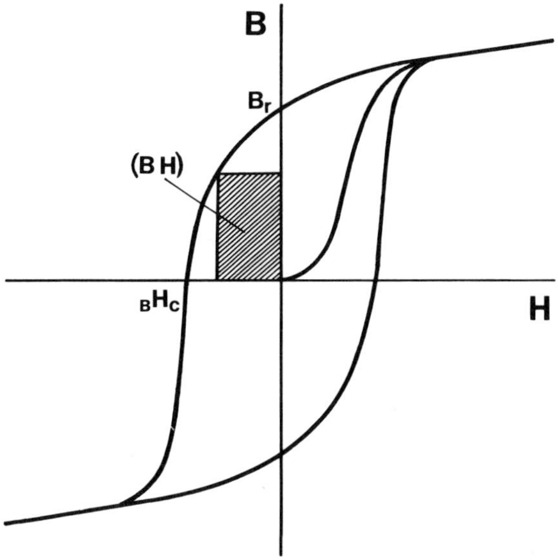

Fig. 1.12 *B–H* hysteresis loop for a permanent magnet material showing definition of (*BH*) energy product.

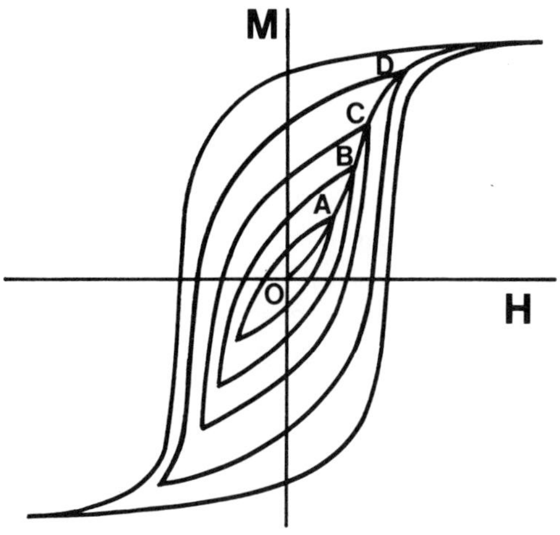

Fig. 1.13 Minor hysteresis loops. These are obtained when the maximum applied field is lower than that required to saturate the material.

When both hard and soft magnetic materials are used in machines and other devices, they may not necessarily operate on the saturation hysteresis loop but on any of the other minor hysteresis loops (Fig. 1.13) unless sufficiently large fields are applied. Furthermore, flux excursions may occur on the main *B–H* and *M–H* loops, producing parasitic loops (Fig. 1.14), which consume energy proportional to the enclosed area just as occurs in traversing the whole hystersis loop, i.e.

$$E = \oint \boldsymbol{H} \cdot \mathrm{d}\boldsymbol{B}$$

(apart from energy losses due to eddy currents).

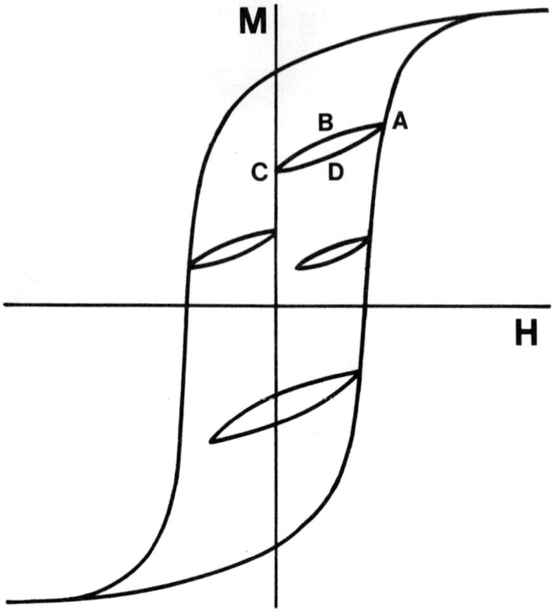

Fig. 1.14 Parasitic hysteresis loops. These are obtained by, for example, reducing the field corresponding to point A through B to C and then increasing again through D to A.

The magnetic properties of various ferromagnetic and ferrimagnetic materials are given in Tables 1.1 and 1.2.

1.1.3 Magnetocrystalline Anisotropy Energy

When a ferromagnetic material is subjected to an applied field, the observed magnetization depends on both the magnitude of the field and the crystallographic direction along which it is applied. The dependence of magnetic properties on crystallographic direction is known as magnetocrystalline anisotropy, and is of fundamental importance in determining the magnetic properties of ferromagnetic materials and hence their suitability for particular applications.

In a sufficiently large field the magnetization will reach the saturation value, which is the same for all crystallographic directions, showing that the individual ionic magnetization vectors have been rotated so as to be parallel to the applied field. The crystallographic direction for which the magnetization reaches saturation in the lowest applied field is known as the easy direction of magnetization or, more precisely, as the easy axis of magnetization. This is the axis along which the magnetization vectors of the domains lie in the absence of an applied field. Magnetization curves for single crystals of iron, nickel and cobalt are shown in Figs 1.15(a), (b) and (c). These curves are representative of a wide range of important ferromagnetic materials. In cobalt, which has the hexagonal close packed crystal structure, there is only one easy axis of magnetization, and this corresponds to the axis of hexagonal symmetry. Uniaxial magnetocrystalline anisotropy is also observed in the important permanent materials

Table 1.1 Magnetic properties of ferromagnetic and ferrimagnetic materials (Soft high-permeability materials).

Material	μ_i	μ_{max}	H_c (A m^{-1})	J_s (T)
3% Si–Fe	9000	40000	12	2.01
Permendur (Fe–Co)	500	6000	160	2.46
Alfenol (Fe–Al)	3000	55000	3	0.80
Sendust (Fe–Al–Si)	—	36000	0.5	0.89
Perminvar (Ni–Fe–Co)	400	2000	100	1.55
Mumetal (Ni–Fe–Cu–Cr)	20000	100000	4	0.75
Supermalloy (Ni–Fe–Mo)	100000	1000000	0.15	0.79
Ni–Zn ferrite	2200 (at 1 kHz)	1600 (at 1 MHz)		0.50
Mn–Zn ferrite	5500 (at 1 kHz)	2500 (at 1 MHz)		0.33
$Fe_{78}B_{13}Si_9$ (amorphous)	—	600000	—	1.56
$(Co,Fe)_{70}(MoSiB)_{30}$ (amorphous)	—	600000	0.4	0.55
$Fe_{66}Co_{18}B_{15}Si_1$ (amorphous)	—	400000	—	1.8
$Fe_{40}Ni_{38}B_{18}Mo_4$ (amorphous)	—	800000	0.6	0.88

Table 1.2 Permanent magnet and recording materials.

Material	B_r (T)	$_BH_c$ (kA m^{-1})	$(BH)_{max}$ (kJ m^{-3})
Alnico (Fe–Co–Ni–Al)	1.30	52	44
Barium ferrite ($BaFe_{12}O_{19}$)	0.35	144	26
$SmCo_5$	0.92	690	200
$Sm(Co–Fe–Cu–Zr)_{7.4}$	1.12	560	240
Nd–Fe–B	1.35	780	320
Cr–Fe–Co	1.56	51	66
Vicalloy II (Fe–Co–V)	1.1	22	14
Cu–Ni–Fe	0.55	42	11
Pt–Co	0.64	400	76
Mn–Al–C	0.6	190	48
γ-Fe_2O_3	0.5	30 (H_c)	—
CrO_2	0.55	48 (H_c)	—

$BaFe_{12}O_{19}$, $SrFe_{12}O_{19}$, $SmCo_5$ and Nd–Fe–B, but in these materials the magnetocrystalline anisotropy energy is very much larger than that in cobalt, i.e. a much larger area is enclosed between the easy-axis magnetization curve, and that in a direction perpendicular to the easy axis.

Magnetization in the various crystallographic directions has an associated energy, and it is the difference in this energy between different directions that determines the magnetic behaviour. The magnetocrystalline anisotropy is described by means of coefficients K_0, K_1 and K_2. For cubic crystals the energy of magnetization in a direction with direction cosines α_1, α_2 and α_3 is given by

$$E = K_0 + K_1(\alpha_1^2\alpha_2^2 + \alpha_2^2\alpha_3^2 + \alpha_1^2\alpha_3^2) + K_2\alpha_1^2\alpha_2^2\alpha_3^2.$$

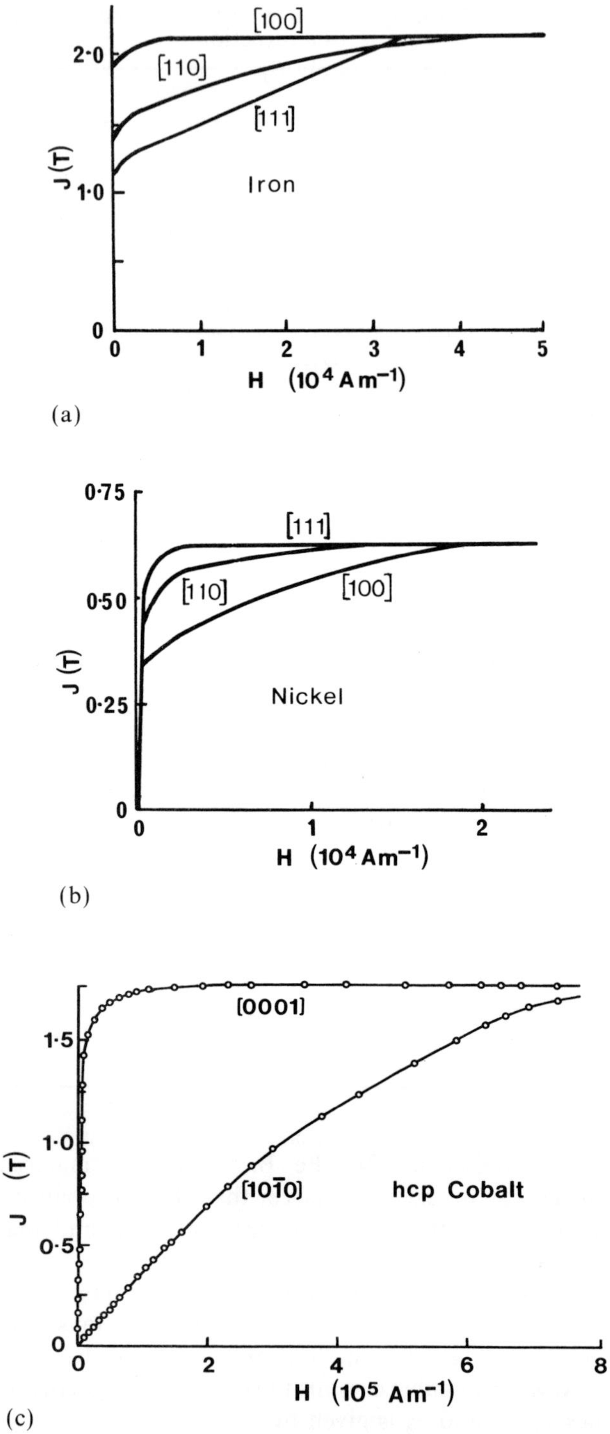

Fig. 1.15 Magnetization curves in various crystallographic directions to illustrate magnetocrystalline anisotropy of (a) iron; (b) nickel; (c) cobalt.

For uniaxial crystals such as hexagonal Co and $SmCo_5$ and tetragonal crystals such as Nd–Fe–Be the energy of magnetization in a direction making an angle θ with the easy axis of magnetization is given approximately by

$$E = K_0 + K_1 \sin^2 \theta.$$

From the magnetization curves for cobalt (Fig. 1.15c) parallel and perpendicular to the easy hexagonal axis (the curves for $SmCo_5$ and Nd–Fe–B are similar in form), we find that

$$E_{\parallel} = K_0 \qquad \text{for } M \text{ parallel to easy axis, i.e. } \theta = 0,$$

$$E_{\perp} = K_0 + K_1 \qquad \text{for } M \text{ perpendicular to easy axis, i.e. } \theta = 90°.$$

Thus the difference in the energy of magnetization per unit volume of material is $E_{\perp} - E_{\parallel}$ i.e. K_1 (J m^{-3}). It can easily be shown that the field required to rotate the magnetization vector 90° away from the easy direction is $2K_1/\mu_0 M_s$ or $2K_1/J_s$. This field is known as the magnetocrystalline anisotropy field H_A, i.e.

$$H_A = \frac{2K_1}{\mu_0 M_s} = \frac{2K_1}{J_s}.$$

Similar expressions can be derived for cubic crystals such as Fe, Ni–Fe, Fe–Co and Si–Fe. Although the above discussion has been concerned with single crystals of the material, the values of K_1 and H_A are of great importance in determining the properties of polycrystalline or grain-oriented materials used in engineering applications. Values of K_1 and H_A vary by many orders of magnitude for different materials.

For permanent magnet materials, we require the magnetocrystalline anisotropy energy to be uniaxial and as large as possible so that there is a very strong preference for magnetization along a particular crystallographic direction, i.e. the easy direction. It is this preference, along with other important effects, that enables permanent magnets to resist demagnetization in reverse fields. The permanent magnet materials $BaFe_{12}O_{19}$, $SrFe_{12}O_{19}$, $SmCo_5$, $Sm_2(Co,Fe,Cu,Zr)_{7.4}$ and Nd–Fe–B have large uniaxial magnetocrystalline anisotropies.

For soft or high-permeability magnetic materials such as Si–Fe transformer core material or the Ni–Fe alloys (Mumetals, Radiometals, Permalloys), we require the magnetocrystalline anisotropy to be as low as possible. This enables changes in magnetization to occur in the lowest possible fields and as rapidly as required – in the case of a transformer core material, at 50 Hz.

1.1.4 Magnetostriction and Magnetoelasticity

When the state of magnetization of a ferromagnetic material is changed by an applied field, there is an accompanying small change in its dimensions, provided that part or all of the magnetization change is due to rotation of the magnetization vector, rather than to domain wall displacement. Although there are several magnetostrictive effects, namely linear, volume and torsional changes, the linear magnetostriction λ is defined as

$$\lambda = \frac{\Delta l}{l},$$

where l is the original length of the specimen and Δl the magnetostrictive change. The effect is anisotropic and very complex, since it is also field-dependent. Typically, the saturation magnetostriction is of order 10^{-5}, although some alloys (e.g. $TbFe_2$) have values of $\lambda_s \approx 200 \times 10^{-5}$. The magnetostrictive strain has an associated magnetoelastic energy, so the two phenomena are closely related, since magnetostriction effectively results in a field-controlled variation in Young's modulus known as the ΔE effect.

The two effects are of most importance in the design of high-permeability alloys, where λ is usually required to be as small as possible. The presence of magnetostriction in hard magnetic materials, i.e. permanent magnets, may be regarded as beneficial, but it is usually of minor significance.

Materials with a large magnetostriction can be used directly in magneto-mechanical transducers – these are discussed in Chapter 2.

1.1.5 Shape Anisotropy Energy

Shape anisotropy energy is a measure of the difference in the energies associated with magnetization in the shortest and longest dimensions of a ferromagnetic body. A long cylinder has a higher magnetostatic energy associated with magnetization perpendicular to its axis than that for magnetization parallel to the axis. This can be understood in a rather over-simplified way by considering the proximities of the associated but oppositely 'charged' magnetic poles that appear on the application of the external field producing the net magnetization. Parallel to the axis, the poles are widely separated and so have a low magnetostatic energy – the long axis is therefore the easy axis of shape

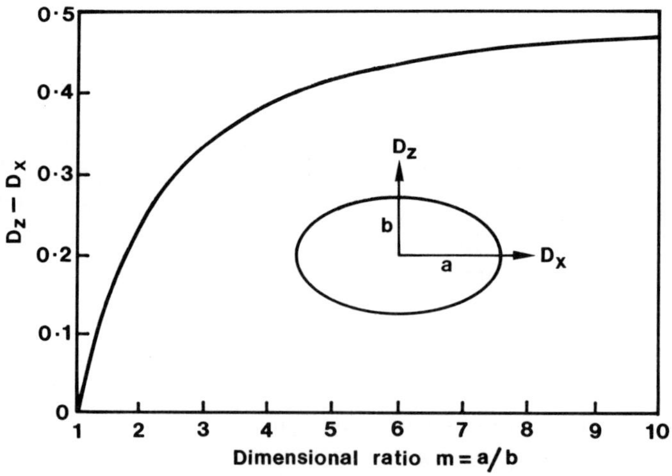

Fig. 1.16 Shape anisotropy in a prolate ellipsoid of revolution showing variation of the difference in the two principal demagnetizing factors $D_z - D_x$ as a function of the axial ratio a/b.

anisotropy, while any axis perpendicular to this is a hard axis of magnetization. The difference, for a magnetization M, in the two energies for a cylinder is given by

$$E=\tfrac{1}{2}\mu_0(D_z-D_x)M^2,$$

where D_x and D_z are the demagnetization factors parallel and perpendicular to the axis of the cylinder. Values of D_z and D_x for any ratio of length to diameter can be calculated from appropriate formulae. Exact values can be obtained only for uniformly magnetized ellipsoids. The variation of D_z-D_x with axial ratio a/b for a prolate ellipsoid of revolution is shown in Fig. 1.16. The magnetization vector therefore always has a preference for the long easy axis. As we shall see later, this property is important in determining the permanent magnet properties of the Alnico alloys and magnetic recording materials such as γ-Fe_2O_3 and CrO_2, for which the coercivity is predominantly due to the shape anisotropy of needle-like (acicular) particles, although in the latter materials there is a contribution due to magnetocrystalline anisotropy.

1.1.6 Magnetic Exchange Energy

Since the spontaneous magnetization of ferromagnetic materials disappears at the Curie temperature, we may assume that the ferromagnetic alignment of the individual ionic magnetic moment vectors is destroyed by the thermal energy kT_c, where k is Boltzmann's constant. The magnetic exchange energy parameter A may be regarded as a measure of the strength of coupling between the ionic magnetic moment vectors separated by the lattice spacing a, so that at the Curie temperature T_c

$$A\approx\frac{kT_c}{a}.$$

For most useful materials A does not vary by more than a factor of about 4.

1.1.7 Domain Wall Energy

The structure of a 180° domain wall (Fig. 1.5) shows how the magnetization reversal between adjacent domains occurs in a series of discrete rotations of the individual ionic magnetic moments. In the ideal ferromagnetic state, corresponding to a minimum in the exchange energy, the spins should be parallel, but in the 180° domain the exchange energy is increased owing to the small angular rotation. At the same time, the ionic magnetic moments are also slightly rotated away from the preferred magnetization axis in each domain, so that there is an increase in the magnetocrystalline anisotropy energy. From a relatively straightforward theoretical argument, it can be shown that the domain wall energy per unit area is given by

$$\gamma=4(AK_1)^{1/2}=4\left(\frac{kT_cK_1}{a}\right)^{1/2}. \tag{1.1}$$

The domain energy is a parameter of considerable importance in the interpretation of magnetic properties, because if there are local variations in the values of A and K_1 within the same material – as there often are at grain boundaries, second phases, inclusions and defects – then γ will depend on the position of the domain wall.

When a field is applied to such materials, the domain walls have to be pushed over the various obstacles or pulled out of domain wall traps (wells), thus giving rise to hysteresis because many of the barriers will be asymmetrical with respect to the direction of movement of the wall. The strength of the field required will therefore be closely related to the observed coercivity H_c, although H_c is a measure of the average domain wall pinning strength. Since the coercivity H_c is defined as the reverse field required to reduce the net magnetization to zero, i.e. the field required to reverse half the magnetization, obstacles with the lowest pinning strengths will be negotiated by the wall first.

1.1.8 Domain Wall Width

From a theoretical argument it can be shown that the width of a 180° domain wall (Fig. 1.5) δ is given by

$$\delta = \pi \left(\frac{A}{K_1} \right)^{1/2} = \pi \left(\frac{kT_c}{aK_1} \right)^{1/2}, \tag{1.2}$$

where the symbols have their usual meanings. Thus, as K_1 increases, so the wall width decreases. In general, therefore, permanent magnets contain narrow domain walls. Since the strength of interaction of a domain wall with regions of differing values of A and K_1 is determined to some extent by the width of the domain wall relative to the width of the defect region, we expect the domain wall width to play an important rôle in determining the coercivity, whether it is controlled by nucleation, pinning, or a mixture of the two.

A wide domain wall in a high-permeability material is not expected to interact strongly with an inclusion, second phase or crystallographic defect that is very much narrower or very much wider than itself, although every attempt is made to remove these obstacles of whatever size from soft high-permeability materials in order to maximize the permeability.

Typical intrinsic magnetic properties for a high permeability material (iron) and a permanent magnet material ($SmCo_5$) are shown in Table 1.3.

Table 1.3 Intrinsic magnetic properties (approximate values) of a typical hard magnetic material ($SmCo_5$) and a typical soft magnetic material (Fe).

Property	$SmCo_5$	Fe	Units
Saturation magnetic polarization J_s	1	2	T
Anisotropy coefficient K_1	10^7	5×10^4	$J\,m^{-3}$
Anisotropy field H_A	2×10^7	5×10^4	$A\,m^{-1}$
Exchange energy parameter A	2×10^{-11}	2×10^{-11}	$J\,m^{-1}$
Domain wall energy γ	5×10^{-2}	5×10^{-3}	$J\,m^{-2}$
Domain wall thickness δ	5×10^{-9}	5×10^{-8}	m

1.1.9 Single-Domain Particles

If a ferromagnetic particle is sufficiently small, it may not have sufficient volume to sustain the presence of a domain wall, or the energy required to construct a domain wall to subdivide the particle into two domains may be greater than the resulting reduction in the magnetostatic energy associated with the uniform magnetization of the spherical particle. Such particles are known as single-domain particles, and, for iron, they have radii less than about 5 nm. Particles as small as this can be produced only by special techniques, such as metallurgical precipitation in an alloy, for example elongated Fe–Co-rich single-domain particles in Alnico permanent magnets (Fe–Co–Ni–Al–Cu–Ti) or acicular γ-Fe_2O_3 particles for magnetic recording.

The magnetic hysteresis properties of ellipsoidal single-domain particles with uniaxial shape anisotropy have been investigated in detail by Stoner and Wohlfarth (1948), who showed that for a prolate ellipsoid of revolution there are only two stable positions of the magnetization vector, i.e. parallel to the long easy axis of magnetization, as shown in Fig. 1.17. The corresponding hysteresis 'loop' is shown in Fig. 1.18. The vertical sections of the 'loop' are shown dotted because they do not represent stable values of the magnetization when the field is parallel to the long x-axis as shown. When the field is applied to the negative x direction, the magnetization vector undergoes a sudden reversal (by coherent rotation of the individual ionic magnetic moments) when the field reaches a critical value H_c known as the coercivity, given by

$$H_c = (D_z - D_x)M_s$$

Strictly speaking, this should be called the intrinsic coercivity to distinguish it from the flux coercivity used in connection with B–H hysteresis loops. When the applied field is perpendicular to the axis of the particle, the magnetization is gradually rotated away from the x axis by the field, and produces a linear magnetization curve with no hysteresis, as shown in Fig. 1.19. If the field is reduced to zero, the magnetization simply returns to the x axis and has no component in the z direction. When the field is applied

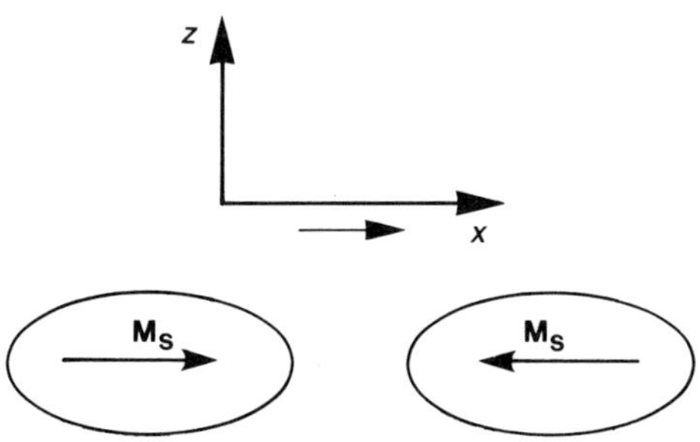

Fig. 1.17 Elongated single-domain particles with uniaxial shape anisotropy. Only two stable states of magnetization are possible, as shown, i.e. parallel and antiparallel to the easy axis.

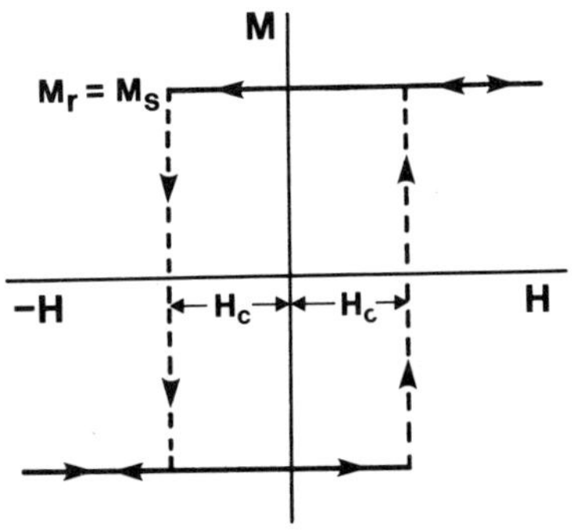

Fig. 1.18 'Rectangular' hysteresis loop for the uniaxial single-domain particle shown in Fig. 1.17.

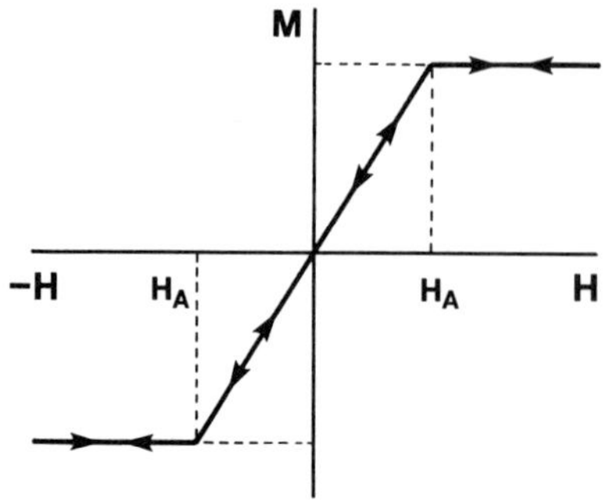

Fig. 1.19 Magnetization curve for a uniaxial single-domain particle when the applied field is perpendicular to the easy axis of magnetization. Note that there is no hysteresis.

at other angles to the x axis, more complex hysteresis loops are obtained, with intermediate values of coercivity and remanence.

The magnetic properties of materials with aligned or random assemblies of single domains can be interpreted to some extent by using the Stoner–Wohlfarth model. The behaviour, however, is complicated by the magnetostatic interactions between particles – not specifically included in the above Stoner–Wohlfarth theory. A further complication arises owing to the possibility that, for certain particle sizes, magnetization reversal in the particles may occur by more complex incoherent mechanisms of magnetization reversal known as curling, buckling, fanning and the chain-of-spheres model (Zijlstra, 1982).

If the single-domain particles are spherical and the material has uniaxial magneto-crystalline anisotropy, Stoner–Wohlfarth theory still applies. In this case the intrinsic coercivity parallel to the easy axis of magnetization is given by

$$H_c = \frac{2K_1}{\mu_0 M_s} = \frac{2K_1}{J_s},$$

and the corresponding hysteresis loop is the same as that shown in Fig. 1.18. Single-domain particles with uniaxial magnetocrystalline anisotropy are not subject to the magnetostatic interaction fields from other particles because the magnetocrystalline anisotropy is a fundamental ionic property of the material. In some single-domain particles, such as acicular γ-Fe_2O_3 or CrO_2 particles, the coercivity is due principally to the shape anisotropy, but there is a significant contribution from the magneto-crystalline anisotropy, so their behaviour is very complex. Much research remains to be done on the magnetic properties of γ-Fe_2O_3 and CrO_2 recording tapes. Since there is usually a distribution of shapes, orientation and magnetic properties of the single-domain particles, the resulting rounded hysteresis loops give the average properties of the particle assembly. Even aligned single-domain particle assemblies have rounded hysteresis loops – although much less so than the random assemblies.

1.1.10 Origin of Hysteresis due to Domain Wall Movement

The origin of hysteresis can be understood from the following simple model. Consider a single crystal in the form of a long rod with a rectangular cross-section, as shown in Fig. 1.20. We assume that the crystal has a uniaxial magnetocrystalline anisotropy and that it contains two domains separated by a 180° wall. If the crystal contains defects

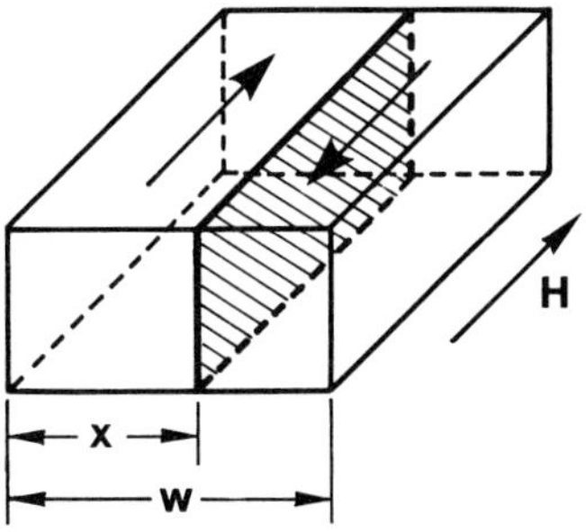

Fig. 1.20 Uniaxial crystal containing a single 180° domain wall.

or other inhomogeneous regions, we should expect the wall energy $\gamma(x)$ to depend on its position in the crystal, because $\gamma(x)$ depends on the local values of the exchange energy A and the magnetocrystalline anisotropy energy K_1, i.e.

$$\gamma = 4(AK_1)^{1/2}.$$

If the crystal is subjected to a magnetic field H parallel to the easy direction of magnetization then the total energy of the crystal per unit volume, when the wall (with unit area) is at a distance x from the edge of the crystal, is given by the equation

$$E(x, H) = \gamma(x) - \mu_0 M_s H x + \mu_0 M_s H(w - x).$$

The magnetostatic energy due to shape anisotropy can be neglected because the length of the crystal is assumed to be large compared with the width w. Thus the stationary values of the energy are given when the following conditions are satisfied:

$$\frac{dE}{dx} = \frac{d\gamma(x)}{dx} - 2\mu_0 M_s H = 0,$$

$$\frac{d^2E}{dx^2} = \frac{d^2\gamma(x)}{dx^2} \gtrless 0.$$

Hence, for a stable equilibrium position of the wall, the field is

$$H = \frac{1}{2\mu_0 M_s} \frac{d\gamma(x)}{dx}.$$

Thus the slope of the $\gamma(x)$ versus x curve, $d\gamma(x)/dx$, can be associated with the applied field or pressure acting on the wall, and the displacement x corresponds to the magnetization. Suppose, for example, that the slope of the domain wall energy versus position curve is as shown in Fig. 1.21. Assume that after saturation the domain wall has just been renucleated at the surface of the crystal corresponding to the point A on the curve. As the field H required for saturation is decreased, the wall moves along the curve to C, where $H = 0$ (i.e. $d\gamma(x)/dx = 0$), corresponding to the remanent magnetization M_r as shown in Fig. 1.21. If the applied field is reversed (i.e. $d\gamma(x)/dx < 0$), the wall continues to move along the curve to the point D where $d^2\gamma(x)/dx^2 = 0$, corresponding

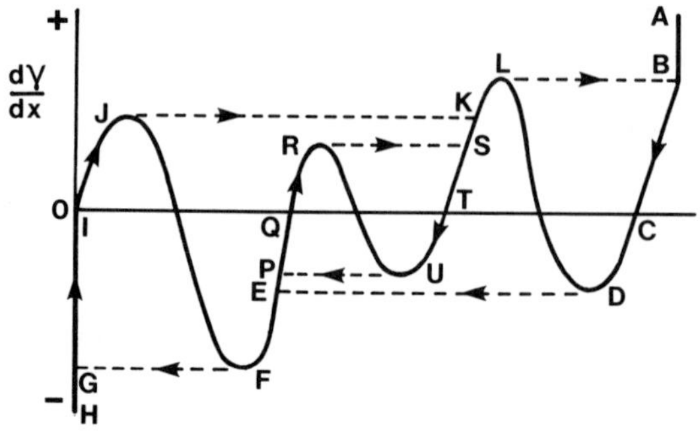

Fig. 1.21 Dependence of the rate of change of domain wall energy with position $d\gamma(x)/dx$ versus position x for the crystal shown in Fig. 1.20.

to a point of inflection in the $\gamma(x)$ versus x curve and to a minimum in the $d\gamma(x)/dx$ versus x curve. The wall is now in an unstable position, and makes a discontinuous jump (i.e. a Barkhausen jump) to the point E, where $d\gamma(x)/dx$ is the same as that at D, but where $d^2\gamma(x)/dx^2 > 0$. This corresponds to a minimum in the energy, and hence to a stable position of the wall. Thus a complete hysteresis loop can be obtained by traversing the whole curve ABCDEFGHIJKLBA. Note that if, when the wall has reached the point E, the applied field is decreased, then the minor hysteresis loop EPQRSTUP can be obtained as shown in Fig. 1.22. In a real material the number of possible hysteresis loops is very large.

Thus we can identify the coercivity with the rate of variation of the wall energy with position, since $d\gamma(x)/dx$ is a measure of the energy barrier that must be surmounted by the domain wall. The principal requirement for a high coercivity is therefore a large and rapid variation of the wall energy with position. Regions of the material in which the wall has a higher or lower energy will act as barriers to domain wall motion or as domain wall traps respectively. Thus, from the major hysteresis loop shown in Fig. 1.22

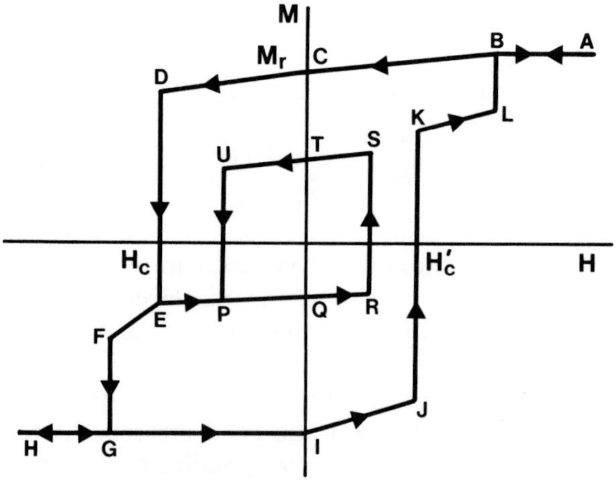

Fig. 1.22 Major and minor hysteresis loops corresponding to relationship between $d\gamma(x)/dx$ versus x shown in Fig. 1.21.

it can be seen that the coercivity can be identified as the field at point D, i.e.

$$[H_c]_D = \frac{1}{2\mu_0 M_s}\left[\frac{d\gamma(x)}{dx}\right]_D, \tag{1.3}$$

and is therefore proportional to the height of the peak in the $d\gamma(x)/dx$ curve at D. Note, however, that for the particular crystal chosen this field reverses more than half the magnetization, so that the net magnetization of the crystal corresponds to that at point E rather than zero. However, in a real material there will be a large number of such single crystal grains with different hysteresis loops, so that we should observe a smooth hysteresis loop in which the observed coercivity and remanence will be averages of all the individual contributions. If the hysteresis curve is observed on a sufficiently fine scale it can be seen that changes in the magnetization occur by a series of small discontinuous jumps known as Barkhausen discontinuities (see e.g. Jiles, 1991). Smooth hysteresis loops are sometimes observed for single crystals because different small regions of the crystal may be slightly different in their hysteric behaviour, particularly if the crystal is chemically or physically inhomogeneous.

In a soft magnetic material we require $d\gamma(x)/dx$ to be as small as possible and to be virtually independent of position. Since the wall energy $\gamma = 4(AK_1)^{1/2}$, we find that the highest-permeability materials are those in which K_1 is very small or nearly zero, so that the coercivities are also as small as possible. As we shall see, it is usually necessary to ensure that the magnetostriction is small so as to minimize the stress-induced anisotropy.

Conversly, in hard magnetic materials we require $d\gamma(x)/dx$ to be as large as possible and to vary rapidly with position. This usually means that K_1 must be as large as possible and that the material be inhomogeneous – often containing two or more phases.

Although the hysteresis loops shown in Fig. 1.22 have been derived from an idealized and simple dependence of the domain wall energy on position, similar loops have been observed for single-crystal particles of $SmCo_5$. These loops show both gradual and discontinuous changes in the magnetization, but it is not usually possible to interpret the behaviour in terms of a single $d\gamma(x)/dx$ versus x curve. This is probably due to the fact that the domain walls do not necessarily negotiate the energy barriers in the same way, and in some cases it may be that more than one domain wall is present. Furthermore, additional domain walls may be nucleated at any point on the hysteresis loop. The properties of permanent magnet materials are discussed in Chapter 5.

1.2 PERMANENT MAGNETS

For a permanent magnet material with uniaxial magnetocrystalline anisotropy K_1 and saturation magnetization M_s, theoretical arguments show that in order to reverse the magnetization, a reverse field equal to the anisotropy field must be applied, i.e. the theoretical coercivity is given by

$$H_c = H_A = \frac{2K_1}{\mu_0 M_s} = \frac{2K_1}{J_s},$$

assuming that magnetization reversal occurs by coherent rotation of the magnetization vectors of the individual ionic magnetic moments. Pure coherent rotation rarely occurs in real materials, so that the best observed values of the coercivity, even for materials such as $SmCo_5$, $Sm_2(Co,Fe,Cu)_{17}$ and Nd–Fe–B, are only about $0.1H_A$. What usually occurs is that in a sufficiently high reverse field domain walls are nucleated in the otherwise saturated permanent magnet, and then magnetization reversal to the coercivity point on the B–H curve occurs by displacement of the domain walls. The ease of nucleation of reverse domains in ferromagnetics can be attributed to the presence of vestigial domains or fragments of domain walls that are not removed during magnetization – ideally to saturation. Even in very carefully grown iron whiskers with a very high degree of crystallographic perfection, it has been shown that domains are still present after a field greater than the anisotropy field has been applied. Thus we must assume that the failure of permanent magnets to exhibit their full potential coercivities can largely be attributed to the presence of crystallographic imperfections, microcracks and surface flaws, all of which act as easy nucleation sites for domain walls.

The fields required to nucleate and displace the walls ultimately determine the value of the coercivity and hence the quality of the permanent magnet. The process is complex, because many domain wall nucleations and displacements will be involved, resulting in relatively smooth rounded B–H hysteresis loops.

The nucleation of domain walls occurs in much lower fields than H_A because they select regions in the material where A and K_1 are lower, thus reducing the energy γ required to construct a domain wall. However, once nucleated, the wall may be displaced to other regions in the magnet where A and K_1 are much lower or much higher. In either case motion of the domain wall under the influence of an external field will be impeded. When γ is lowered in a region of low A and K_1, this will act as a trap, an energy well or a pinning site for the wall as shown by the wells in Fig. 1.23 whereas when A and K_1 are higher, the region will appear as an energy barrier to the wall as shown by the barriers in Fig. 1.23. In either case a significant increase in the applied field will be required, either to drag the domain wall out of the energy well or to push it over the energy barrier, thus changing the overall magnetization state of the magnet.

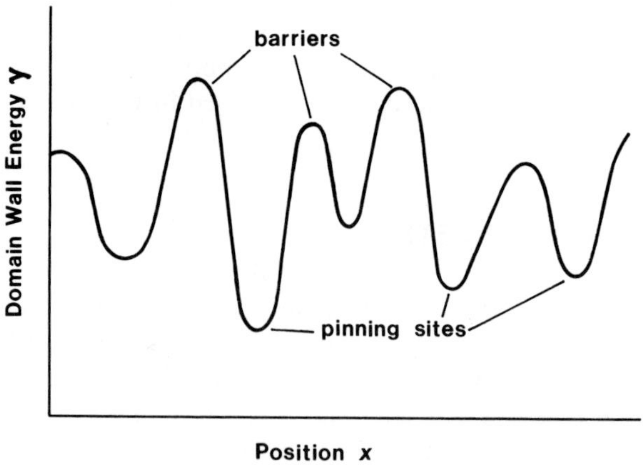

Fig. 1.23 Domain wall energy $\gamma(x)$ as a function of position x in a permanent magnet material, showing domain wall pinning sites and barriers to domain wall motion.

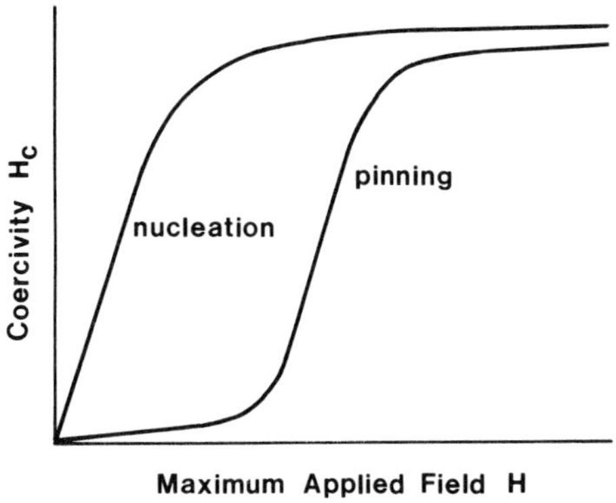

Fig. 1.24 Dependence of the coercivity H_c on the maximum applied field for permanent magnets where the magnetic properties are determined principally by domain wall nucleation and domain wall pinning.

Regions with higher or lower values of A and K_1 may, of course, be due to grain boundaries, second phases, inclusions and crystallographic defects. This explains why permanent magnet alloys generally have more complex compositions and metallurgical microstructures so as to introduce a large and rapid variation of domain wall energy as a function of position in the material. The domain walls therefore have to overcome the many pinning sites and energy barriers, and ideally do so with the largest possible hysteresis.

For a planar defect of width D and magnetic parameters J_s, K_2 and A_2 in a matrix phase with parameters J_s, K_1 and A_1, the coercivity due to domain pinning is given by (Zijlstra, 1982)

$$H_c = 0.2\left(\frac{2K_1}{J_s}\right)\frac{D}{\delta}\left(\frac{A_1}{A_2} - \frac{K_2}{K_1}\right),$$

where δ is the width of the domain wall in the matrix phase.

In some materials both nucleation and pinning mechanisms operate, but in others one or other of these dominates. The two mechanisms can be distinguished by plotting a graph of the variation of the observed coercivity as a function of the maximum applied magnetizing field as shown in Fig. 1.24.

1.2.1 Requirements for Permanent Magnet Materials

These are as follows:

(1) high saturation magnetization M_s;
(2) high remanence M_r;
(3) ideally rectangular $M-H$ hysteresis loop with $M_r = M_s$ and $H_c \geqslant \frac{1}{2}M_s$;
(4) very high uniaxial magnetocrystalline anisotropy energy K_1;
(5) high coercivity due to domain wall nucleation, domain wall pinning or a combination of the two, requiring as large and rapid a variation of the domain wall energy with position in the material as possible, i.e. there must be significant structural and/or composition fluctuations in the material;
(6) high maximum energy product $(BH)_{max}$;
(7) high Curie temperature;

(8) good temperature stability;
(9) corrosion resistance;
(10) mechanical strength;
(11) machinability;
(12) cheapness.

A considerable amount of research effort is being devoted to the search for better permanent magnets, since they have a very wide range of applications with a large potential world market, estimated in 1990 to be worth about £1 000 000 000 per annum.

A derivation of the $(BH)_{max}$ quality criterion for permanent magnets and the factors determining its maximum possible value are given in Chapter 5.

1.3 SOFT OR HIGH-PERMEABILITY MAGNETIC MATERIALS

For high permeability we require materials with very high saturation magnetization and very low magnetocrystalline anisotropy energy K_1, preferably with many easy directions of magnetization (eight $\langle 111 \rangle$ directions in the case of many Ni–Fe alloys). For materials with typical Curie temperatures and hence magnetic exchange energies (Table 1.1) this means that the domain wall energy γ will be low (equation (1.1)), thus facilitating the nucleation and displacement of domain walls in the lowest possible applied fields in both the forward and reverse directions of magnetization. A low value of K_1 also implies that the domain walls are very wide (equation (1.2)), so that they are not therefore expected to interact strongly with grain boundaries, inclusions or crystallographic defects, since the widths of these will usually be much smaller than the domain wall width δ. Ideally, high-permeability materials should be as homogeneous as possible, certainly single-phase and free from inclusions, impurities (especially carbon and sulfur), crystallographic defects and internal and external stresses. All these factors imply the requirement that the domain wall energy $\gamma(x)$ should not depend strongly on its position x within the material and that the $\gamma(x)$ versus x curve should be as flat and smooth as possible. In the presence of an applied field the domains will move very easily, so that large changes in B will be produced in the lowest possible fields, thus giving a high permeability $\mu = B/H$.

Since the energy consumed in traversing a hysteresis loop is proportional to the enclosed area, the low coercivity also minimizes these losses, while the high saturation magnetization ensures that as much magnetic flux as possible is transported in the magnetic circuit, as is required for example in transformer and inductor cores.

The presence of stress produces an effective additional anisotropy and further impedes the motion of domain walls. This occurs because there is an interaction between mechanical stress and the magnetization of the material known as the magnetoelastic interaction, which manifests itself in the phenomenon of magnetostriction. When a material undergoes a change in its state of magnetization involving some rotation of the magnetization vectors, there is an accompanying small change in the dimensions of the material of order 10^{-5} of the original. The change may be an increase or a decrease in dimensions, and is known as magnetostriction. Changes in the dimensions due to stress also produce (conversely) changes in the magnetization state and hence the domain structure.

In transformer cores the state of magnetization, i.e. the direction of magnetization, changes 50 times per second, and so is accompanied by 2×50 mechanical changes per second, giving rise to an audible magnetostrictive hum. Considerable research effort has been expended in order to reduce this unwanted hum to a minimum. The magnetostrictive hum from transformer substations can be very irritating, especially at night.

Since high-permeability conducting alloys are used for transformer and inductor cores that require AC excitation, it is also important to minimize energy losses due to eddy currents induced by flux changes in the core alloys. This is achieved to some extent by careful choice of composition consistent with the magnetic requirements. The addition of silicon to iron reduces the magnetocrystalline anisotropy and also increases the electrical resistivity, which reduces eddy currents and hence heat losses. For very high frequencies it is necessary to use the insulating high-permeability ceramic materials Ni–Zn and Mn–Zn ferrites. The materials are mixtures of the ferrites $NiFe_2O_4$ and $MnFe_2O_4$ with $ZnFe_2O_4$. A wide range of properties can be obtained by careful choice of composition. Ferrites with more complex compositions are used at microwave frequencies.

1.3.1 High-Permeability Amorphous Alloys

Very-high-permeability materials can also be produced from amorphous alloys such as $Fe_{40}Ni_{40}P_{14}B_6$, $Fe_{80}B_{20}$ and $Fe_{66}Ci_{18}B_{15}Si_1$. Since these materials are amorphous, i.e. they are glasses produced by very rapid quenching from the melt, they have no regular crystalline structure and hence no magnetocrystalline anisotropy energy, and hence no well-defined easy directions of magnetization. The magnetic anisotropy coefficient K_u is not zero, however, as might be expected, because a small induced anisotropy is produced in the thin ribbons of the materials as they are very rapidly quenched onto a rotating cooled drum. Since the induced anisotropy is low, the domain wall energy is low and domain walls very wide and irregular. Thus in an applied field the domain walls move through the alloy with little impedance and so give rise to exceptionally high permeabilities. The permeability can usually be increased still further by annealing in order to remove residual stresses that, via magnetoelestic effects, impede domain wall motion. Preferentially higher permeabilities can be obtained by annealing in a magnetic field.

Amorphous ferromagnetic materials can be used for small transformer and inductor cores, but there is some doubt as to whether they will be suitable for use in high-power transformers. The maximum thickness of the amorphous ribbons is limited by the requirement that the cooling rate has to be of order $10^6\ {}^\circ C\, s^{-1}$ in order to quench-in the amorphous 'structure'. The presence of the glass-forming elements P, B, C, Si etc. in the alloys has the further advantage of increasing their electrical resistivities, so reducing eddy current losses.

High-permeability amorphous alloys are very useful for detecting small currents and magnetic fields, where their large response can easily be amplified and measured.

Amorphous alloys have interesting magnetoelastic properties, and are finding applications in magnetomechanical transducers.

1.3.2 Requirements for High-Permeability Materials

These are as follows:

(1) high saturation magnetization M_s;
(2) Very low magnetocrystalline anisotropy energy K_1, or $K_u \rightarrow 0$ for amorphous alloys;
(3) very low coercivity, so that domain walls are easily nucleated and displaced; materials should be as homogeneous as possible and free from second phases, inclusions, impurities, stress and crystallographic defects; the low coercivity ensures that hysteresis energy losses are small;
(4) high Curie temperature;
(5) high electrical resistivity to minimize energy losses due to eddy currents;
(6) good temperature stability;
(7) corrosion resistance;
(8) mechanical strength;
(9) machinability;
(10) cheapness.

For further information on magnetism and ferromagnetic materials the reader is referred to the bibliography, especially to the books by Crangle (1991), Jiles (1991), Jakubovics (1987) and Cullity (1972) for excellent introductions to the subject.

BIBLIOGRAPHY

Anderson, J. C. (1969). *Magnetism and Magnetic Materials* (Chapman and Hall, London).
Bates, L. F. (1961) *Modern Magnetism*, 4th edn (Cambridge University Press, London)
Bozorth, R. M. (1951). *Ferromagnetism* (Van Nostrand, Princeton).
Bradley, F. N. (1971). *Materials for Magnetic Function* (Hayden, New York).
Brailsford, F. (1966) *Physical Principles of Magnetism* (Van Nostrand, London).
Burke, H. E. (1986). *Handbook of Magnetic Phenomena* (Van Nostrand–Reinhold, New York).
Carey, R. and Isaac, E. D. (1966). *Magnetic Domains and Techniques for their Observation* (English Universities Press, London).
Chen, C. W. (1977). *Magnetism and Metallurgy of Soft Magnetic Materials* (North-Holland, Amsterdam).
Chikazumi, S. (1964). *Physics of Magnetism* (Wiley, New York).
Chikazumi, S. and Charap, S. H. (1978). *Physics of Magnetism* (Krieger, Florida).
Craik, D. J. (1971). *Structures and Properties of Magnetic Materials* (Pion, London).
Craik D. J. (ed.) (1975). *Magnetic Oxides*, Parts 1 and 2. (Wiley, London).
Craik, D. J. and Tebble, R. S. (1965). *Ferromagnetism and Ferromagnetic Domains* (North-Holland, Amsterdam).
Crangle, J. (1971). *Magnetism in Solids* (Edward Arnold, London).
Crangle, J. (1991). *Solid State Magnetism* (Edward Arnold, Sevenoaks).
Cullity, B. D. (1972). *An Introduction to Magnetic Materials* (Addison-Wesley, London).
Evetts, J. (ed) (1992) *Concise Encyclopedia of Magnetic and Superconducting Materials* (Pergamon Press, Oxford)
Hadfield, D. (ed.) (1962). *Permanent Magnets and Magnetism* (Iliffe Books, London).
Heck, C. (1974). *Magnetic Materials and their Applications* (Butterworths, London).

Hoselitz, K. (1952). *Ferromagnetic Properties of Metals and Alloys* (Clarendon Press, Oxford).
Jakubovics, J. P. (1987). *Magnetism and Magnetic Materials* (The Institute of Metals, London).
Jiles, D. (1991). *Introduction to Magnetism and Magnetic Materials* (Chapman and Hall, London).
Kalvius, G. M. and Tebble, R. S. (eds) (1979). *Experimental Magnetism* (Wiley, Chichester).
McCaig, M. and Clegg, A. G. (1987). *Permanent Magnets in Theory and Practice*, 2nd edn (Pentech Press, London).
Martin, D. H. (1967). *Magnetism in Solids* (Iliffe, London).
Morrish, A. H. (1965). *Physical Principles of Magnetism* (Wiley, New York).
Nesbitt, E. A. and Wernick, J. H. (1973). *Rare-Earth Permanent Magnets* (Academic Press, New York).
Parker, R. J. (1990). *Advances in Permanent Magnetism* (Wiley-Interscience, New York).
Smart, J. S. (1966). *Effective Field Theories of Magnetism* (Saunders, Philadelphia).
Smit, J. (1971). *Magnetic Properties of Materials* (Wiley, New York).
Smit, J. and Wijn, H. P. J. (1959). *Ferrites* (Philips Technical Library, Eindhoven).
Snelling, E. C. (1988). *Ferrites*, 2nd edn (Iliffe, London).
Standley, K. J. (1972). *Oxide Magnetic Materials* (Clarendon Press, Oxford).
Stoner, E. C. and Wohlfarth, E. P. (1948). *Phil. Trans. R. Soc. Lond.* **A240**, 599.
Tebble, R. S. and Craik, D. J. (1969). *Magnetic Materials* (Wiley-Interscience, London).
Thompson, J. E. (1968). *The Magnetic Properties of Materials* (Newnes, Feltham, Middlesex).
Vonsovskii, S. V. (1976). *Magnetism*, Vols 1 and 2 (Wiley, New York).
Watson, J. K. (1980). *Applications of Magnetism* (Wiley-Interscience, New York).
Williams, D. E. G. (1966). *The Magnetic Properties of Matter* (Longmans, London).
Wohlfarth, E. P. (ed) (1980a,b and 1982). *Ferromagnetic Materials: A Handbook on the Properties of Magnetically Ordered Substances*, Vols 1, 2 and 3 (North-Holland, Amsterdam).
Wohlfarth, E. P. and Buschow, K. H. J. (eds) (1988 and 1989). *Ferromagnetic Materials: A Handbook on the Properties of Magnetically Ordered Substances*, Vols 4 and 5 (Elsevier, Amsterdam).
Zijlstra, H. (1967). *Experimental Methods in Magnetism*, Vols 1 and 2 (North-Holland, Amsterdam).
Zijlstra, H. (1982). 'Permanent magnets: Theory', *Ferromagnetic Materials: A Handbook on the Properties of Magnetically Ordered Substances*, ed. E. P. Wohlfarth and K. H. J. Buschow, Vol. 3 (Elsevier, Amsterdam).

Chapter 2

Soft Magnetic Alloys

2.1 INTRODUCTION

Soft or high-permeability magnetic materials are defined as those in which very large changes in the magnetic flux density can be produced by very small fields, sometimes as low as $8\ \mathrm{A\,m^{-1}}$. The most convenient parameter for describing a soft magnetic material is the permeability $\mu = B/H$, where B is the flux density produced by the applied field H. The relative permeability $\mu_r = \mu/\mu_0$, where $\mu_0 = 4\pi \times 10^{-7}\ \mathrm{H\,m^{-1}}$ is the permeability of free space; μ_r is therefore a pure number. In most discussions, including those in this book, the term 'permeability' is often used instead of 'relative permeability' but it is quite clear from the context that the dimensionless number quoted is really the relative permeability. Thus a 'permeability' of 1000 is really a permeability $\mu = 1000\mu_0 = 1000 \times 4\pi \times 10^{-7}\ \mathrm{H\,m^{-1}} = 4\pi \times 10^{-4}\ \mathrm{H\,m^{-1}}$. All the permeabilities in textual, tabular and graphical form in this book (and indeed in virtually all other publications) are presented as the relative permeabilities or as permeabilities in units of μ_0. Since $B = \mu_0(M + H)$, and H is usually small compared with M, the flux density B is due almost entirely to the magnetization of the material. Thus, for high permeabilities, we require materials with high saturation magnetizations and very low coercivities, so that large changes in the magnetization, occurring either by domain wall displacement or rotation of the magnetization vector, can be obtained with very low applied fields. Ideally $\mu = \infty$, corresponding to saturation of the material by an applied field $H \to 0$. Since the coercivity must be as small as possible, soft magnetic materials with high permeabilities have very narrow hysteresis loops. The M–H and B–H loops for a typical soft magnetic material are shown in Figs 2.1(a, b), and since the contribution to the flux density B from the magnetic field is usually very small, the loops are almost identical. It is usually more convenient to describe the properties of soft magnetic materials by their initial B versus H and μ_r versus H curves. Examples of these curves for three representative materials are shown in Fig. 2.2. The principal applications of soft magnetic materials are in alternating current machines and devices such as transformers, generators, motors, inductor cores and magnetic amplifiers, in which the hysteresis loops

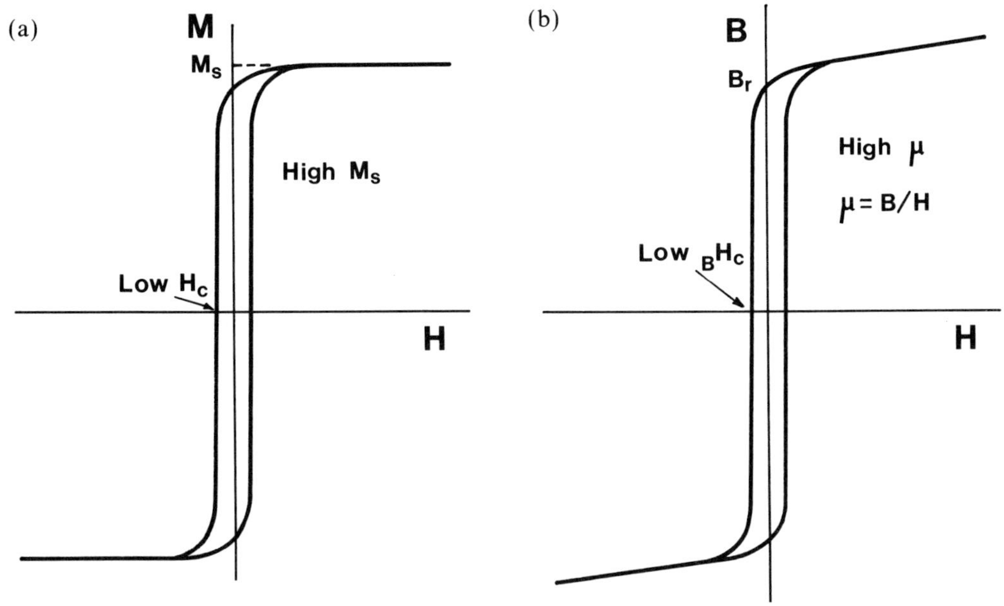

Fig. 2.1 (a) Typical *M–H* hysteresis loop for a soft magnetic material with high permeability. (b) the *B–H* hysteresis loop for the same material. Note that for soft magnetic materials, there is little difference between the *M–H* and *B–H* loops because the contribution from the field *H* to the total flux density *B* is small compared with that due to the magnetization *M*.

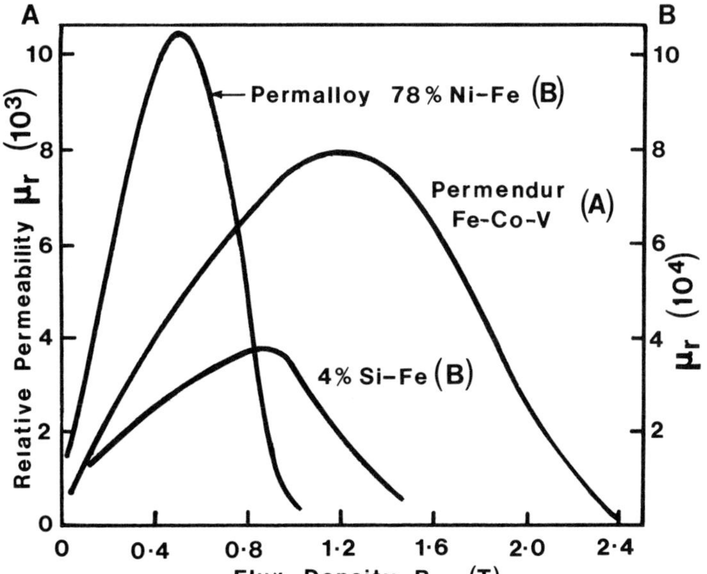

Fig. 2.2 Variation of permeability with flux density for Permendur, 48% Fe–50% Co, –2% V, 4% Si–Fe and 78% Ni–Fe Permalloy.

of the core material may be traversed many times per second, depending on the frequency used. Since the energy consumed W_h is proportional to the area enclosed by the hysteresis loop,

$$W_h = \oint \boldsymbol{H} \cdot d\boldsymbol{B},$$

the coercivity must be as low as possible. The hysteresis loss could of course be reduced by decreasing the height of the loop, i.e. by using a material with a relatively low saturation magnetization, but this is self-defeating, since the function of the material is to provide as much flux as possible with the smallest applied field and quantity of material.

Apart from the hysteresis losses, there are also energy losses due to induced eddy currents, so the electrical resistivity is also an important parameter in the selection of a high-permeability material for a particular application. The eddy current power loss P_e per unit volume of material is approximately

$$P_e = \eta \frac{2\pi^2 f^2 B_m^2 a^2}{3\rho} \quad (\mathrm{J\,s^{-1}\,m^{-3}}),$$

where a is the thickness of the material, B_m is the flux density in the material near the surface, f is the frequency, ρ is the electrical resistivity and η is the loss anomaly factor, which takes account of the additional eddy current loss due to the complex movement of domain walls during the magnetization cycle. This excess loss is discussed in more detail later in this chapter. The losses therefore increase very rapidly with increasing frequency. For exceptionally high frequencies, e.g. $f > 1$ kHz, the insulating or semi-conducting oxides known as ferrites are used. These materials are discussed in Chapter 4. For the present, we shall confine our attention to high-permeability metals and alloys.

2.2 FACTORS DETERMINING THE PERMEABILITY OF METALS AND ALLOYS

High permeabilities can be obtained only if the coercivity is as low as possible, and since this is a highly structure-sensitive magnetic parameter, it depends on both intrinsic and extrinsic properties. We can therefore divide our discussion into two sections:

(a) the effects of the fundamental magnetic properties such as saturation magnetization, anisotropy and magnetostriction;
(b) the effects of metallurgical structure at both the atomic and microstructural levels, e.g. crystallographic defects, atomic order, impurities, the presence of second phases and their dependence on heat treatment.

Both (a) and (b) depend strongly on composition, so that this must be chosen to optimize all the factors simultaneously. Since changes in soft high-permeability alloys occur by domain wall displacements, we require the domain wall energy $\gamma = 4(AK_1)^{1/2}$ to vary as little as possible with its position x in the alloy: ideally the variation of $\gamma(x)$ with x should be a horizontal line. Clearly in real alloys there will inevitably be impedance to domain wall movement as described in (a) and (b) above.

2.3 EFFECTS OF FUNDAMENTAL MAGNETIC PROPERTIES ON PERMEABILITY

2.3.1 Magnetocrystalline Anisotropy

The fundamental requirements for a high-permeability material are that the saturation magnetization and Curie temperature should be as high as possible and that the coercivity should be as low as possible. The first requirement is most likely to be satisfied by the transition metals Fe, Co and Ni, or their alloys, which also have relatively high Curie temperatures. The requirement of very low coercivity is usually satisfied by choosing materials that have the lowest possible values of magnetocrystalline anisotropy and magnetostriction so as to ensure that changes in the magnetization due to domain wall displacement or magnetization rotation occur in the lowest possible applied fields. Thus, by varying the composition, we can optimize the above magnetic parameters. Consider for example the very important Ni–Fe alloys. The dependence of the magnetocrystalline anisotropy constant K_1 and the magnetostriction constants $\lambda_{[100]}$ and $\lambda_{[111]}$ on composition are shown in Figs 2.3 and 2.4. From these figures, we can see that K_1 is zero at 70% nickel, while the magnetostriction is low over the range 78–83% Ni. Thus we should expect that alloys with compositions in the range 70–83% will have potentially very high permeabilities, and this is indeed confirmed experimentally. The alloy 78% Ni–17% Fe–5% Mo, known as 'Supermalloy', has a maximum permeability of order 10^6 and a coercivity of about 0.15 A m^{-1}, and is certainly one of the highest-permeability materials known. Unfortunately, the high nickel content reduces the saturation magnetization to 0.79 T, but for many applications this is of secondary importance to the high permeability.

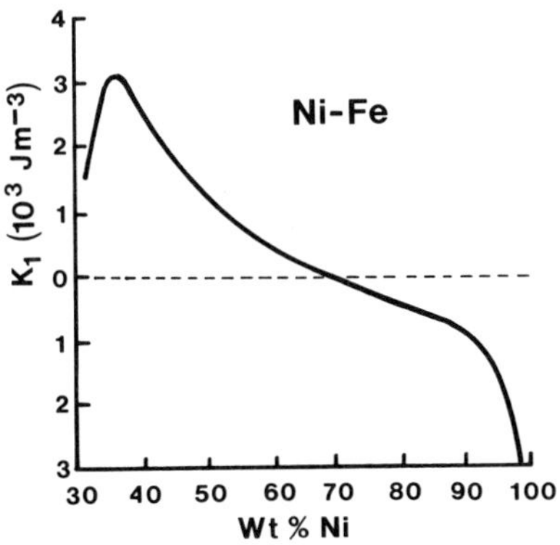

Fig. 2.3 Dependence of the magnetocrystalline anisotropy constant K_1 on composition in the Ni–Fe system. Note that K_1 is zero at about 70% Ni.

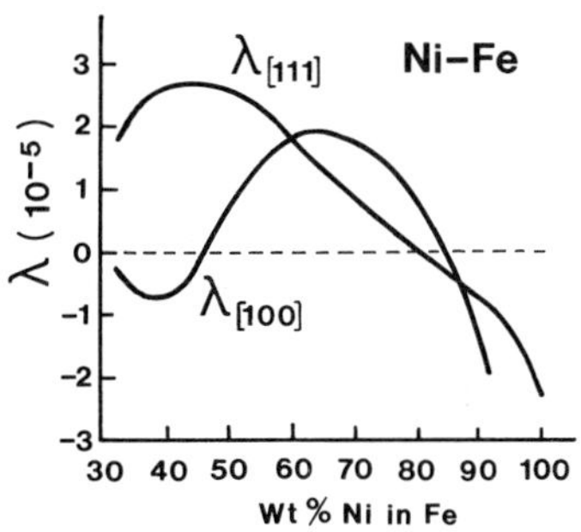

Fig. 2.4 Dependence of the magnetostriction constants $\lambda_{[100]}$ and $\lambda_{[111]}$ on composition in the Ni–Fe system.

Since nearly all high-permeability alloys have a cubic crystal structure, the magnetocrystalline anisotropy energy E can be expressed as

$$E = K_0 + K_1(\alpha_1^2\alpha_2^2 + \alpha_2^2\alpha_3^2 + \alpha_1^2\alpha_3^2) + K_2\alpha_1^2\alpha_2^2\alpha_3^2,$$

with $K_2 \ll K_1$. We can estimate the initial permeability, assuming coherent rotational processes, from the anisotropy field H_A as follows:

$$H_A = \frac{2K_1}{\mu_0 M_s} \quad (\text{A m}^{-1}),$$

so that the rotational susceptibility χ_{rot} is

$$\chi_{rot} = \frac{M_s}{H_A} = \frac{\mu_0 M_s^2}{2K_1}.$$

Hence

$$\mu_{rot} = \mu_0\left(1 + \frac{\mu_0 M_s^2}{2K_1}\right) \quad \text{or} \quad \mu_r(\text{rot}) = 1 + \frac{\mu_0 M_s^2}{2K_1}$$

For the alloy 78% Ni–22% Fe, $K_1 \approx -5 \times 10^2$ J m^{-3} (Fig. 2.3) and $M_s = 835 \times 10^3$ A m^{-1} ($J_s = 1.05$ T), so that the relative rotational permeability $\mu_r(\text{rot}) \approx 800$ and we may regard this as an estimate of the initial permeability. Since the experimental values for this alloy are at least an order of magnitude larger than this estimate, i.e. $\mu_r \approx 8000$, we may conclude that other easier mechanisms of magnetization are operative. Chikazumi (1952) has suggested that in high-permeability materials in which the magnetocrystalline anisotropy is small, incoherent rotational magnetization processes are likely to be more important than simple domain wall displacements. However, since the domain walls are very wide (of order 1 μm because of the extremely low values of the magnetocrystalline anisotropy energy), the displacement of these walls combined with incoherent rotational processes is the most probable mechanism of magnetization change.

2.3.2 Magnetostriction and Magnetoelasticity

In order to simplify our discussion of these effects, let us assume for the present that the magnetocrystalline anisotropy energy is negligibly small compared with the magnetoelastic energy, i.e. $K_1 \ll \lambda_s\sigma$. Further, in representing the magnetostriction by λ_s, we have also assumed that it is isotropic. However, this is quite realistic, since we shall usually be dealing with polycrystalline or grain-orientated materials rather than single crystals.

Rotational changes in magnetization are accompanied by changes in dimensions, i.e. there is an elongation of the material if the magnetostriction is positive or a contraction if the magnetostriction is negative. In the presence of either internal or external stresses there is a preferred orientation of the domain magnetization vectors that depends on the sign of the product $\lambda_s\sigma$. Since σ may be a tensile or compressive stress, there are four possible combinations of λ_s and σ as shown in Table 2.1.

Table 2.1 Magnetoelastic anisotropy

λ_s	σ	$\lambda_s\sigma$	Domain magnetization preference
+	+	+	Parallel to σ
−	+	−	Perpendicular to σ
+	−	−	Perpendicular to σ
−	−	+	Parallel to σ

Thus, from the table, the effect of the stress is to introduce an effective uniaxial magnetoelastic anisotropy so that the domain magnetization vectors have a preference for the axis parallel to σ when $\lambda_s\sigma$ is positive and perpendicular to this axis when $\lambda_s\sigma$ is negative. In the Ni–Fe alloys with a nickel constant $<80\%$ the magnetostriction constants $\lambda_{[100]}$ and $\lambda_{[111]}$ (Fig. 2.4) are both positive, so that we can assume that a polycrystalline alloy will be approximately isotropic. The polycrystalline saturation magnetostriction constant λ_s is

$$\lambda_s = \tfrac{2}{5}\lambda_{[100]} + \tfrac{3}{5}\lambda_{[111]}.$$

We should therefore expect that the presence of stress in Ni–Fe will in general produce a preferred magnetization direction parallel to the stress direction. The effect is illustrated in Fig. 2.5(a–c), which show the hysteresis loops for a stressed and unstressed 68% Ni–Fe alloy usually referred to as 68 Permalloy. Thus the tensile stress increases the permeability considerably, but only in the stress direction. Note that the coercivity is increased slightly, so that the applied field must be slightly greater than this if the very high permeability is to be obtained. By contrast, the permeability of pure nickel, which has a negative magnetostriction, is decreased when it is subjected to a tensile stress.

However, when there are no externally applied stresses, the existence of a significant magnetostriction (either positive or negative) will in general decrease the permeability owing to randomly orientated internal microstresses. The resulting magnetoelastic energies will not, therefore, produce an overall uniaxial anisotropy unless there is a resultant uniaxial stress. There may of course be many localized regions in which the anisotropy is uniaxial, and these will in general impede domain wall movement and magnetization rotation. We may therefore regard a polycrystalline material as consisting of a large number of randomly orientated local regions with uniaxial magnetoelastic anisotropy. During magnetization, the applied field will make angles between 0° and 180° with these local magnetization vectors in localized uniaxial regions, so that the field will have to rotate the magnetization vector against the uniaxial magnetoelastic anisotropy energy.

An estimate of the rotational permeability associated with a uniaxial magnetoelastic anisotropy energy can be obtained as follows. Suppose, for example, in the most unfavourable situation there is a stress σ perpendicular to the intended direction of magnetization. Assuming that the magnetostriction λ_s is positive, the sum E of the magnetoelastic energy and the magnetic energy is

$$E = \tfrac{3}{2}\lambda_s\sigma \sin^2\theta - \mu_0 H M_s \cos(90° - \theta)$$

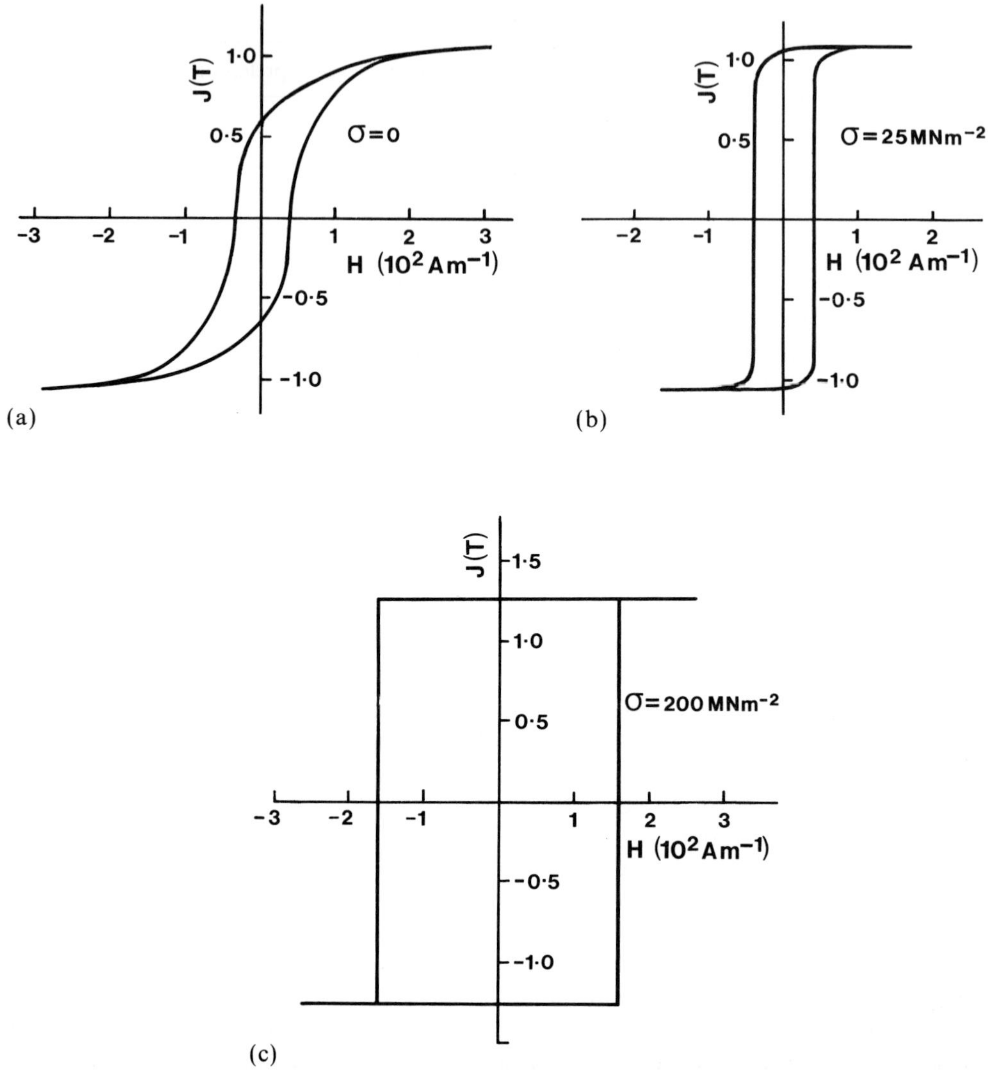

Fig. 2.5 Hysteresis loops for stressed and unstressed 68 Permalloy (68% Ni–Fe).

when the magnetization vector makes an angle θ with the stress axis and the magnetic field H is perpendicular to this as shown in Fig. 2.6. In equilibrium, this energy must be a minimum so that

$$\frac{\partial E}{\partial \theta} = 3\lambda_s \sigma \sin\theta \cos\theta - \mu_0 H M_s \cos\theta = 0;$$

$$H = \frac{3\lambda_s \sigma \sin\theta}{\mu_0 M_s}.$$

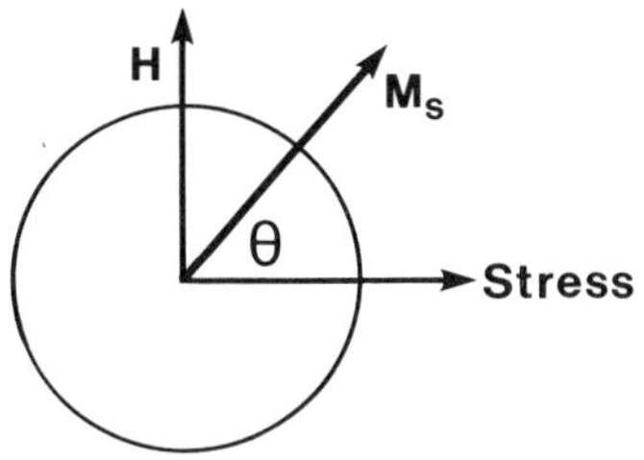

Fig. 2.6 Relationship between stress axis, applied field H, magnetization M_s and the angle θ.

Now the magnetization parallel to the field is $M_s \sin\theta$, so that the field required to saturate the material in the field direction is that for which $\sin\theta = 1$, i.e. $\theta = 90°$. This field is the anisotropy field

$$H_A = \frac{3\lambda_s\sigma}{\mu_0 M_s}.$$

The rotational susceptibility is

$$\chi_{rot} = \frac{M_s}{H_A} = \frac{\mu_0 M_s^2}{3\lambda_s\sigma},$$

and the rotational permeability is therefore

$$\mu_{rot} = \mu_0\left(1 + \frac{\mu_0 M_s^2}{3\lambda_s\sigma}\right) \quad \text{or} \quad \mu_r(\text{rot}) = 1 + \frac{\mu_0 M_s^2}{3\lambda_s\sigma}. \tag{2.1}$$

The permeability is proportional to the square of the saturation magnetization and inversely proportional to the product $|\lambda_s\sigma|$. We therefore require $|\lambda_s\sigma|$ to be as small as possible so that the magnetization vector can be rotated away from the preferred axis (whether this is parallel or perpendicular to the stress axis, see Table 2.1) by the lowest possible fields. In a polycrystalline material there will of course be a large number of randomly stressed regions. This means that the field required to cause large changes in the magnetization by purely rotational processes will be rather less than this, because the angles between the field direction and the various stress axes will be in the range 0°–90°. The above theory is of course highly simplified, but it does at least show the relative importance of the fundamental inevitable presence of a low but finite magneto-elastic anisotropy energy, which also affects the equilibrium direction of the magnetization vector and hence the permeability. Furthermore, the magnetostriction constants themselves depend on the stress, so that a rigorous determination of the preferred magnetization axes even in a single domain would be extremely difficult. The calculation of the permeability of a polycrystalline material is therefore even more difficult.

We conclude that, in order to obtain very high permeabilities, it is essential that the material has very low or zero values for K_1 and λ_s. However, in certain alloys such as 68 Permalloy, the positive magnetostriction can be used to advantage provided that the alloy is stressed in tension in the intended direction of magnetization.

2.4 MAGNETIC PROPERTIES OF HIGH-PERMEABILITY Ni–Fe ALLOYS: PERMALLOYS

Since the permeabilities for the Ni–Fe alloys are generally very much higher than those predicted by (2.1), we now consider how this arises. For the alloy 78% Ni–Fe, $M_s = 835 \times 10^3$ A m^{-1} ($J_s = 1.05$ T) and $\lambda_s = 5 \times 10^{-6}$. Since the initial relative permeability is about 8000, this implies that the internal stress $\sigma \approx 10^7$ N m^{-2}, which is approximately 10^{-5} of the yield strength of the alloy. Since stresses as low as this are unlikely to be present throughout the whole volume of the material, we may conclude that purely coherent rotation of the magnetization is unlikely to give such high initial permeabilities. Chikazumi (1952) has suggested that movement of the very wide domain walls in combination with incoherent rotational processes is the most probable mechanism of magnetization change. Both these processes will occur in fields very much less than the anisotropy fields $H_A = 2K_1/\mu_0 M_s$ or $3\lambda_s\sigma/\mu_0 M_s$. Similar mechanisms are thought to be operative in many other high-permeability materials, such as Permendur (50% Co–Fe), Perminvar (43% Ni–34% Fe–Co), and in the Ni–Zn and Mn–Zn ferrites.

2.4.1 Effects of Microstructure on Permeability

(*i*) *Heat treatment and chemical purity*

In general, when a domain wall moves through a material, its motion is impeded by metallurgical and chemical heterogeneities in the material. The impedance arises because the domain wall energy γ varies as a function of position owing to local variations in the parameters A, K_1, λ_s and the stress σ. The approximate wall coercivity H_c is given by

$$H_c \approx \frac{1}{2\mu_0 M_s}\left(\frac{\partial \gamma}{\partial x}\right)_{\max},$$

and this will be small provided that the maximum variation of the wall energy with position, $\partial\gamma/\partial x$, is small. Since the wall energy γ depends on many parameters, such as magnetocrystalline anisotropy energy, exchange energy and strain, the above equation can be written in the more general form

$$H_c = \frac{1}{2\mu_0 M_s}\left[\sum_i \left(\frac{\partial \gamma}{\partial P_i}\frac{dP_i}{dx}\right)\right]_{\max},$$

where P_i are the parameters determining the wall energy. Note that in the presence of a stress σ the domain wall energy γ is given approximately by

$$\gamma = 4[A(K_1 + \lambda_s\sigma)]^{1/2},$$

and the width

$$\delta = \pi\left(\frac{A}{K_1 + \lambda_s\sigma}\right)^{1/2}.$$

Since K_1 and λ_s are very small or nearly zero for high-permeability materials, the domain walls have very low energies and are very wide; they are therefore unlikely to interact strongly with obstacles that are small compared with the domain wall width δ. We should therefore expect that careful annealing of the alloy will tend to reduce the number and amplitude of the metallurgical and chemical inhomogeneities, and hence increase the permeability. The impedance to domain wall motion and rotation of the magnetization vectors will also be reduced if all non-magnetic inclusions are removed, since subsidiary domain structures tend to form on inclusions, thus pinning the main domain walls. Relatively large fields are then required to detach the walls to continue the magnetization process. The fields required to displace domain walls are therefore very sensitive to all kinds of structural irregularities, whereas the rotational component of the permeability depends principally on the values of K_1 and $\lambda_s\sigma$. Thus, for really high permeabilities, we require very pure and homogeneous alloys.

The effects of heat treatment on the initial and maximum permeabilities for various Ni–Fe alloys (Bozorth, 1953) are shown in Fig. 2.7. The four heat treatments used are as follows:

(1) annealing at 900–950°C for 1 h; furnace cooled at a maximum of 100°C h^{-1};
(2) annealing at 900–950°C for 1 h; furnace cooled slowly and then reheated to 600°C, which is above the ordering temperature of about 500°C, followed by cooling on a copper plate.
(3) baked: annealing at 900–950°C then maintained at 450°C for about 20 h.
(4) double treatment: (2) followed by quenching.

From Fig. 2.7, it is clear that there are two peaks in the permeability: one at about 78% Ni and the other at about 45% Ni. Since alloys near these compositions have very low values of K_1 and λ_s (Figs 2.3 and 2.4), the high permeabilities are to be expected. However, we also note that the double heat treatment followed by a quench gives very much higher permeabilities, particularly for compositions close to Ni_3Fe, where the driving force for the ordering reaction is greatest.

Note that all Ni–Fe alloys with an Ni content $>11\%$ are face-centred cubic. The appearance of an increased magnetocrystalline anisotropy in the ordered structures can be understood if we consider the ordered arrangement of Ni and Fe ions along the [100], [110] and [111] directions. Thus, if the development of long- and short-range order is to be avoided, the alloys must be quenched from above the ordering temperature rather than slowly cooled.

There is also an additional anisotropy due to ion pair ordering: as we shall see below, it is sometimes advantageous to heat-treat the alloys in a magnetic field.

(ii) Magnetic annealing and thermomagnetic anisotropy in the presence of an external field

When Ni–Fe alloys with compositions in the range 35–85% Ni are annealed below the Curie temperature in the presence of a magnetic field, they develop an additional uniaxial

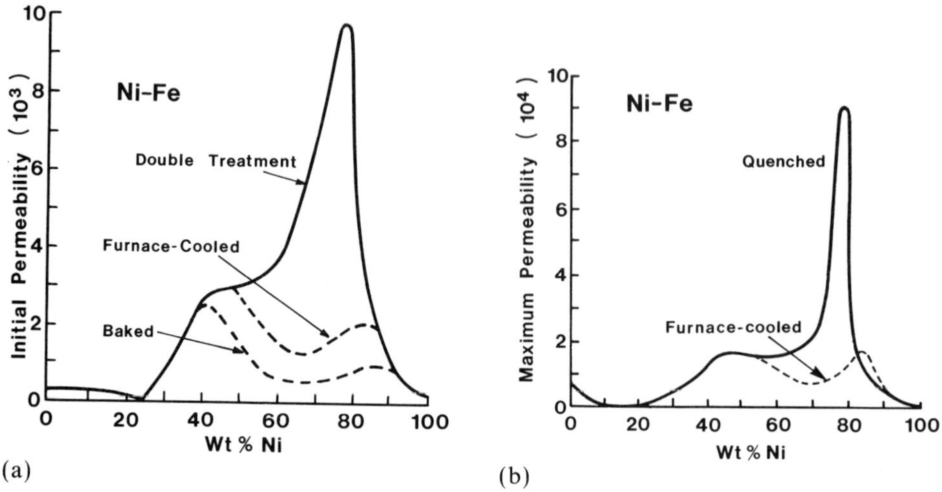

Fig. 2.7 Effects of heat treatment on (a) the initial and (b) the maximum permeability for various Ni–Fe alloys. The various heat treatments are described in the text.

magnetocrystalline anisotropy, with the easy axis of magnetization parallel to the field direction. This magnetic annealing usually improves the rectangularity of the hysteresis loops, and hence increases the permeability considerably. The hysteresis loops for a 65% Ni–Fe alloy annealed with and without an applied field are shown in Fig. 2.8(a). The dependence of the maximum permeability on the annealing temperature is shown in Fig. 2.8(b).

The origin of this thermomagnetic anisotropy has been attributed to the short-range directional ordering of ionic pairs. Since the possible ionic pairing in Ni–Fe alloys is Ni–Ni, Fe–Fe or Ni–Fe, magnetic annealing below the Curie temperature will tend to align those pairs with the lowest coupling energy in the field direction. Rapid cooling to a low temperature will then freeze this locally ordered pair structure so that the easy direction of magnetization is parallel to the axes of the favoured pairs, and therefore parallel to the applied field. Theoretical discussions of this directional ordering of ion pairs during magnetic annealing have been given by Néel (1954), Taniguchi and Yamamoto (1954), Taniguchi (1955) and Chikazumi and Oomura (1955). The basis of these theories is that the coupling energy between a pair of atoms depends on the angle between the local magnetic flux and the axis of the ion pair. The annealing temperature must be below the Curie temperature, but high enough for efficient diffusion so that preferential alignment of the pairs can occur after reasonably short periods of a few hours. The degree of alignment of the pairs will of course depend on the composition of the alloy, the effective values of the diffusion coefficients, and the time and temperature of the anneal. A high degree of directional order after thermomagnetic heat treatment is produced by the local magnetization vector, which aligns the ion pairs. The function of the applied field, which must be sufficient to saturate the material, is to ensure that the material is a single domain with its magnetization vector parallel to the external field. Since the Ni–Fe alloys can be saturated by fields of order 10 A m^{-1}, the contribution to the local internal magnetic flux is negligibly small compared with that due to magnetic polarization J_s which is about 0.8–1.6 T. When the specimen is removed

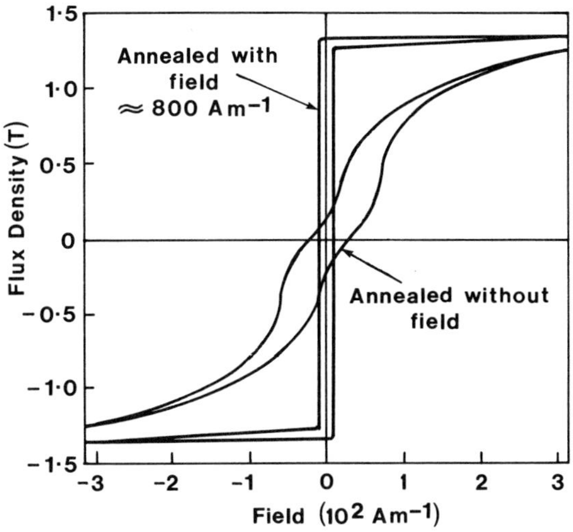

(a)

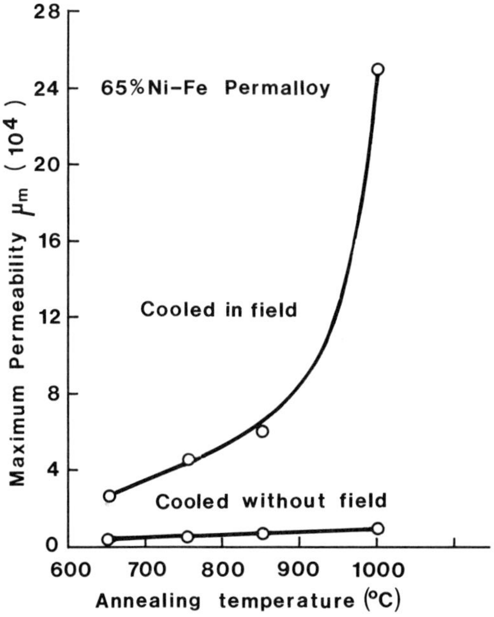

(b)

Fig. 2.8 (a) Hysteresis loops for 65% Ni–Fe Permalloy annealed with and without an applied field. (b) Effects on maximum permeability of 65% Ni–Fe Permalloy of deformation followed by annealing at various temperatures for 1 h with and without a magnetic field.

from the magnetic annealing furnace, the domain structure of the demagnetized state has a strong preference for its magnetization vectors to be parallel to the thermomagnetic field. This increases the rectangularity of the hysteresis loop and also decreases the magnetostriction; both effects give increased permeabilities.

(*iii*) *Annealing below the Curie temperature in the absence of an applied field or in a very small applied field*

Since the preferential ion pair alignment is in the direction of the domain magnetization, we find that annealing below the Curie temperature in the absence of an external magnetic field also induces a thermomagnetic anisotropy. However, since the alloy is in a demagnetized state, there is no preferred direction of magnetization, because the domain magnetization vectors have various orientations. After this annealing treatment, the hysteresis loops are often constricted or wasp-waisted. This phenomenon is known as the 'Perminvar effect', since it is most pronounced in the Perminvar (Fe–Ni–Co) alloys. The appearance of these constricted loops after slow cooling in the absence of a magnetic field is a useful effect, since it often suggests that the alloy will respond to magnetic annealing in the presence of an applied field, thus enabling a rectangular, high-permeability loop to be obtained. However, it should be emphasized that the constricted loop is observed only if the maximum applied field is small; if the material is saturated, the hysteresis loop has a 'normal' appearance, and the constricted loop can be regained only if the alloy is given another heat treatment. Alloys that show the Perminvar effect also have similar properties to the Perminvar alloys, i.e. constant permeability over a wide range of applied fields.

The following alloys have been shown to respond to magnetic annealing:

Fe + (65–85%)Ni;
Fe + (45–60%)Co;
Co + (30–85%)Ni;
Ni(20–60%) + (15–35%)Fe, balance Co.

Typical hysteresis loops for Perminvar (43% Ni–34% Fe–23% Co) annealed without and with an applied field are shown in Fig. 2.9. The uniaxial thermomagnetic anisotropy is of course superimposed on the normal magnetocrystalline anisotropy (although this is usually very low or nearly zero), so that the domain structure and behaviour is that which corresponds to a minimum in the total energy. Since the magnetocrystalline

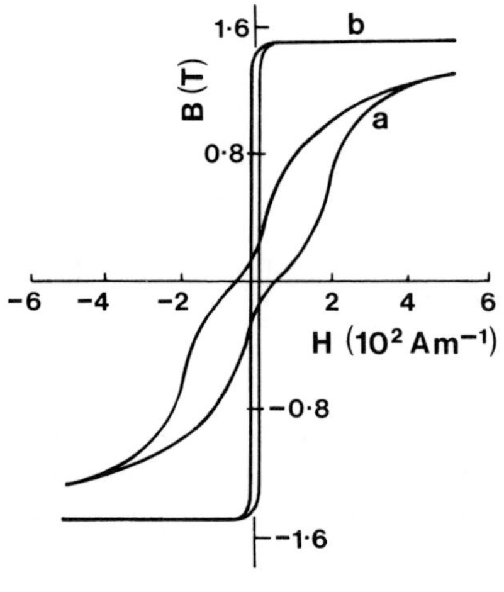

Fig. 2.9 Typical hysteresis loops for Perminvar (43% Ni–34% Fe–23% Co): (a) annealed without an applied field; (b) annealed with an applied field (after Williams and Goertz, 1952).

anisotropy energy is so low, it is often very difficult to observe clear equilibrium domain structures in the demagnetized state because the domains are usually very small and are enclosed by very wide walls (the wall width $\delta = \pi(A/K)^{1/2}$). Clearer structures can usually be obtained by applying a small non-saturating field.

(*iv*) *Slip-induced anisotropy and rolling textures*

A large uniaxial anisotropy can also be obtained in many alloys by rolling or drawing, Chikazumi *et al.* (1957); Chikazumi (1958). Since the accompanying plastic deformation usually occurs by crystallographic slip (on the $\{111\}\langle 110\rangle$ system for fcc alloys), the isotropically ordered alloy can be converted into a directionally ordered alloy through the formation of Fe–Fe and Ni–Ni pairs across the (111) slip plane. This slip-induced or roll magnetic anisotropy is an extremely complex phenomenon, since the induced anisotropy in polycrystalline material is entirely different from that in single crystals, and even in the latter it is very strongly dependent on the crystallographic planes and directions chosen for rolling. For (110) [001] rolling in Ni_3Fe the easy axis of the induced anisotropy is in the (110) rolling plane and perpendicular to the rolling direction [001] (Fig. 2.10). Since the magnetocrystalline anisotropy is so low in the unrolled crystal, no clear domain structures can be observed. However, in the rolled crystal the slip-induced anisotropy is sufficient to give very clear domain structures with well-defined magnetization vectors. The magnitude of the induced anisotropy depends on the degree of rolling, but reaches a maximum after about 75% reduction in the sheet thickness.

The roll magnetic anisotropy of polycrystalline Ni_3Fe has its average easy axis parallel to the rolling direction, although domain observations show that the easy axes differ from grain to grain. In addition to the slip-induced magnetic anisotropy, the alloys also develop a crystallographic texture due to the deformation during rolling. This can sometimes be used to great advantage, as we shall see below.

(*v*) *Recrystallization textures*

Since the Ni–Fe alloys with 75% Ni content have very low magnetocrystalline anisotropy, no improvement in their magnetic properties can be obtained by rolling to produce anisotropy and a recrystallization texture. However, Ni–Fe alloys with a Ni content of about 50% have a positive magnetocrystalline anisotropy constant K_1 and $\langle 100\rangle$ easy directions of magnetization, so that their magnetic properties can be markedly improved by the production of a rolling texture. These alloys are commercially important because the high iron content gives them a high saturation magnetic polarization of 1.6 T. After severe cold working, i.e. rolling, and an anneal, all the face-centred cubic Ni–Fe alloys develop a $\{100\}\langle 100\rangle$ primary crystallization texture. Since the easy

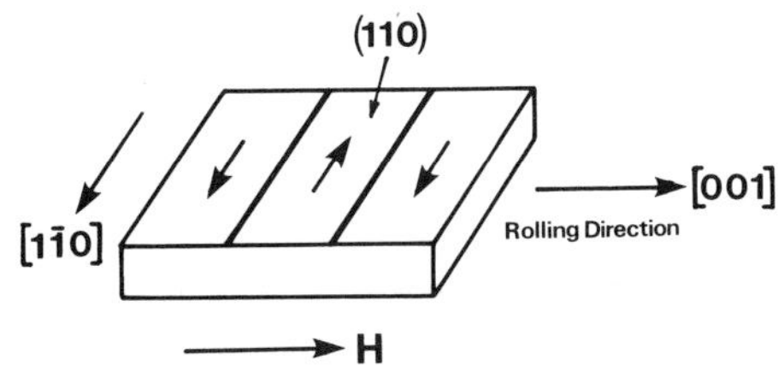

Fig. 2.10 Definition of (110)[001] rolling texture for face-centred cubic Ni_3Fe.

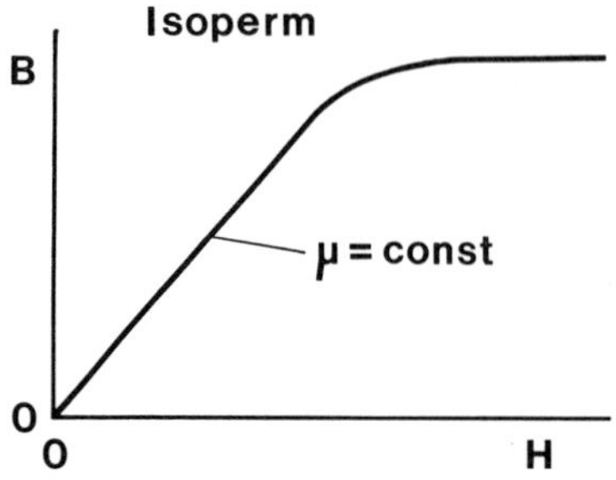

Fig. 2.11 Magnetization curve for material with constant permeability up to the saturating field. Materials with these properties are known as Isoperms.

directions of magnetization in 50% Ni–Fe are ⟨100⟩, the plane of the rolled sheet contains two of these, e.g. [100] and [010]. The 50% Ni–Fe alloy has the advantage that it combines a high permeability and a rectangular hysteresis loop with a high saturation magnetic polarization $J_s = 1.6$ T.

The essential steps in the production of the {100}⟨100⟩ texture are cold rolling (90% reduction) followed by an anneal in pure hydrogen at 900–1000°C. The resulting material is expensive, but its special properties make it suitable for a wide range of applications, particularly in devices where the material is magnetized to saturation. The high J_s means that a smaller volume of material is required for a given flux. The material is sold under various trade names, e.g. Squaremu 49, Orthonik, Deltamax and Orthonol.

If annealed 50% Ni–Fe with the {100}⟨100⟩ cube texture is given a further cold rolling to about 50% reduction, it acquires a large slip-induced anisotropy. This results in a preferred direction of magnetization, which is perpendicular to the rolling direction so that magnetization by an external field parallel to the rolling direction is relatively difficult. However, since magnetization can occur only by rotation of the domain magnetization vectors, the resulting magnetization curve is linear up to saturation, thus giving a constant permeability up to the saturation field, as shown in Fig. 2.11. The material is known as Isoperm. It has very low hysteresis losses, and is used as a high-frequency inductor core material in telecommunications.

Ordinary polycrystalline 50 wt% Ni–Fe Permalloy has a very much lower permeability than 78 or 79% Ni Permalloy, because of its high value of K_1. However, if 50 wt% Ni–Fe is subjected to prolonged annealing at 1000°C in pure hydrogen, its maximum relative permeability may be increased to about 7×10^4, a value comparable to that of 78% Permalloy. The alloy, known as Hypernik, combines a very high permeability with a very high saturation magnetic polarization. The high μ and J_s mean that the external fields and the volume of the material required are as low as possible. It is therefore particularly suitable for use in devices where the material is magnetized to saturation.

2.4.2 High-Permeability Ni–Fe Alloys with Additional Alloying Metals

As we have seen, the Ni–Fe system forms a series of very versatile high-permeability alloys, nearly all of which are sensitive to composition, heat treatment, slip-induced anisotropy and thermomagnetic treatment. The fundamental requirements for high initial and maximum permeabilities are low or zero values of K_1 and λ_s. These

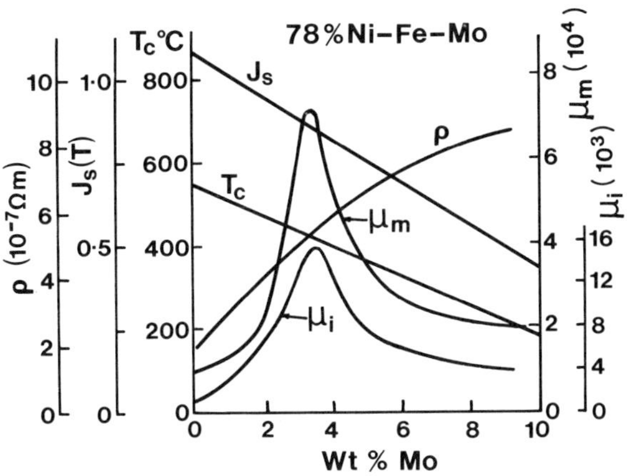

Fig. 2.12 Magnetic properties and resistivity of 78% Ni–Fe–Mo alloys as a function of Mo content (after Bradley, 1971).

requirements are usually satisfied by choosing compositions close to 78% Ni. However, it has been found that the addition of a few per cent Mo has two beneficial effects:

(1) it increases the initial permeability because the presence of Mo inhibits the development of the undesirable short- and long-range crystallographic order;
(2) the molybdenum ions act as electron-scattering centres and hence increase the electrical resistivity, thus reducing the eddy current losses.

The addition of the Mo decreases the saturation magnetization, but this is outweighed by the suppression of the ordering reaction, which simplifies the heat treatment required for the high permeability. The dependence of the magnetic properties and the resistivity of Ni–Fe alloys on the Mo content is shown in Fig. 2.12.

The addition of a few per cent Cu is also beneficial. It does not appear to affect the magnetic properties directly, but it increases the range of compositions over which high permeabilities can be obtained, and, like Mo, it also simplifies the heat treatment. The copper also improves the ductility and working properties of the alloys, so that the alloy 77 wt% Ni–16% Fe–5% Cu–2% Cr can be rolled to very thin sheets, of order 10 μm, thus minimizing eddy current losses. The improved ductility is also useful in producing magnetic shielding boxes in which the high permeability of the material preferentially 'conducts' or diverts any unwanted magnetic flux away from the central part of the box. These separate beneficial effects of adding Mo or Cu can be combined in alloys such as 78.7% Ni–15.3% Fe–1.4% Mo–2.4% Cu–1% Cr–1.2% V. The properties of high-permeability alloys based on the Ni–Fe system are given in Table 2.2. Although most of the compositions are rather complex, the values of K_1 and λ_s are all very low in spite of the additional alloying metals. General rules for achieving this rather surprising result have been given by Enoch and Fudge (1966) and Rassmann and Hofmann (1968). If C_i is the atomic percentage concentration of the added metal of atomic species i then a high-permeability alloy with composition

$$\sum_i C_i + (14.5 \pm 1.5)\%\ \text{Fe} + \text{balance Ni}$$

Table 2.2 Magnetic properties of high-permeability metals and alloys.

Material	Composition (wt%)	μ_i (μ_0)	μ_m (μ_0)	H_c (A m^{-1})	J_s (T)
Iron	Commercial 99 Fe	200	6 000	70	2.16
Iron	Pure 99.9 Fe	25 000	350 000	0.8	2.16
Silicon–iron	96 Fe–4 Si	500	7 000	40	1.95
Silicon–iron (110)[001]	97 Fe–3 Si	9 000	40 000	12	2.01
Silicon–iron {100}⟨100⟩	97 Fe–3 Si	—	100 000	6	2.01
Mild steel	Fe–0.1% C–0.1% Si –0.4% Mn	800	1 100	200	—
Hypernik	50 Fe–50 Ni	4 000	70 000	4	1.60
Deltamax {100}⟨100⟩	50 Fe–50 Ni	500	200 000	16	1.55
Isoperm {100}⟨100⟩	50 Fe–50 Ni	90	100	480	1.60
78 Permalloy	78 Ni–22 Fe	4 000	100 000	4	1.05
Supermalloy	79 Ni–16 Fe–5 Mo	100 000	1 000 000	0.15	0.79
Mumetal	77 Ni–16 Fe–5 Cu–2 Cr	20 000	100 000	4	0.75
Hyperco	64 Fe–35 Co–0.5 Cr	650	10 000	80	2.42
Permendur	50 Fe–50 Co	500	6 000	160	2.46
2V-Permendur	49 Fe–49 Co–2 V	800	4 000	160	2.45
Supermendur	49 Fe–49 Co–2 V	—	60 000	16	2.40
25 Perminvar	45 Ni–30 Fe–25 Co	400	2 000	100	1.55
7 Perminvar	70 Ni–23 Fe–7 Co	850	4 000	50	1.25
Perminvar (magnetically annealed)	43 Ni–34 Fe–23 Co	—	400 000	2.4	1.50
Alperm or Alfenol	84 Fe–16 Al	3 000	55 000	3.2	0.80
Alfer	87 Fe–13 Al	700	3 700	53	1.20
Aluminium–iron	96.5 Fe–3.5 Al	500	19 000	24	1.90
Sendust	85 Fe–10 Al–5 Al	36 000	120 000	1.6	0.89

will be obtained, provided that

$$\sum_i C_i Z_i = (19.5 \pm 1.5)\%,$$

where Z_i is the number of valence electrons in the added metals, e.g. $Z_i=1$ for Cu and $Z_i=6$ for Mo. A large number of high-permeability alloys satisfy this condition reasonably well, so that the relation appears to have some general validity. Enoch and Fudge (1966) found that for high permeability the optimum cooling rate R from 600°C depends on the non-magnetic content as follows:

$$R \approx 10^8 \exp(-0.42Z)$$

where R is in °C h^{-1}.

Thus the optimum cooling rate decreases with increasing non-magnetic content; this is of course due to the suppression of atomic ordering. The maximum permeability, of order 10^6, is observed for an alloy with composition 79% Ni–16% Fe–5% Mo known as Supermalloy.

A review of the magnetic properties of Ni–Fe alloys has been published by Chin and Wernick (1980); see also the book by Chen (1977).

2.4.3 Nickel–Iron Permalloy Powders

Nickel dust can be prepared by the carbonyl process, and if iron and nickel dusts are prepared simultaneously then various Ni–Fe alloys can be obtained after suitable annealing at about 500°C. However, various atomizing techniques are usually used to prepare Ni–Fe and Fe–Ni–Mo Permalloy dusts, which can be insulated with a variety of coatings such as silicone lacquers, Teflon and Makrolon.

Molybdenum Permalloys close to the composition 81% Ni–Fe–2% Mo are used in powdered form in high-Q filters, loading coils and resonant circuits. The initial relative permeabilities range from about 15 to 20 000, and the powder cores can be operated at frequencies up to about 300 kHz. These powder cores are superior to the ferrites because they have a high saturation magnetic polarization and hence give a better inductive stability, especially when used in high-DC-bias conditions. However, because of their relatively low resistivities, they are not suitable for use at high frequencies (>1 MHz), where the insulating ferrites are preferred.

The very important Sendust powders (Fe–Al–Si) are discussed later.

2.5 HIGH-PERMEABILITY, HIGH SATURATION Fe–Co ALLOYS

From the Fe–Co phase diagram, we know that Co is soluble in Fe up to Fe–75% Co and that the alloys are all bcc at room temperature. At compositions close to Fe–50% Co and below 730°C, they can form a simple ordered bcc superlattice. The maximum saturation magnetization at 0 K occurs at about Fe–35 wt% Co, although the maximum room temperature value occurs at about Fe–30 wt% Co. Another interesting feature of these alloys is that the magnetocrystalline anisotropy constant K_1 passes through zero at about Fe–45% Co, so we might expect an alloy of this composition to have a very high permeability. However, it is found that both the initial and maximum permeabilities have their maximum values at about Fe–50% Co. This alloy has considerable commercial importance, and is known as Permendur. The magnetization curve and the variation of the permeability as a function of flux density is shown in Fig. 2.2.

The permeabilities of Fe–Co alloys show considerable variation with composition and heat treatment, as shown in Fig. 2.13. The rapid increase in the permeability at Fe–45 wt% Co can be associated with the very low magnetocrystalline anisotropy K_1. Fe–Co alloys also have the advantage of a very high saturation magnetic polarization (Chen, 1961), which is very useful in machines and devices involving attractive forces. Unfortunately, the use of Fe–Co alloys is limited by difficulties in mechanical working, but their workability can be improved by the addition of the small quantities of vanadium – usually about 2–4%. Major *et al.* (1989) showed that additions of Ta or Nb enabled the Fe–Co alloy to be cold-worked and achieved a higher saturation magnetic polarization than was possible with 2–4% V.

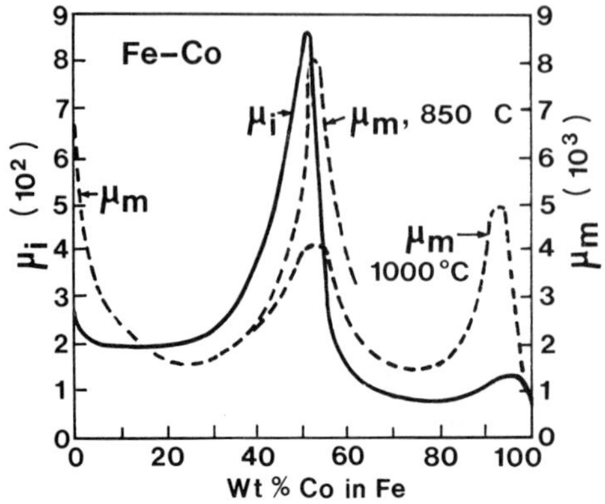

Fig. 2.13 Initial and maximum permeabilities of Fe–Co alloys as functions of composition and heat-treatment temperature.

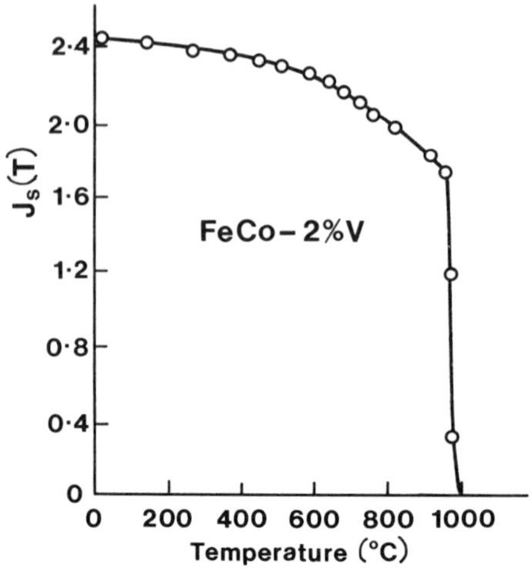

Fig. 2.14 Temperature variation of saturation magnetic polarization of Supermendur (49 wt% Fe–49% Co–2% V) (after Chen, 1961).

The above alloys are sensitive to magnetic annealing, after which treatment the alloy 49% Fe–49% Co–2% V is known as Supermendur. The magnetic anneal increases the relative permeability from about 5000 to 60 000. Supermendur also has a very high Curie temperature and a relatively slow decrease in the saturaton magnetic polarization with increasing temperature (Fig. 2.14). Iron–cobalt and iron–cobalt–vanadium alloys are used for transductor cores, transformers, and the diaphragms in telephone earpieces.

2.6 HIGH-PERMEABILITY Ni–Fe–Co ALLOYS

The magnetic properties of Ni–Fe–Co alloys have been investigated by Williams and Goertz (1952). The most important member of this system is the alloy 45 wt% Ni–30% Fe–25% Co, known as Perminvar, which shows great sensitivity to magnetic annealing and is characterized by a high permeability that is constant over a wide range of applied field values. In order to obtain these properties, the alloy must be annealed for about 24 h at a temperature in the range 400–450°C. According to Taniguchi (1955), the domain walls are stabilized by the usual directional pair ordering process; the field that produces the uniaxial anisotropy is of course the domain magnetization vector itself. In low applied fields the permeability is constant, but once the field is sufficient to remove the walls from their pinning sites, the permeability becomes a function of the applied field.

If the alloys are annealed below the Curie temperature in the presence of a saturating external field, the resulting hysteresis loops have a high degree of rectangularity. The high cost of the alloys limits their application, but it has been found that the alloy 70 wt% Ni–23% Fe–7% Co has similar properties and is considerably cheaper since it has a lower cobalt content.

2.7 HIGH-PERMEABILITY IRON AND FERRITIC IRON

Although the magnetocrystalline anisotropy and magnetostriction of iron are relatively large, permeabilities of about 2×10^5 have been obtained for very carefully purified (impurities <0.01%) and annealed single crystals. The coercivities associated with these very high permeabilities are generally of order $1\ \mathrm{A\ m^{-1}}$ or less. However, these are exceptional values and are due to the complete absence of domain wall pinning inclusions such as iron carbides and nitrides. The main impurities in commercially 'pure' iron are C, Mn, Si, P, S, N and O, but C, O, S and N have by far the greatest effects on the magnetic properties. These impurities may be present as ions in interstitial sites or as non-ferromagnetic (or less-ferromagnetic) inclusions in the form of carbides, nitrides, sulfides and phosphides, which impede domain wall motion and therefore decrease the permeability. Commercial irons may have permeabilities in the range 10 000–20 000, but these can be increased to about 100 000 by annealing in vacuo or in hydrogen, a process that removes most of the C, S, N and O. Permeabilities as high as 1 000 000 can be obtained by prolonged annealing in hydrogen at 1200–1500°C. Unfortunately, this process is very expensive.

The importance of purity is shown by the initial flux density curves of the two iron specimens that are compared in Fig. 2.15. Cioffi (1934) has shown that when iron is exceptionally pure (impurities <0.001%), permeabilities as high as 280 000 can be obtained after very careful heat treatment. A maximum permeability of 1 040 000 was recorded (Cioffi *et al.*, 1937) for a single crystal of iron in the form of a picture frame with the edges parallel to the easy ⟨100⟩ directions. Williams (1937) later improved on this by obtaining a maximum permeability of 1 380 000 in the [100] direction of a single crystal of 3.8% Si–Fe.

However, well-annealed iron or rather low-carbon steels (0.05–0.15% C with other impurities) may have maximum permeabilities in the range 1000–5000 and core losses of about $12\ \mathrm{W\ kg^{-1}}$ at 1.5 T. Since low-carbon steels have high saturation magnetic polarizations (or about 2 T), these permeabilities are quite sufficient for many applications, particularly in small electric motors, which are manufactured in very large quantities. Although the core loss of $12\ \mathrm{W\ kg^{-1}}$ is about 10 times that for grain-oriented

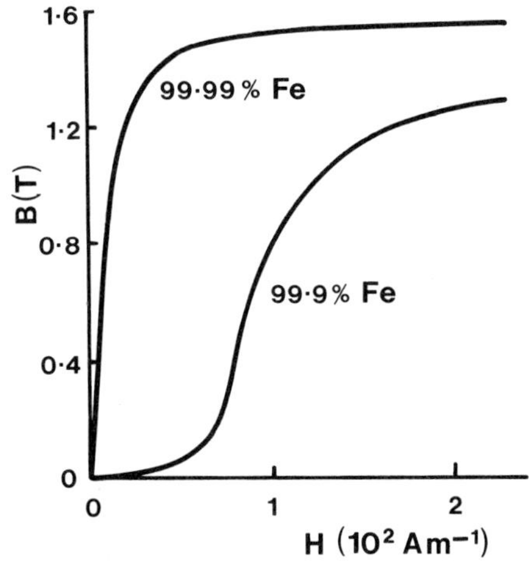

Fig. 2.15 Initial flux density curves for two iron specimens: (a) 99.9% pure Fe; (b) 99.99% pure Fe.

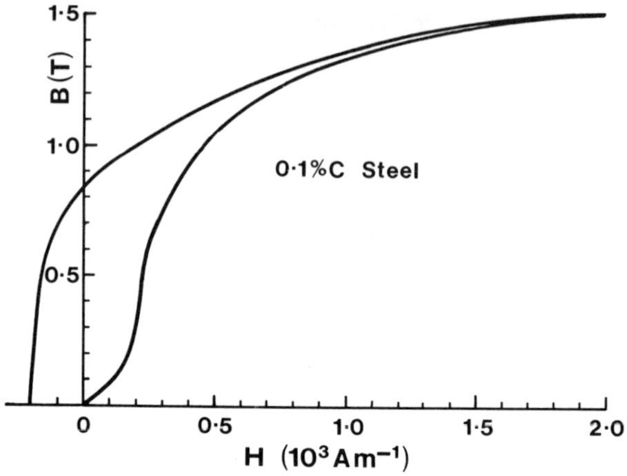

Fig. 2.16 Initial and demagnetization curves for a typical motor steel containing 0.1% C.

Si–Fe, this is a relatively unimportant consideration because the cost of manufacture is more important than the efficiency of the motor. The initial and demagnetization curves for a typical motor steel containing 0.1% C are shown in Fig. 2.16. Low-carbon steel is used in the frames of rotating machines, pole pieces, pole supports in motors, and in many other machines and devices. Higher permeabilities can be obtained by reducing the carbon, phosphorous and sulfur content, but this increases the cost of the material very significantly. The permeability of iron decreases very markedly with increasing carbon content. Iron containing 0.3–0.8% P is also a very useful high-permeability alloy with a high saturation magnetic polarization.

The main disadvantages of iron as a soft magnetic material are that it has a relatively low electrical resistivity and mechanical strength. The magnetic properties of various grades of iron are given in Table 2.3. The first three grades of iron are used in telephone relays and switches, contactors, electromagnets and many other DC machines and

Table 2.3 Magnetic properties of various grades of iron.

Material	H_c (A m^{-1})	μ_i (μ_0)	μ_{max} (μ_0)
Ingot (99.8% Fe)	112	10	1 000
Armco	80	200	7 000
Commercially pure	20–100	200–500	3 500–20 000
Carbonyl iron powder	6	3 000	20 000
Vacuum-melted	25	—	21 000
Electrolytic	7	1 000	26 000
Electrolytic annealed	18	—	41 500
Vacuum-smelted and hydrogen-annealed	3	—	88 400
Purified Armco (99.95% Fe)	4	10 000	227 000
Vacuum-annealed	—	14 000	280 000
Single-crystal	—	—	680 000
Single-crystal, magnetically annealed	12	—	1 430 000

Saturation magnetic polarization $J_s = 2.15$ T at 20°C, except for ingot with $J_s = 2.05$ T.

devices. For further details the reader is referred to the books by Tebble and Craik (1969), Bradley (1971), Heck (1974), Chen (1977) and Boll (1979).

2.7.1 Iron Powder

Iron in powder form with particle sizes in the range 10–100 μm can be compacted to form core materials with relatively low hysteresis and eddy current losses. The cores are manufactured by compressing the powder together with an insulating material, a binder and a lubricant. They can be used at frequencies up to about 100 kHz. Carbonyl iron is often used for this purpose. It is so-called because it is prepared by vaporizing iron pentacarbonyl, which is liquid at room temperature:

$$Fe(CO)_5 \rightarrow Fe + 5CO$$

The reaction is rarely complete, and is accompanied by the reaction

$$2CO \rightarrow CO_2 + C$$

so that the resulting powders may contain up to about 0.8% C. The carbon content can be decreased by heat treatment in a reducing atmosphere to produce very high purity iron powder.

Electrolytically produced iron, which also has very high purity, can also be milled to produce suitable fine powders.

The initial permeabilities of the cores are very sensitive to the amount of carbon (or nitrogen) remaining in the iron, its state of dispersion within the particles, their insulating coating, the packing fraction of the particles within the binder and the moulding pressure. Although the particles are multi-domain, the permeabilities of the finished cores are relatively low, typically 10–100.

Carbonyl iron can be sintered and worked into sheet or strip. The coercivity of the strip (about 8 A m^{-1}) can be further reduced if it is stretched to a critical degree and recrystallized by annealing.

2.7.2 Ferritic Stainless Steels, Fe–Cr Alloys

The ferritic (bcc) stainless steels that are ferromagnetic usually contain 12–18 wt% Cr and 0.1–0.2 wt% C. Although the alloys have relatively low permeabilities, their saturation magnetic polarizations are relatively high. Typical magnetic properties are summarized in Table 2.4.

Table 2.4 Magnetic properties of ferritic stainless steels (Chen, 1977).

Cr (%)	C (%)	Other	Condition	μ_{max} (μ_0)	H_c (A m^{-1})	B_r (T)	B_{max} (T)
12	0.15	—	Annealed	670	610	1.06	1.62
12	0.15	—	Hardened	70	4760	0.65	1.25
17	0.12	—	Annealed	960	488	1.15	1.62
17	0.12	—	Hardened	278	784	0.51	1.46
16	0.20	2 Ni	Hardened	—	1840	0.77	1.53

The ferritic stainless steels are used principally for magnetic cores, solenoid switches, pole pieces and relays where their high corrosion resistance is required.

2.8 NON-ORIENTED SILICON–IRON ALLOYS

Were it not for the very high cost of producing high-purity iron, its high saturation magnetic polarization and high permeability would make it one of the best materials for application in electric motors, generators, transformers, electromagnets, relays and other electromagnetic devices. However, iron has largely been superseded by silicon–iron alloys in both the polycrystalline and oriented forms for nearly all of the above applications, although it is still used in less demanding applications.

In all applications low hysteresis and eddy current losses are a prime requirement, but these have to be achieved at minimum cost. The adition of silicon (1–4% Si) reduces the magnetocrystalline anisotropy and the coercivity, and hence increases the permeability with the added advantage of increasing the resistivity and hence reducing the eddy current losses. Even better properties can be obtained if the silicon content is increased to about 6%, but this decreases the ductility of the alloys, making it difficult to roll ingots into the thin sheets or laminations that are required in most applications. Unfortunately, the addition of silicon reduces the saturation magnetic polarization, but this is by far offset by the overall improvement in the magnetic properties.

The alloys are decarbonized in moist hydrogen at about 820°C in order to optimize the magnetic properties. This decreases the carbon content to less than about 0.001% by weight, which is below the threshold where a deterioration in the magnetic properties can occur during the lifetime of a transformer due to the precipitation of iron carbides. The latter increase the coercivity by acting as domain wall pinning sites or impedances to easy domain wall motion, which reduces the permeability. Transformer core ageing can be virtually eliminated by the addition of aluminium or titanium, which enables any excess carbon or nitrogen to form stable harmless compounds (Littman, 1971).

Better melting and casting techniques are resulting in much cleaner steels with very much lower levels of carbon, nitrogen and sulfur, thus reducing still further the energy losses due to hysteresis. Even in the best grades of non-oriented silicon steels, hysteresis losses may account for as much as 60–70% of the total losses. These losses can be minimized by controlling the grain size; the optimum grain size in 3.2% Si–Fe is about 0.15 mm. Grain sizes larger than this also result in increased eddy current losses, but these can be reduced by the addition of aluminium or manganese.

Nearly all non-oriented silicon–iron alloys are coated with a thin insulating layer, which should be of uniform thickness and have good surface adherence. The coating may be just a thin oxide film or a complex organic or inorganic compound. Modern coatings are designed to react with the steel surface, which reduces hysteresis losses and increases the permeability. Further improvements in coatings are likely to be made when a better understanding of the complex interactions between the coating and the steel, together with their effects on domain size and domain wall dynamics, is obtained.

The non-oriented silicon steels are available in standard thicknesses of 0.35 mm, 0.5 mm and 0.65 mm in widths up to 1070 mm. Total losses for a high quality steel of thickness 0.35 mm at 1.5 T and 50 Hz amount to about 2.5 W kg^{-1}, whereas the losses in steels for cheap domestic motors are about 10–20 W kg^{-1}.

Table 2.5 Magnetic properties of Si–Fe alloys (after Adams, 1962).

Si (%)	μ_{20} (μ_0)[c]	μ_{max} (A m^{-1})	H_c (A m^{-1})	B (T)[b]	Core loss at 60 Hz (W kg^{-1} at 1.0 Tesla)
0.94	860	14 780	35	1.64	1.5
2.81	1720	17 440	15	1.53	1.3
4.80	1410	17 160	16	1.52	1.0
6.40	1720	28 860	12	1.40	0.9
6.40[a]	1390	66 940	16	1.46	0.8

[a] Magnetic anneal in 136 A m^{-1}.
[b] Measured in a field of 2.4 kA m^{-1}.
[c] Measured in a field of 1.6 KA m^{-1}.

Apart from the removal or careful control of impurities, the final cold rolling process determines the losses and the permeability, although the first hot rolling must be carefully controlled to ensure the production of the required randomly oriented grain texture. The magnetic properties of various silicon–iron alloys are shown in Table 2.5.

Although the coercivities of the alloys are low, the random orientation of the grains in the polycrystalline sheet means that fields much greater than the coercivity must be applied in order to produce saturation, because when all the domain wall motion is completed, the domain magnetization vectors must ultimately be rotated away from the easy directions of magnetization, i.e. ⟨100⟩. This is a considerable disadvantage, since in most applications we require the material to saturate in the lowest possible applied fields in order to provide as much magnetic flux as possible for the smallest volume of material.

As we shall see, a great improvement in the magnetic properties can be obtained if the alloys are grain-oriented.

2.9 MICROCRYSTALLINE SILICON–IRON ALLOYS

These alloys are produced by very rapid solidification similar to that used for amorphous alloys (see Chapter 3). No glass-forming elements are included, so that a ductile microcrystalline strip is produced, and, by optimizing the grain size and strip thickness, good magnetic properties can be obtained. A 6% Si–Fe alloy with thicknesses between 0.06 and 1 mm has been shown by Narita and Yamashiro (1980) to be ductile and to have lower losses than conventional grain-oriented 3.2% Si–Fe. A 0.08 mm thick ribbon containing 4.5% Si had almost the same loss as an amorphous alloy, while at 1.7 T and 50 Hz its loss was only 0.56 W kg^{-1}, which is as good as the best laboratory-processed grain-oriented alloy. A 6% Si–1% B–Fe alloy also has very useful soft magnetic properties.

2.10 GRAIN-ORIENTED SILICON–IRON WITH THE (110)[001] TEXTURE

Grain-oriented 3.25% Si–Fe with the (110)[001] texture is produced in very large quantities, and is among the most important of all magnetic materials since it is used as the core material in electrical power transformers. In view of this, its magnetic and metallurgical properties will be considered in detail.

2.10.1 Requirements for Transformer Core Materials

A transformer consists essentially of two electrically conducting coils of wire, known as the primary and secondary, wound onto a closed ring of high-permeability material that is electrically insulated from the coils. The function of the core is to transport the maximum magnetic flux from the primary coil to the secondary coil with the minimum alternating magnetizing current in the primary. The fundamental requirements for an efficient transformer core material are that it should have a high saturation magnetic polarization and high magnetic permeability, preferably in combination with a high electrical resistivity so as to minimize eddy current losses.

(*i*) *Hysteresis losses*

When a material is magnetized through a complete hysteresis loop, the energy consumed in one cycle, W_h, is equal to the area enclosed by the loop, i.e.

$$W_h = \oint \boldsymbol{H} \cdot \mathrm{d}\boldsymbol{B} \quad (\mathrm{J\,m^{-3}}),$$

$$\boldsymbol{B} = \mu_0(\boldsymbol{M} + \boldsymbol{H}), \qquad \mathrm{d}\boldsymbol{B} = \mu_0\,\mathrm{d}\boldsymbol{M} + \mu_0\,\mathrm{d}\boldsymbol{H},$$

$$W_h = \oint \mu_0 \boldsymbol{H} \cdot \mathrm{d}\boldsymbol{M} + \oint \mu_0 \boldsymbol{H} \cdot \mathrm{d}\boldsymbol{H} \quad (\mathrm{J\,m^{-3}}).$$

Since a graph of H versus H is a straight line, it cannot enclose an area and is not therefore hysteretic, so that

$$\oint \mu_0 \boldsymbol{H} \cdot \mathrm{d}\boldsymbol{H} = 0,$$

$$W_h = \oint \boldsymbol{H} \cdot \mathrm{d}\boldsymbol{B} = \oint \mu_0 \boldsymbol{H} \cdot \mathrm{d}\boldsymbol{M}.$$

Thus the area enclosed by the B–H loop is the same as that for the M–H loop. Since this loss is usually measured in DC or slowly varying fields, it is often termed the static hysteresis loss, although this term has limited validity if we consider the effects of cyclical magnetization at a frequency f, when the shape and area of the loop depends in a complex way on the frequency and waveform. However, the shape and area of the static hysteresis loop is a very good guide as to the suitability of the material for a transformer core. Hysteresis loops that enclose the minimum area are of course the most suitable, but they must have a large saturation magnetic polarization so that they can supply the maximum magnetic flux for the minimum volume of material. If these two requirements are to be satisfied simultaneously, the coercivity must be as low as possible.

Thus we require a material with a high permeability, a high saturation magnetization, a low coercivity, and hence a hysteresis loop that encloses the smallest possible area.

(*ii*) *Eddy current losses*

Since the core material is a conductor of electricity and is subjected to an alternating field, the associated flux changes in the material induce eddy currents. These eddy currents are driven through the finite resistance of the core material, and therefore constitute a net loss of energy, which appears as heat (I^2R, in J s^{-1}) in the core.

Consider a section of the core material with the dimensions shown in Fig. 2.17(a). Let the flux density B in the material at time t be given by

$$B = B_m \sin \omega t,$$

where ω is the angular frequency and B_m is the maximum flux density – which is not necessarily the saturation flux density $B_s = J_s + \mu_0 H$. Since the flux changes are perpendicular to the plane of the cross-section in Fig. 2.17(a), the eddy current paths are shown by the arrowed lines. One such path is shown on a larger scale in Fig. 2.17(b). For simplicity, we neglect the curved ends of the eddy current loops and assume that the current is flowing in two thin flat strips of width δx and depth equal to unity. If the resistivity of the Si–Fe core is ρ, the resistance of the circuit is approximately

$$R = \frac{2l}{1\,\delta x}\rho = \frac{2\rho l}{\delta x} \quad (\Omega).$$

The total magnetic flux enclosed by the eddy current loop is

$$\varphi = 2xlB_m \sin \omega t,$$

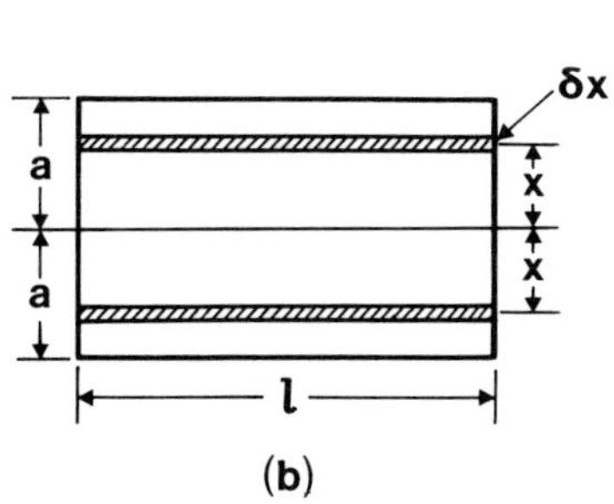

Fig. 2.17 (a) Cross-section of a transformer core material, showing directions of eddy currents. (b) Simplified eddy current paths (shaded areas) in a transformer core material.

since the enclosed area is $2xl$ and the flux density at time t is $B_m \sin \omega t$. From the Faraday–Lenz law of electromagnetic induction, the induced emf due to the changing flux is

$$V = -\frac{d\varphi}{dt} = -2xl\omega B_m \cos \omega t.$$

The Ohmic power loss is therefore

$$I^2R = \frac{V^2}{R} = \frac{2x^2 l\omega^2 \, \delta x \, B_m^2 \cos^2 \omega t}{\rho}.$$

The mean power loss

$$\delta P = l\omega^2 B_m^2 x^2 \, \delta x / \rho,$$

since the mean value of $\cos^2 \omega t$ is $\frac{1}{2}$. Now δP is the power loss for one elemental eddy current loop, so that for the whole block the power loss is

$$P = \int_0^a \frac{l\omega^2 B_m^2 x^2}{\rho} \, dx = \frac{l\omega^2 B_m^2 a^3}{3\rho}.$$

Since the volume of the block is $2al \times 1 = 2al$, the power loss per unit volume is $P/2al = P_e$:

$$\frac{P}{2al} = P_e = \frac{\omega^2 B_m^2 a^2}{6\rho} = \frac{2\pi^2 f^2 B_m^2 a^2}{3\rho}. \tag{2.2}$$

This result must be regarded as an approximation, since we have made a number of assumptions, namely

(1) that the flux density is not affected by eddy currents – this will be true only if the block is very thin and the frequency f very low;
(2) that the permeability of the core material is constant throughout the cycle and uniform over the whole volume of the material.

Since we know that Si–Fe contains domains, the material cannot be completely magnetically homogeneous, so that the response to the field at particular points in the material will depend on the local domain structure and orientation. Thus the eddy current losses depend to some extent on the permeability and hysteresis properties of the material. Similarly the hysteretic behaviour depends on changes in magnetization produced by the eddy currents. The complexity of the behaviour can be appreciated when we remember that domains and hence domain walls are present at every point on the hysteresis loop, except of course at the extreme ends of the loop, where the material is saturated. Thus it is not possible to separate hysteresis and eddy current losses. The net result of this complex behaviour is that the energy losses are very much greater (by a factor of about 1.5–2.0) than the sum of those due to pure hysteresis and eddy current effects. However, the discussions above do at least provide a valuable guide to the most important parameters that must be considered, particularly at normal power frequencies. The principal conclusions from (2.2) are that the eddy current losses are

(1) proportional to the square of the frequency, f^2;
(2) proportional to the square of the block thickness, a^2;
(3) inversely proportional to the resistivity, ρ^{-1}.

Since the frequency f will be fixed by the power frequency, the eddy current losses will be reduced if the resistivity is as high as possible and the core is laminated such that the sheets are as thin as is convenient. A further reduction can be obtained if the sheets are electrically insulated from each other, since this will eliminate eddy currents perpendicular to the laminations. If a block of thickness $2a$ is divided into N laminations of thickness $2a/N$, the approximate eddy current loss is

$$P = l\omega^2 B_{\mathrm{m}}^2 \left(\frac{a}{N}\right)^3 \frac{N}{3\rho}$$

$$= \frac{4\pi^2 f^2 l B_{\mathrm{m}}^2 a^3}{3N^2\rho}.$$

That is,

$$P = \frac{2\pi^2 f^2 B_{\mathrm{m}}^2 a^2}{3N^2\rho} \quad \text{per unit volume.}$$

Thus lamination of the core reduces the eddy current losses by a factor $1/N^2$. It should be emphasized that the eddy currents themselves set up magnetic fields that oppose the applied field (Lenz's law) and thus cause the flux density to decrease progressively from the surface of the lamination to the centre (O'Kelly, 1975). The degree of flux penetration is a rather complex function of the sheet thickness and the electromagnetic properties of the core material. The variation of the flux with position for various sheet thicknesses is shown in Fig. 2.18, from which we can see that if the sheet is too thick, a large volume of the material in the centre is not being used efficiently for the transportation of magnetic flux. We must therefore choose the thickness so as to maximize the efficient use of the material subject to ease of rolling, mechanical handling and stability. The energy losses could of course be reduced still further by decreasing the maximum flux density B_{m}, but this would be self-defeating, since the function of the material is to provide as much flux as possible for the smallest possible applied field produced by the primary coil.

The requirements for an efficient transformer core material (alloy) are as follows:

(1) high saturation magnetic polarization J_{s};
(2) high permeability μ;
(3) low magnetocrystalline anisotropy energy K_1;
(4) high electrical resistivity ρ, to minimize eddy currents;

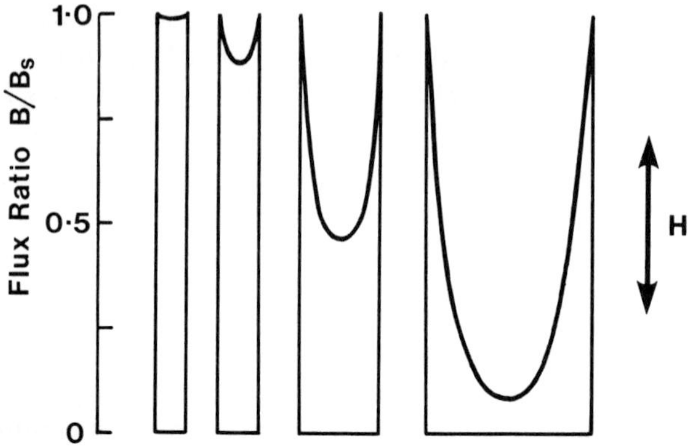

Fig. 2.18 Flux penetration in transformer core materials as a function of sheet thickness. The flux penetration is plotted as a fraction of the maximum value at the surface of the sheets.

(5) low hysteresis loss, i.e. narrow hysteresis loop with very low coercivity H_c;
(6) low magnetostriction coefficient λ_s;
(7) single-phase material, to ensure magnetic homogeneity;
(8) high degree of purity, in particular no non-ferromagnetic non-metallic elements such as C, S, P or N;
(9) ductility, for cold rolling at room temperature;
(10) rolling texture such that grains contain at least one easy direction of magnetization in a common preferred direction, preferably the rolling direction;
(11) relatively large grain size, so as to minimize grain boundary area, although the grains must not be too large: 1–5 mm is satisfactory for Si–Fe;
(12) stable properties over a reasonable temperature range;
(13) ability to retain an insulating coating;
(14) reasonably high thermal conductivity;
(15) low susceptibility to magnetic and metallurgical ageing during a long operational lifetime.

We shall now show in detail how these requirements are satisfied by (110)[001] grain-oriented Si–Fe alloys with an Si content in the range 3–4%, which can be produced economically in large quantities for use in power transformers.

2.10.2 Grain-Oriented Si–Fe with the (110)[001] Texture: Saturation Magnetization, Magnetocrystalline Anisotropy Energy and Resistivity

The addition of a few weight per cent silicon to iron has two beneficial effects:

(1) it decreases the magnetocrystalline anisotropy energy K_1 linearly (Fig. 2.19);
(2) it increases the electrical resistivity linearly (Fig. 2.20).

There is unfortunately a decrease in the saturation magnetic polarization J_s, but this is far outweighed by the improvement in properties described by (1) and (2). The lower

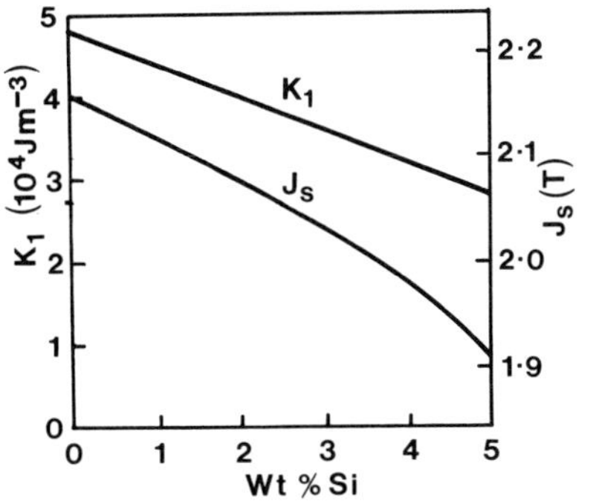

Fig. 2.19 Dependence of the magneto-crystalline anisotropy K_1 and the saturation magnetic polarization J_s on silicon content.

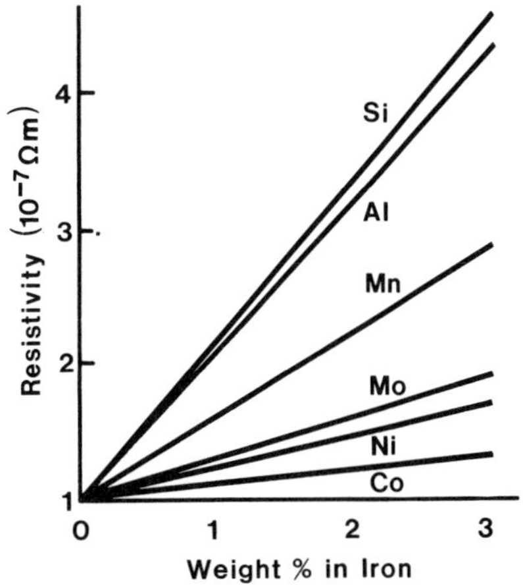

Fig. 2.20 Dependence of the electrical resistivity of iron on the silicon, aluminium, manganese, molybdenum, nickel and cobalt content.

value of K_1 decreases the domain wall energy, increases the domain wall width and increases the rotational permeability. The coercivities of the Si–Fe alloys are therefore considerably reduced, thus decreasing the hysteresis loop area and hence the associated losses.

2.10.3 Phase Structure and Effects of Carbon

The iron-rich side of the Si–Fe phase diagram (Crangle, 1954) is shown in Fig. 2.21(a). The $\alpha+\gamma$ and γ loops extend to about 2.25% Si. Since we require the material to be single-phase up to the melting point, we must choose an alloy with a silicon content greater than 3%. Alloys near this composition are bcc and can be recrystallized at high temperatures without phase changes occurring on cooling. This is a very important property, because the presence of the non-ferromagnetic fcc γ phase acts as a barrier to domain wall motion, and therefore increases the coercivity and decreases domain wall mobility. Another important feature of the Si–Fe system is that the presence of carbon as an impurity (as little as 0.07 wt% C) shifts the nose of the $\alpha+\gamma$ loop to about 6 wt% Si (Fig. 2.21b). Alloys containing about 6 wt% Si are brittle at room temperature, so it is important to reduce the carbon content to below 0.01 wt%. At this composition the $\alpha+\gamma$ loop remains unchanged. There are two other important reasons for minimizing the carbon content. First, the reduction in the saturation magnetic polarization for each added per cent carbon is about ten times greater than that for each added per cent silicon. Secondly, the carbon may be precipitated as iron carbides, which are either non-ferromagnetic or, as in the case of orthorhombic cementite Fe_3C, have ferromagnetic properties different from the matrix. The precipitates will therefore act as domain wall pinning sites and hence impede domain wall motion, resulting in an increased coercivity and hysteresis loss. In general, the carbon content of the non-oriented Si–Fe is about 0.03%, whereas in grain-oriented Si–Fe it must be about 0.01%, and would be reduced

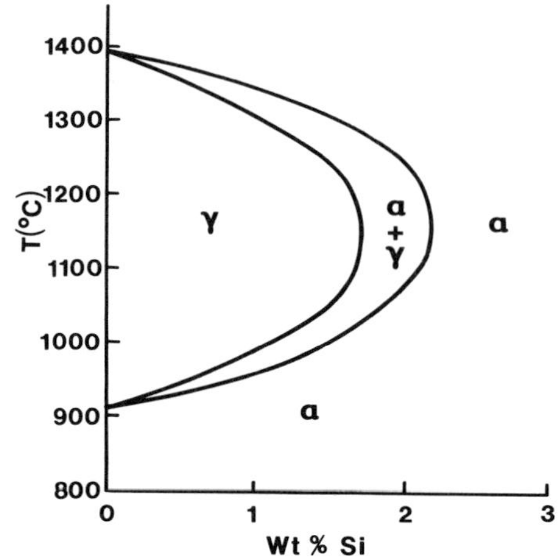

(a)

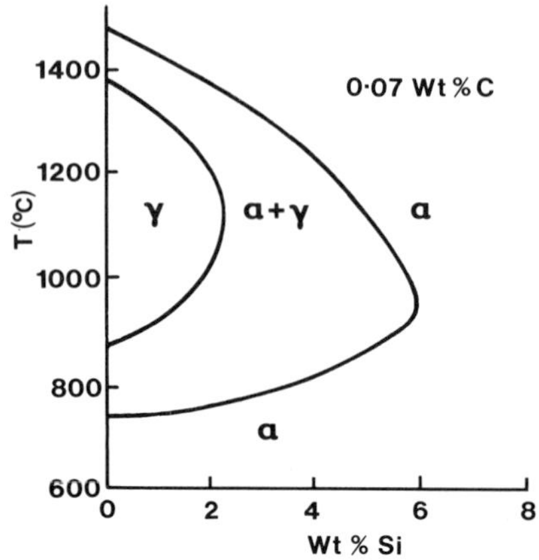

(b)

Fig. 2.21 (a) The iron-rich portion of the Si–Fe equilibrium phase diagram (Crangle, 1954). (b) The effect of 0.07 wt% C on the $\alpha+\gamma$ loop.

still further if it were economically feasible to do so. For the same reasons, it is essential to minimize the sulfur content.

2.10.4 Ductility

Since the alloys are to be rolled at room temperature, they must be sufficiently ductile and have a ductile–brittle temperature below room temperature. Referring to Fig. 2.22, we can see that an alloy containing a little more than 3% Si will satisfy this condition and still have reasonable ductility. The percentage elongation in a tensile test is taken as a measure of ductility.

2.10.5 Magnetostriction and Transformer Hum

The magnetostriction of Si–Fe alloys depends not only on the composition as shown in Fig. 2.23, but also on the stress and the flux density. However, for polycrystalline materials with an Si content of about 5% the saturation magnetostriction coefficient

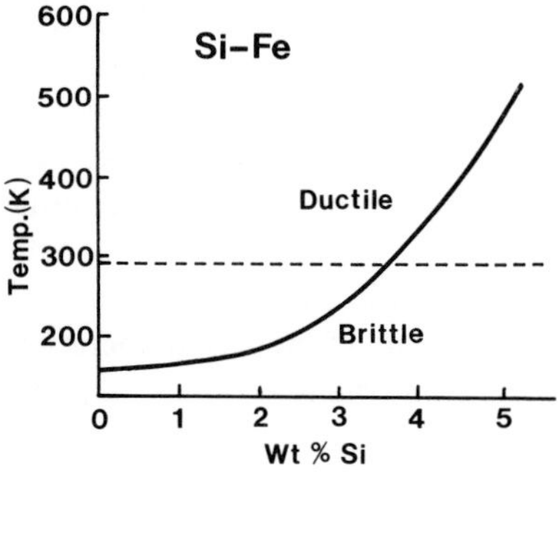

Fig. 2.22 The ductile–brittle transition temperature in Si–Fe alloys as a function of composition. Room temperature is indicated by the broken line.

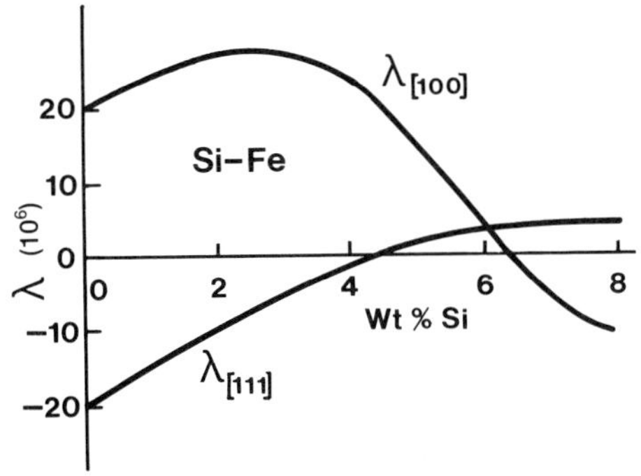

Fig. 2.23 Magnetostriction $\lambda_{[100]}$ and $\lambda_{[111]}$ as a function of Si content in the Si–Fe system.

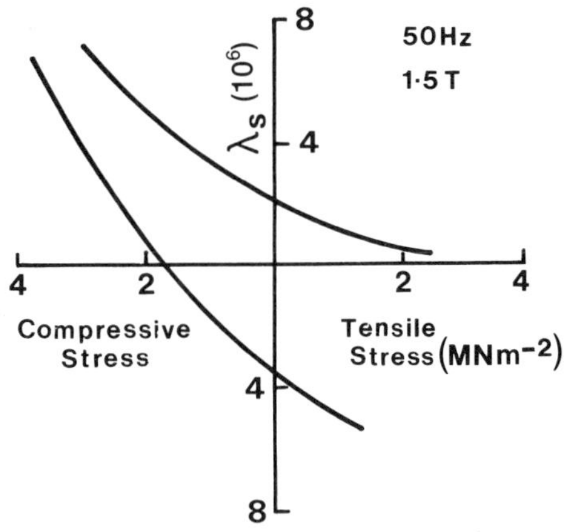

Fig. 2.24 Dependence of the saturation magnetostriction constant λ_s on the applied stress σ for grain-orientated 3.2% Si–Fe with the (110)[001] texture.

$\lambda_s \approx 5 \times 10^{-6}$. In the (110)[001] grain-oriented material the behaviour is complicated, but the effective magnetostriction λ_s parallel to the rolling direction [001] is very small, of order 10^{-6}, particularly when the material is under tensile stress. The dependence of λ_s on stress is shown in Fig. 2.24, from which we can see the importance of avoiding compressive stresses.

During operation, the material experiences cyclical magnetization, so that its dimensions change owing to magnetostriction. The core material therefore vibrates, producing sound of frequency twice that of the applied alternating field (2 × frequency because the core material expands and contracts twice every cycle). This sound is of course the familiar transformer hum, which is a source of irritation to those who live near transformer substations. A great deal of effort has been made to minimize the hum by ensuring that the transformer core is assembled so as to avoid the occurrence of compressive stresses. The lower limit of this undesirable hum is determined by the fundamental magnetostriction constant, which can only be minimized once all the other requirements have been satisfied. Fortunately the magnetostriction is very low at a composition of about 3 wt% Si–Fe. Since magnetostriction changes occur only when the domain magnetization vectors are rotated by the applied field, the magnetostriction will be reduced if the material consists mainly of well-aligned domains with 180° walls parallel to the [001] easy direction. The (110) plane of the grain-oriented material contains only one easy direction, i.e. [001] (see Fig. 2.25), so that this condition will usually be satisfied for well-oriented grains.

2.10.6 Effects of Grain Size on Hysteresis and Eddy Current Losses

Increasing the grain size is generally beneficial, but for grain sizes in the range 1–5 mm Littmann (1967) has shown that the hysteresis losses begin to increase slowly with grain size. When the grain size is very large, the domains tend to be large and increase the losses, even when the material is relatively free of defects. This is due to the fact that if

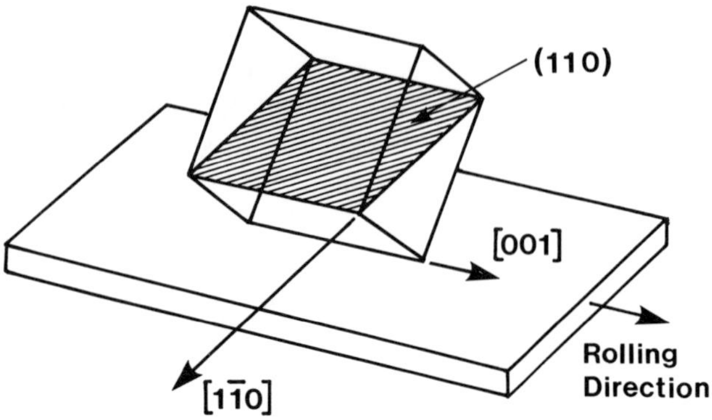

Fig. 2.25 Orientation relationship for the (110)[001] texture.

the domains are large, a given change in magnetization means that the small number of domain walls present have to move further, and this wall movement itself generates micro-eddy currents. However, if the grain size is too small, there are more walls and an increased grain boundary area, which impedes their motion and hence increases the hysteresis losses. After a great deal of research, it has been found that a grain size of 1–5 mm minimizes losses due to both hysteresis and micro-eddy currents. Dragoshansky *et al.* (1972), Shilling *et al.* (1978) and Nozawa *et al.* (1978) have shown that the eddy current losses in single crystals of Si–Fe are higher than those in the grain-oriented material. Thus it appears that the grain boundaries act as nucleation sites for domain walls.

2.10.7 Effects of Insulation on the Magnetic Properties

The principal function of the insulating coating on the laminations is of course to minimize the eddy currents that flow perpendicular to the plane of lamination. A coating of amorphous magnesium silicate and phosphate has a smaller coefficient of thermal contraction than the Si–Fe, and therefore produces a tensile stress when it is cooled from the final high-temperature anneal. According to Fig. 2.24, this reduces the magnetostriction, and since the stress is predominantly longitudinal, the alloy tends to form 180° domain walls parallel to the rolling direction, thus increasing the permeability.

2.10.8 Effects of Temperature and Core Assembly

In large transformers it is essential that the cooling rate of the core be sufficient to remove heat generated at the centre, otherwise compressive stresses will be developed owing to differential thermal expansion, thus causing an increase in the magnetostrictive hum.

Compressive stresses may also be produced if the laminations are not perfectly flat, so that when they are assembled, the 'wavy' sections will be compressed as the core is bolted together. Since it is practically impossible to ensure mechanical perfection, the laminations also vibrate slightly owing to electromagnetic forces between them. However, if the core is assembled very carefully so as to avoid stressing the laminations, the hum can be reduced to an acceptable level.

2.10.9 Commercial Production of (110)[001] Grain-Oriented 3.2% Si–Fe

The ideal orientation relationship for the (110)[001] texture is shown in Fig. 2.25. For 3.2% Si–Fe the deviation from the ideal orientation is usually not more than about 10° for the (110) plane and 7° for the [001] direction, which should ideally be parallel to the rolling direction and hence the direction of magnetization in the transformer core.

Although precise details of the process are not readily available, the following summary gives the most important steps. The basic process was first developed by Goss (1935), and the (110)[001] texture is often referred to as the Goss texture. It is a curious coincidence that Goss is an acronym for the material, i.e. Grain-Oriented Silicon Steel!

1. The starting material has a composition Fe–3.2% Si–0.03% C–(0.06–0.10%) Mn–0.02% S.
2. The cast ingot is hot-rolled at about 1300°C to a thickness of about 1.5–2.5 mm.
3. The oxide scale is then removed by acid – a process known as pickling.
4. The material is then cold-rolled to a final thickness of about 0.25–0.35 mm in two approximately equal steps, with an intermediate anneal at about 800°C–1000°C so as to soften the sheet sufficiently for the second cold roll.
5. The material is then decarbonized at about 800°C in moist hydrogen and nitrogen so as to reduce the carbon content to about 0.003% C, and also to cause primary recrystallization, i.e. the growth of new strain-free grains.
6. The material is then coated with MgO (usually in the form of $Mg(OH_2)$), which combines with the Si in the surface layers of the material to form an insulating glassy layer of magnesium silicate, $MgSiO_3$, with the additional beneficial effects described above.
7. The sheet is annealed at 1100–1200°C in dry hydrogen and nitrogen to form the (110)[001] texture by secondary recrystallization. This anneal also removes sulfur from any remaining MnS by reducing it to Mn and H_2S gas. The manganese thus liberated goes into solid solution with iron, thus increasing the resistivity. The presence of MnS in the alloy is important because it inhibits excessive grain growth during the primary recrystallization, and is now considered to be essential to the formation of a good (110)[001] texture. It is essential to remove the MnS particles, since if they are capable of inhibiting grain growth, they could also inhibit domain wall movement. Finally, the alloy is given a flattening anneal at 800°C in hydrogen and nitrogen, when excess MgO is removed. A thin phosphate coating is applied, which reacts with the magnesium silicate to form a strong highly insulating coating.

The secondary (110)[001] recrystallization texture, unlike the primary texture, is very strong with relatively little scatter about the ideal orientation. The most accurate method of determining the quality of the grain orientation is by the Laue back-reflection X-ray technique. Since the grain size of (110)[001] Si–Fe is relatively large, this technique is quite straightforward, though tedious. A back-reflection Laue photograph of one grain

Fig. 2.26 Back reflection X-ray Laue photograph of a single grain in (110)[001] textured 3.2% Si–Fe. The grain is misoriented by about 3°.

is shown in Fig. 2.26, for which the (110) plane makes an angle of 3° with the surface and the [001] direction.

2.10.10 Domain Structures

The domain structures in 3.2% Si–Fe with the (110)[001] texture can be observed by means of the well-known magnetic colloid technique. The nature of the domain structures gives a good indication of the degree of orientation that has been achieved. Such domain structures have been published by Paxton and Nilan (1955) and McCurrie *et al.* (1975).

The domain structures in two grains of (110)[001] grain-oriented 3.2% Si–Fe are shown in Fig. 2.27, and are very close to the ideal orientation, so that both contain simple widely spaced 180° domain walls.

The domain structures can more conveniently be observed by means of a viewer in which the domain structure is made visible by the concentration of the ferrofluid particles, namely colloidal magnetite (Fe_3O_4) suspended in water. The viewer, which is available commercially, consists of a layer of ferrofluid sandwiched between two glass plates. The domains are observed by simply placing the viewer on the surface of the material. In (110)[001] grain-oriented Si–Fe the domains are easily resolvable with the naked eye.

2.10.11 Torque Curves for the (110)[001] Texture

The degree of grain orientation can also be determined by measuring the torque curves on a sample of the (110)[001] material containing a reasonably large number of grains (about 100) and comparing this with the torque measurements on a single crystal with the (110) orientation as shown in Fig. 2.28. The torque curve from the grain-oriented material is the algebraic sum of the torques due to individual grains, so that in principle

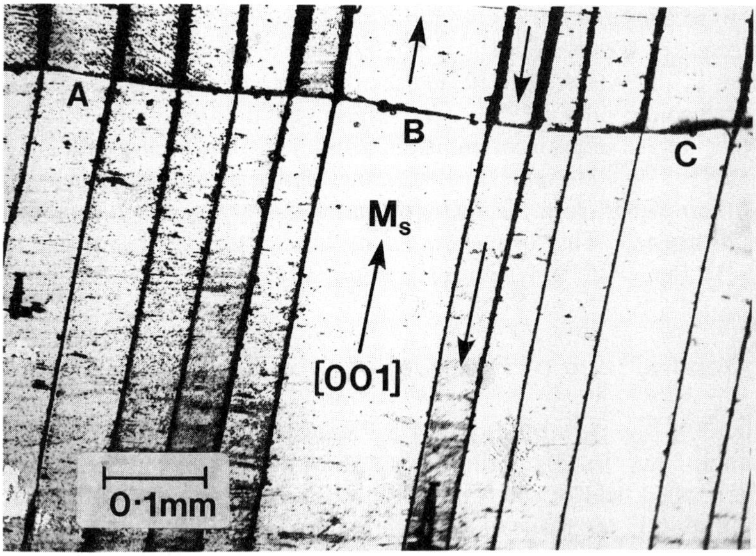

Fig. 2.27 Simple 180° wall domain structure in 3.2% Si–Fe with the (110)[001] texture. Note that while the domain walls are continuous across the grain boundary ABC, there is a slight change in angle due to the change in the easy direction of magnetization.

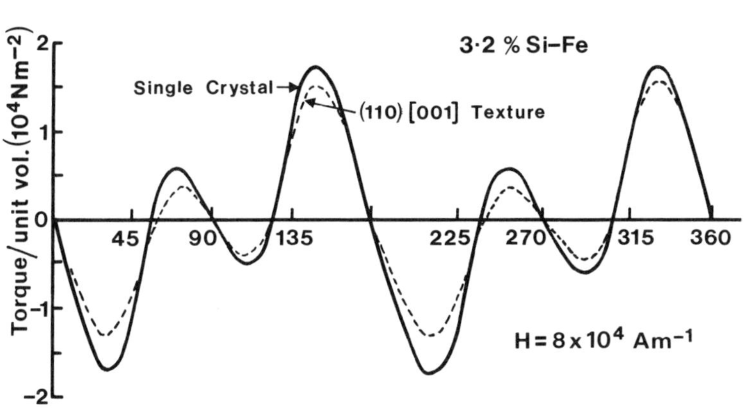

Fig. 2.28 Comparison of torque curves for single-crystal and (110)[001] textured 3.2% Si–Fe (after Graham, 1969).

it is possible to calculate a theoretical torque curve corresponding to the range of misorientation angles α and β. Unfortunately, this is rather difficult for the (110) orientation, since it involves summing a large number of terms, each of which contains a K_1 and K_2 dependence. The torque Γ in the (110) plane of a single crystal of Si–Fe is given by

$$\Gamma=\tfrac{1}{8}K_1(2\sin 2\theta+3\sin 4\theta)+\tfrac{1}{64}K_2(\sin 2\theta+4\sin 4\theta-3\sin 6\theta),$$

so it is easy to see that the full expression for the grain-orientated material will be very complex. However, Westerstrand *et al.* (1975) have shown that for 3.2 wt% Si–Fe, $K_2=0$, so that the term in K_2 can be neglected. General expressions for the torques are given by Dunn and Walter (1959). The experimental torque curve is a very useful indicator of the degree of grain orientation, although it has been pointed out by Dunn and Walter (1959) that attempts to determine textures from torque curves are often unreliable if the torques from the textured material are very much smaller than those from the single crystals. The torque curves for various textures have been investigated by Swift (1973), Benford (1974) and Swift *et al.* (1975).

2.10.12 Dependence of Permeability on Texture

The dependence of the permeability at 1.5 T of grain-oriented (110)[001] Si–Fe as a function of angle from the rolling direction is shown in Fig. 2.29, together with the flux density in an applied field of 800 A m^{-1}. Since the (110)[001] textured material has a single easy direction in the plane of the sheet, large power transformers are constructed so that all the sections contain the [001] direction parallel to the direction of magnetization. For smaller transformers construction of the cores from individual sections would be much too expensive, and as a compromise they are made of E–I pieces. The flux is therefore parallel to the easy direction in all sections of the core, except the upright section of the E, where the flux is in the [110] direction and is therefore perpendicular to the easy direction. In order to compensate for this, the magnetic reluctance of this section must be decreased by increasing the cross-sectional area so that the total flux transport is maintained at the required level. We can see why this is effective by remembering that the magnetic reluctance $\mathbb{R}$ of a circuit is defined as

$$\mathbb{R}=\frac{l}{\mu A},$$

where l is the effective length of the magnetic material, μ is the permeability and A is the cross-sectional area. Since μ is determined by the properties of the material, $\mathbb{R}$ can be decreased if l is decreased or A is increased.

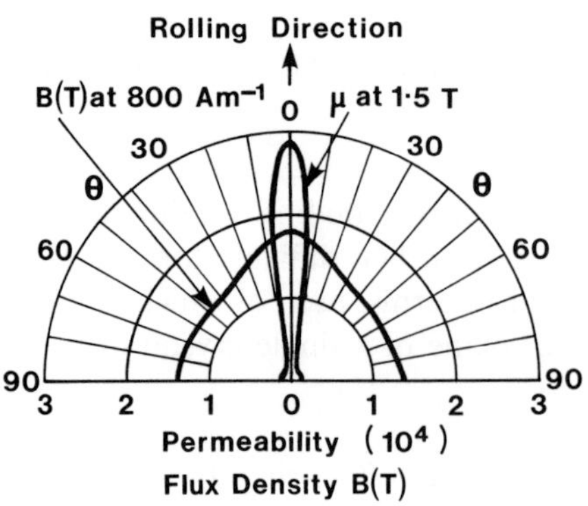

Fig. 2.29 Dependence of the relative permeability of grain-oriented 3.2% Si–Fe at a flux density of 1.5 T on the angle θ from the rolling direction, $\theta=0$. The angular variation of the flux density in a field of 800 A m^{-1} is also shown.

2.11 GRAIN-ORIENTED Si–Fe WITH THE (001)[100] TEXTURE

Unlike the Goss-textured material, the (001)[100] texture (Assmus *et al.*, 1957) contains two easy directions of magnetization in the rolling plane, e.g. [100] and [010], so that it can be magnetized with equal facility in two orthogonal directions. The orientation relationships are illustrated in Fig. 2.30. This texture is much more difficult to produce than the Goss texture, and requires very careful control of the necessary impurities and the annealing atmosphere. Unfortunately these requirements make the material very expensive to produce, so that it is not available commercially, although it is used occasionally for very special applications.

A comparison of the flux densities in a field of 800 A m^{-1} for the (110)[001] and (001)[100] textures as a function of the average deviation θ of the easy direction of magnetization [001] from the rolling direction is shown in Fig. 2.31.

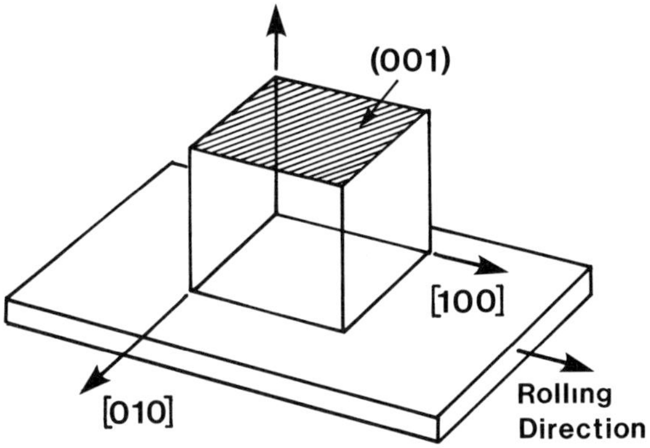

Fig. 2.30 Orientation relationships for the (001)[100] texture.

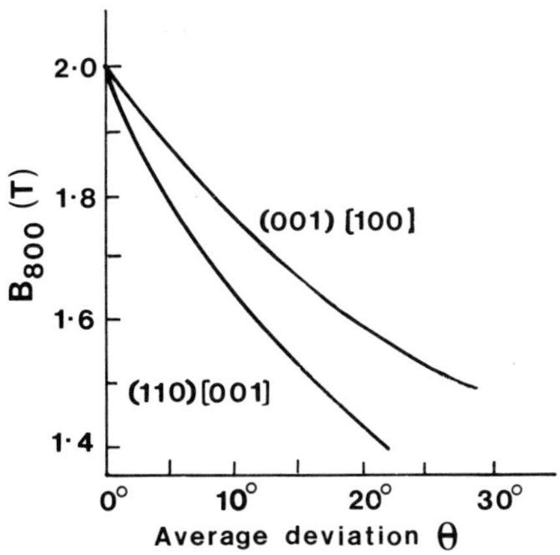

Fig. 2.31 Flux density at 800 A m^{-1} for (110)[001] and (001)[100] grain-oriented 3.2% Si–Fe as a function of the average deviation of the [001] easy direction of magnetization from the rolling direction (after Littmann, 1967).

2.11.1 Torque Curves for the (001)[100] Texture

The interpretation of torque curves for textured polycrystalline materials is in general very difficult, and even in relatively simple cases is subject to considerable error. However, as a simple example (Graham, 1969), consider a polycrystalline material with a (001)[100] texture in which the deviations from the perfect mixture occur only as rotations in the plane of the specimen, i.e. about the normal to the surface of the specimen. As the torque curve is measured, the magnetization vector in each grain rotates in the plane, so that for a material with cubic magnetocrystalline anisotropy, the torque on an individual grain is given by

$$\Gamma_i = -\tfrac{1}{2}K_1 V_i \sin 4(\theta+\beta),$$

where K_1 is the first magnetocrystalline anisotropy constant, θ is the angle between the magnetization vector and the rolling direction, β is the angular deviation of the [100] direction from the rolling direction, and V_i is the volume of the grain. The torque curve for the whole polycrystalline specimen is the sum of the torque curves of the individual grains:

$$\frac{\Gamma}{V} = \sum_i \left[-\tfrac{1}{2}K_1 V_i \sin 4(\theta+\beta_i)\right] \Big/ \sum_i V_i,$$

where V is the total volume of the specimen. If it is assumed that the grains have equal volumes uniformly distributed from $-\beta$ to $+\beta$ then the above sums can be replaced by integrals over the angle from $-\beta$ to $+\beta$, giving

$$\frac{\Gamma}{V} = -\tfrac{1}{2}K_1 \sin 4\theta \left(\frac{\sin 4\beta}{4\beta}\right).$$

Thus the torque curve of a polycrystalline specimen has exactly the same shape as that for a single crystal with the (001) orientation, but reduced in magnitude by a factor $(\sin 4\beta)/4\beta$, which is always less than one. In real materials the texture is usually much more complex, so that the analysis is much more difficult. Dunn and Walter (1959) gave general equations that can be applied to particular cases, but they emphasize the uncertainties in trying to deduce information about the crystal texture from torque curves, particularly where the torques are small compared with the torques from the corresponding single crystal.

The torque curves produced by various textures have been investigated by Swift (1973), Benford (1974) and Swift *et al.* (1975).

2.12 ENERGY LOSSES IN TRANSFORMER CORES

The energy losses in transformer cores are often measured by means of an Epstein test frame in the form of a square made from the transformer core laminations (Bozorth, 1951). Each side of the square has primary and secondary windings. The Epstein test is divided into two parts: (1) determination of the static hysteresis loss; and

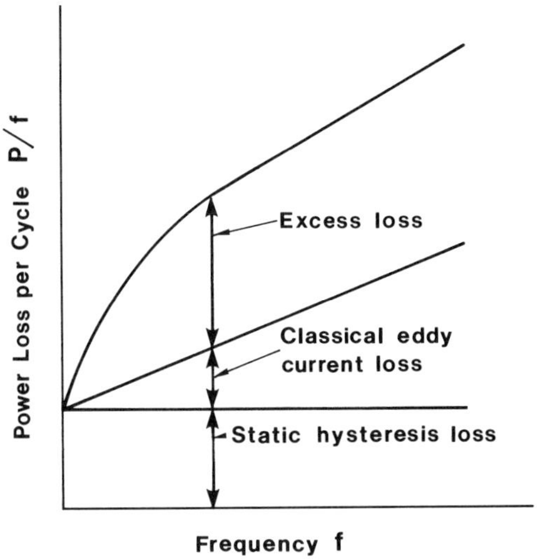

Fig. 2.32 Separation of transformer core power losses as a function of frequency. The total power loss per cycle is the sum of the three components.

(2) determination of the eddy current loss. The hysteresis loss is determined from the area of the B–H loop measured by a galvanometer, fluxmeter or with a very slowly operated hysteresis loop plotter. This loss is assumed, tentatively, to be independent of frequency. The power loss due to eddy currents in the core material is measured directly by means of a wattmeter, although minor corrections are usually necessary. The core losses are usually measured at $B = 1.5$ T.

If we assume for the moment that the total power loss P_T can be separated into two components, namely hysteresis losses P_h and eddy current losses P_{ec}, then

$$\begin{aligned} P_T &= P_h + P_{ec} \\ &= W_{hf} + Kf^2, \end{aligned}$$

where f is the frequency and K is a constant. The total power loss per cycle is

$$P_T/f = W_h + Kf.$$

Thus we expect P_T/f to vary linearly with frequency and to be equal to W_h at zero frequency. The typical energy loss per cycle of a Si–Fe transformer core material is shown in Fig. 2.32, where it has been assumed that the power losses in the core can be separated. From this figure, we can see immediately that the total energy loss per cycle is very much greater (a factor of nearly 2) than that predicted by simple summation of the hysteresis and classical eddy current losses. The additional losses are often referred to as the 'excess' losses.

2.12.1 Eddy Current Losses due to Domain Wall Motion

In the derivation of the classical expression for the eddy current loss it was assumed that the permeability μ is a constant – which is certainly not the case. Furthermore, if the applied field H varies sinusoidally with time, the variation of B is not in general

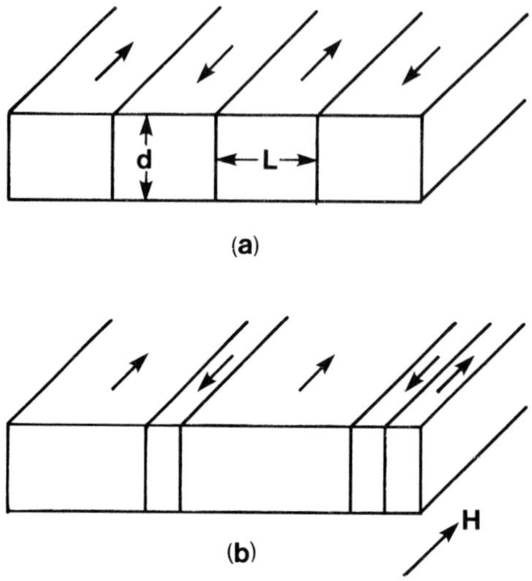

Fig. 2.33 Magnetization by simple domain wall displacement in a thin ferromagnetic sheet: (a) the demagnetized state; (b) the partially magnetized state.

sinusoidal. The fact that the transformer core material has a domain structure that varies in a complex way with H and position in the sheet complicates the behaviour still further. The net result is that (2.2) underestimates the eddy current losses by a factor of about 2.

It is now generally accepted that the so-called 'anomalous' or 'excess' eddy current losses associated with cyclic magnetization or transformer core laminations are attributable to 'micro' eddy currents induced by the motion of individual domains walls. A detailed explanation of this important effect has been attempted by a number of authors. The best known and most universally employed model is that due to Pry and Bean (1958), although this is very similar to the model given by Lee (1958) and Carr (1959), who also considered the behaviour of plane domain walls perpendicular to a ferromagnetic sheet, as shown in Fig. 2.33. According to Pry and Bean (1958), the ratio of the power loss due to domain wall motion and the classical eddy current loss is

$$\frac{P}{P_c}=\frac{48L}{\pi^3 d}\sum_{\substack{n=1\\ n\text{ odd}}}^{\infty}\frac{1}{n}\coth\frac{\dot{n}\pi L}{2d}, \tag{2.3}$$

assuming that (a) the material is excited at an angular frequency ω in such a way that the flux density at time t is $B=B_m\cos\omega t$, and (b) that $B_m/B_s \ll 1$, where B_m is the maximum flux density and B_s is the saturation flux density. For large values of L/d, i.e. large domains, (2.3) reduces to

$$\frac{P}{P_c}=1.628\frac{L}{d}.$$

The variation of P/P_c as given by (2.3) is shown in Fig. 2.34 as a function of L/d for $B_m/B_s=1$ and $B_m/B_s \ll 1$. It is clear that P/P_c increases rapidly with increasing L/d. The relationship between domain spacing, grain size, sheet thickness and power loss

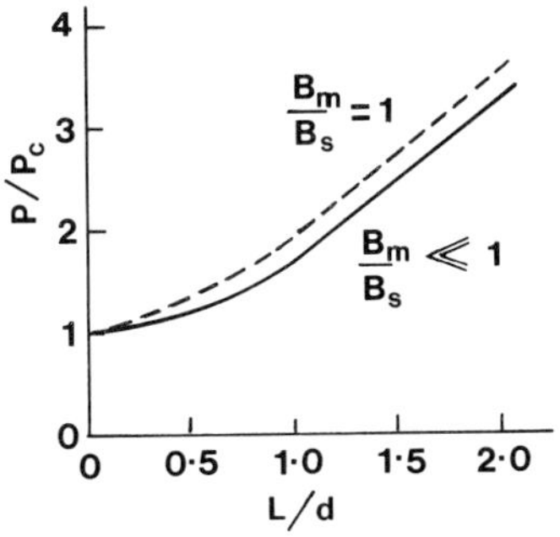

Fig. 2.34 Ratio of domain wall eddy current power loss to the classical eddy current power loss P/P_c as a function of the domain wall spacing L relative to the sheet thickness d (after Pry and Bean, 1958).

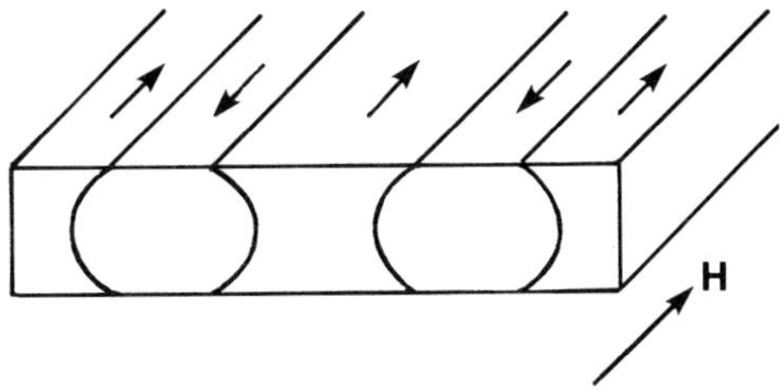

Fig. 2.35 Domain wall bowing in the presence of an applied field H.

has been the subject of many investigations – see e.g. the paper by Overshott and Thompson (1970).

In spite of the simplicity of the Pry and Bean model, it is in reasonable agreement with the behaviour of many transformer core materials. One of its principal limitations, however, is that it assumes that the domain walls remain rigid throughout the magnetization cycle. It has been suggested by Lee (1960) and Overshott *et al.* (1966) that magnetization changes occur by domain wall bowing (Fig. 2.35) as well as domain wall displacement (Bishop, 1976a). The theory of wall bowing has been investigated by Bishop (1978), Carr (1976) and Boon and Robey (1969). A more general and theoretical discussion of the AC magnetization of ferromagnetic sheets has been given by Bishop and Lee (1963), Lee and Lynch (1959), Lee (1960) and Polyvanov (1952).

Since the wall bowing occurs in the interior of the material, the phenomenon has not yet been confirmed by observation, although Hill and Overshott (1978) have shown that if the wall profiles are assumed to be parabolic, there is a good correlation between the measured flux density and that determined from the surface domain structure (Overshott *et al.*, 1966). The observation of surface domain structure by the Kerr magneto-optic effect has been used extensively in the study of eddy current and hysteresis losses in Si–Fe transformer core materials. The published papers on Si–Fe are too numerous to cite individually; the reader is therefore referred to the reviews by Littman (1967, 1971), Shilling and Houze (1974), Taguchi *et al.* (1976), Fiedler (1977) and Moses (1990). A summary of the use of domain observations in the understanding and improvement of transformer core materials has been published by Overshott (1976).

Bishop (1976a, b) has suggested that losses in (110)[001] Si–Fe could be due to a new process called domain wall ruckling. The possibility of this process arises because the 180° domain walls in (110)[001] Si–Fe have minimum energy when they are inclined at 32° to the sheet normal, so that hybrid or ruckled walls with alternate segments inclined at ±32° may occur. The movement of these walls also produces micro-eddy currents, but Bishop (1976a, b) has shown that the overall eddy current drag should be reduced to between 41 and 91% of that for a rigid planar wall.

Apart from the material parameters that have already been discussed, there are several other factors with significant effects on the overall performance of power transformers. These include the losses introduced during the assembly of the transformer, e.g. the stresses resulting from bolting the laminations and the non-uniform flux distributions that occur at 'T' joints or other 90° joints in the core. The latter effects have been investigated in considerable detail. See e.g. Joslin *et al.* (1970), Moses *et al.*(1972), Moses and Thomas (1973a,b) and Moses (1974).

2.13 GRAIN-ORIENTATED Si–Fe WITH AN IMPROVED (110)[001] TEXTURE

One of the most important advances in the technology of soft magnetic materials is the development of an improved (110)[001] texture in Si–Fe by Taguchi and Sakakura (1969), Yamamoto *et al.* (1972) and Taguchi *et al.* (1974). These authors have shown that if AlN (approximately 0.06 wt%) is added to the conventional Goss composition, the AlN acts as a very effective primary grain growth inhibitor and produces a higher degree of grain orientation. The average deviation of the easy [001] axis from the rolling direction is approximately 3°, whereas in the conventional Goss material it is approximately 7°. The improved texture increases the permeability of the alloy considerably. It is usual to specify the permeability in terms of the flux density at a given field strength – usually 800 A m^{-1}. For the new Hi-B material (the full name is Orientcore Hi-B) the flux density at 800 A m^{-1} is typically in the range 1.9–1.98 T so that the relative permeability $\mu_r \approx 2000$. This is not the maximum permeability, but is often used as a figure of merit that is convenient for the comparison of different materials. For polycrystalline and conventional grain-oriented Si–Fe the core losses are usually measured at 1.5 T and at 50 or 60 Hz. The losses are usually denoted by $W_{1.5/50}$ or $P_{1.5/50}$. However, for the new Hi-B material, measurement of the losses at a flux density of 1.7 T is more appropriate. (The name Hi-B was chosen to indicate that the new core material operates at higher flux densities than the conventional Goss material.) The core losses $W_{1.7/50}$ as a function of the flux density at 800 A m^{-1} for over 200 specimens of Hi-B Si–Fe are shown in Fig. 2.36 (Sato *et al.*, 1978). The minimum core loss of 0.95 W kg^{-1} occurs for a specimen with a flux density of 1.95 T at 800 A m^{-1},

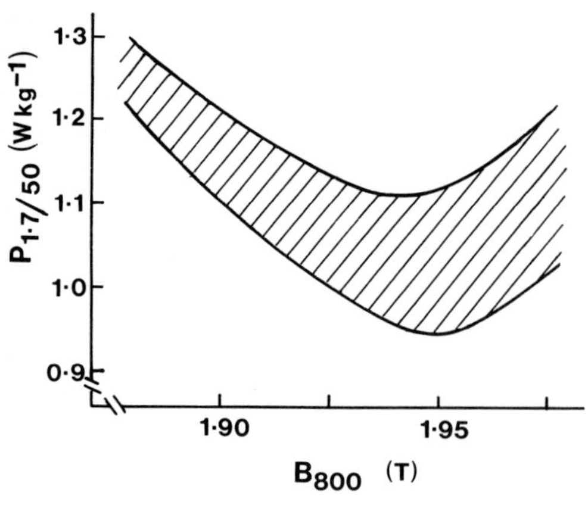

Fig. 2.36 Transformer core losses $P_{1.7/50}$ at 1.7 T and 50 Hz for various specimens of Hi-B Si–Fe with the (110)[001] texture. The specimens are characterized in terms of their flux densities in an applied field of 800 A m^{-1} (after Sato *et al.*, 1978).

corresponding to a relative permeability of about 1950. Although there is considerable scatter in the values, it is now well established that the core losses for the Hi-B material are about 20% lower than those for the Goss material. Another important advantage of the new Hi-B material is that the exciting power *VA* (W) is also considerably reduced, by 20–25%.

Although the addition of AlN increases the sharpness of the secondary (110)[001] texture, its effectiveness as a primary grain growth inhibitor is doubly beneficial because it allows the sheet material to be produced by a single-step cold roll with a large reduction (about 80%), which itself favours a sharp (110)[001] texture. Usually, a large final cold reduction increases the tendency for growth of the primary grains.

Another important factor contributing to the improved properties of the Hi-B material is that the new surface coating produces a tensile stress on the core material. The new insulating coating consists of two layers: an inorganic material on a glass film, which is mainly 24% SiO_2–21% Al_2O_3–16% MgO. The higher tensile stress has two important effects: (1) it decreases the magnetostriction coefficient, and hence the magnetostrictive hum (noise reduced by 2–7 dB (A scale)); (2) it decreases the domain wall spacing and favours the formation of 180° walls parallel to the rolling direction, i.e. the direction of magnetization. The latter effect results in a significant reduction in the eddy current loss.

If we assume that the total core losses can be separated into two components – hysteresis losses and eddy current losses – then the physical features determining the losses can be summarized as follows:

Hysteresis losses
(1) crystal orientation;
(2) purity (inclusions, precipitates);
(3) internal stress;
(4) surface condition;
Eddy current losses
(1) electrical resistivity (silicon content);
(2) sheet thickness;
(3) domain size and grain size;
(4) elastic tensile stress (surface coating).

Sato *et al.* (1978) have shown that the surface condition of the core material itself is also very important in determining the core loss. Specimens of uncoated Hi-B material that had been deliberately scratched at intervals of 10 mm showed a substantially reduced core loss, particularly at high flux densities with an applied tensile stress (1.5 kg mm^{-2}).

The improvement in the soft magnetic properties of the Hi-B (110)[001] Si–Fe is due principally to the sharper crystal orientation and to the higher tensile stress produced by the new surface coating. A comparison of the standard and improved (Hi-B) (110)[001] Si–Fe is shown in Table 2.6. The magnetic properties of typical specimens of Hi-B Si–Fe as functions of sheet thickness are shown in Table 2.7.

The spectacular success of Hi-B Si–Fe has stimulated a great deal of research by other workers, who have obtained similar properties by slightly different processes. Goto *et al.* (1975) produced a material known as RG-H Si–Fe with similar properties to Hi-B Si–Fe except that these authors used Sb with MnS or MnSe as the primary grain growth inhibitors. The insulating coating used (Shimanaka *et al.*, 1975) consisted

Table 2.6 Comparison between the improved and old Goss-textured 3% Si–Fe laminations of 0.30 mm thickness (after Chen, 1978).

	Standard laminations	Improved laminations
Processing:		
Final reduction by cold rolling	80% in two steps, each about 40%, separated by an anneal at 800–1000°C in a reducing atmosphere	85–93% in one step
Heat treatments after final reduction	At 800°C for a few minutes in wet hydrogen, then at 1200°C in dry hydrogen	At 850°C for 2 min in wet hydrogen, then at 1200°C for 20 h in dry hydrogen and nitrogen
Grain-growth inhibitor	MnS	AlN plus MnS
Surface coating		
Structure and composition	One layer of glass (mainly 35% SiO_2 and 17% MgO)	Two layers – an inorganic deposited on a glass (mainly 24% SiO_2, 21% Al_2O_3 and 16% MgO) film
Induced tensile stress in the Si–Fe sheet	0.1–0.2 kg mm^{-2}	0.3–0.5 kg mm^{-2}
Grain structure:		
Grain diameter	2–4 mm	7–9 mm
Average deviation of the [001] axis from the rolling direction	7°	3°
Magnetic properties:		
Permeability at 800 A m^{-1}	1820	1920
Flux density at 800 A m^{-1}	1.82 T	1.92 T
Linear magnetostriction with surface coating intact	-0.66×10^{-6}	-0.30×10^{-6}
Total energy loss at 60 Hz		
(i) $B=1.5$ T	1.04 W kg^{-1}	0.81 W kg^{-1}
(ii) $B=1.7$ T	—	1.45 W kg^{-1}

Table 2.7 Magnetic properties of Hi-B high-permeability Si–Fe with the (110)[001] texture (after Taguchi *et al.*, 1976).

Property (units)[a]	Thickness (mm)		
	0.35	0.30	0.28
B_{800} (T)	1.965	1.958	1.960
P_t (kg^{-1}) at 1.7 T, 50 Hz	1.13	0.98	0.95
P_t (kg^{-1}) at 1.7 T, 60 Hz	1.5	1.3	1.25
P_h (kg^{-1}) at 1.7 T, 50 Hz	0.202	0.210	0.221
P_h (kg^{-1}) at 1.7 T, 60 Hz	0.26	0.29	0.29
λ ($\times10^{-6}$) at 1.5 T	−0.5	−0.5	−0.6
λ ($\times10^{-6}$) at 1.9 T	−0.2	−0.2	−0.3

[a] $B_{800}\equiv$ flux density at 800 A m^{-1}, $P_t\equiv$ total loss in W kg^{-1}, $P_h\equiv$ hysteresis loss in W kg^{-1}, $\lambda\equiv$ magnetostriction coefficient.

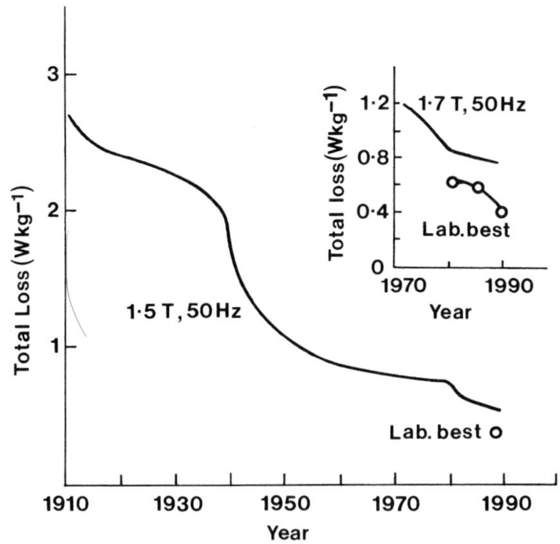

Fig. 2.37 Development of high permeability Si–Fe alloys. The core loss at 1.7 T and 1.5 T at 50 Hz is plotted as a function of calendar year (after Moses, 1990).

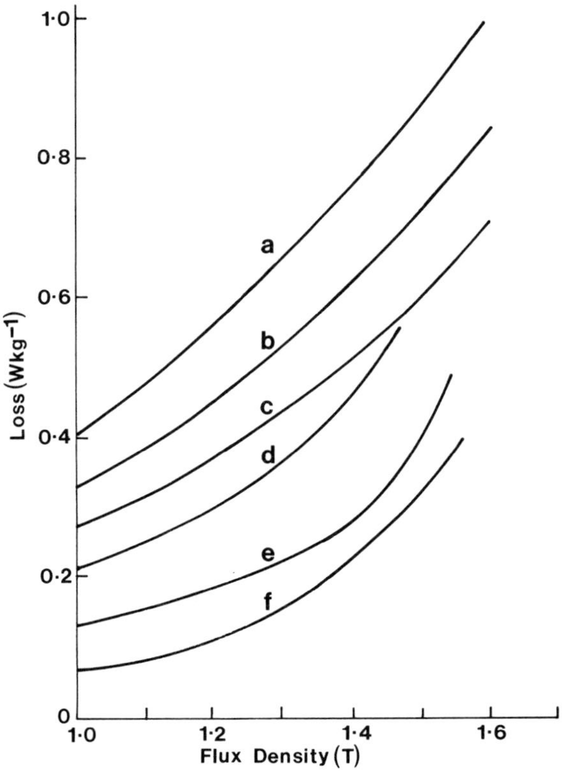

Fig. 2.38 Core losses as a function of frequency for various materials: (a) 0.3 mm thick conventional grain-oriented (110)[001] 3.2% Si–Fe; (b) 0.33 mm thick special high-permeability grain-oriented (110)[001] 3.2% Si–Fe; (c) 0.23 mm thick domain-refined (110)[001] 3.2% Si–Fe; (d) 6.5% Si–Fe semicrystalline; (e) Powercore; (f) Metglas 2605-S2 (after Moses, 1990).

Table 2.8 Typical applications of electrical steels (after Moses, 1990).

Application	Non-oriented steel			Grain-oriented steel	
	Silicon-free	Low-silicon	High-silicon	Conventional	High-permeability
Small motors	←	→			
Lamp ballasts	←	→			
Medium AC motors	←	—	→		
Welding transformers	←	—	→		
Audio transformers		←	—	→	
Small power transformers		←	—	→	
Large rotating machines			←	→	
Medium generators		←	—	→	
Distribution transformers				←	→
Power transformers				←	→

of colloidal silica, magnesium phosphate and a small quantity of chromium trioxide. This coating also produces a high tensile stress parallel to the rolling direction, with a consequent reduction in magnetostrictive hum and eddy current losses.

The reduction of core losses in Si–Fe alloys as a function of calendar year is shown in Fig. 2.37. The total core losses as a function of flux density for various materials are compared in Fig. 2.38. Typical applications for various types of Si–Fe are shown in Table 2.8.

2.14 DOMAIN REFINEMENT OF GRAIN-ORIENTED SILICON–IRON

Further improvements to the high permeability and low losses in Si–Fe alloys continue to be made. The misorientation angle in the Hi-B alloy has been reduced to about 3°, and a significant reduction in the losses can also be achieved by reducing the thickness of the laminations, but this increases the difficulties in handling the material on an industrial scale.

As has already been mentioned, the number of domains and their spacing within the grains is of vital importance in determining the eddy current losses. Important improvements have been made by the process known as domain refining. This can be achieved by three techniques:

(1) scratching (Sato *et al.*, 1978; Nozawa *et al.*, 1979; Pfutzner and Zehetbauer, 1982; Fukawa and Yamamoto, 1982);
(2) laser irradiation (Iuchi *et al.*, 1982; Ichijima *et al.*, 1982; and Nakamura *et al.*, 1987;
(3) spark ablation (Beckley *et al.*, 1985).

All of these techniques introduce stress lines or spots on the surface of the lamination perpendicular to the rolling direction, which results in the nucleation of domain walls parallel to the [001] direction with a spacing determined by the spacing of the stress lines.

a

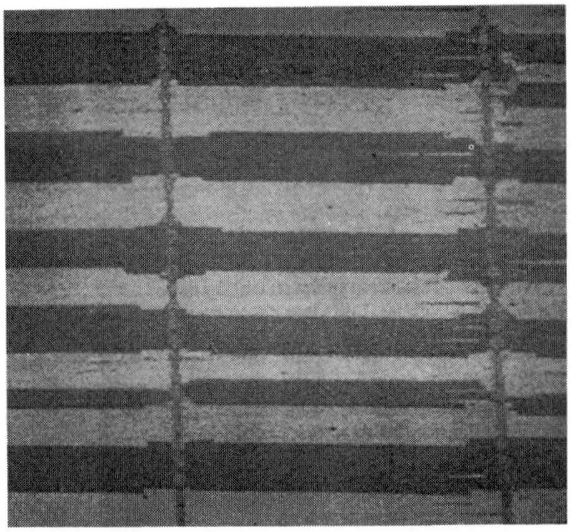

b

Fig. 2.39 Domain refining by laser, i.e. reduction of domain size. Magnification $\times 100$ (courtesy of Nippon Steel Corporation, Kitakyushu, Japan).

Careful choice of the domain wall spacing reduces the micro-eddy currents produced by the wall displacement during magnetization.

The effect of laser irradiation on the domain spacing in grain-oriented Si–Fe is shown in Fig. 2.39.

2.15 ROTATIONAL HYSTERESIS LOSS

When changes in magnetization occur by rotation of the magnetization vector, i.e. when the applied field vector is at an angle to the magnetization vector, losses occur due to rotational hysteresis. As the magnitude and direction of the applied field changes, the magnetization vector is rotated against magnetocrystalline anisotropy forces in an attempt to follow the field. This produces a torque Γ given by

$$\Gamma = \mu_0 |\boldsymbol{M} \times \boldsymbol{H}| = \mu_0 MH \sin\theta,$$

where θ is the angle between the field $\boldsymbol{H}$ and the magnetization vector $\boldsymbol{M}$. The rotation of the magnetization through the material is of course not spontaneously reversible, so that energy is consumed in the process, and is known as the rotational hysteresis loss.

The angle θ between the vectors $\boldsymbol{M}$ and $\boldsymbol{H}$ may vary considerably in different parts of the laminations. Rotational hysteresis is obviously of particular importance in rotating machines, but also occurs at junctions and corners in transformer cores. Although there is no satisfactory theory, it is clear that rotational hysteresis will be determined by processes of domain wall pinning and impedance similar to those that occur in alternating hysteresis. Rotational hysteresis losses are generally much larger than alternating hysteresis losses, and have certain features that are quite distinct. The rotational and alternating hysteresis losses for a specimen of 3.13% Si–Fe are shown in Fig. 2.40. The rotational losses in a material are usually 1–2 times the alternating losses, depending on the frequency and degree of magnetization, although the rotational hysteresis loss is of course zero in a saturating field, because in this case the angle between the applied field and the magnetization vector is zero.

When polycrystalline or grain oriented Si–Fe alloys are used in electric motors, generators or transformers, they are subjected to rotating magnetic fields that vary in magnitude – often in a complex way. Rotational hysteresis in a rotating machine stator core may account for 50% of the total losses. In general, rotational effects include eddy current losses as well as hysteresis losses, which depend on the same factors as alternating losses. For low or medium flux densities, the rotational losses may be a factor of three greater than the alternating losses at the same flux densities, although the rotational losses are significantly reduced if the material is isotropic. The dependence of the rotational power losses on the flux density in various alloys is shown in Fig. 2.41.

Ultimately, a cost–performance balance has to be struck for particular applications of the materials. There is still of course considerable scope for improvement in high-permeability alloys for large-tonnage applications, especially as the environmental and economic effects of energy consumption and wastage become ever more important.

Rotational hysteresis in soft magnetic materials has been investigated by Brailsford (1938, 1939, 1966), Kornetzki and Lucas (1955), Archenhold *et al.* (1960), Boon and

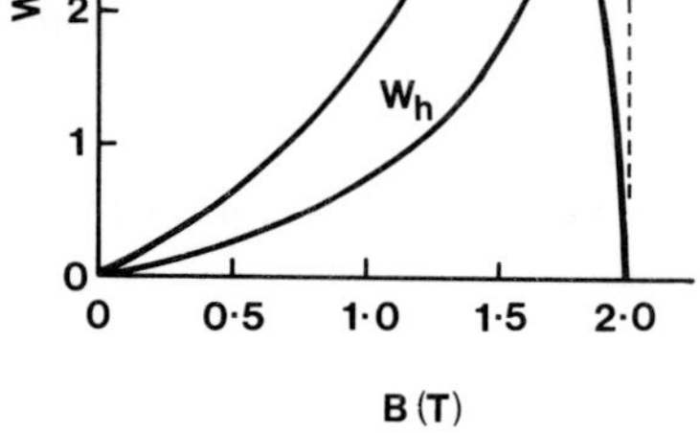

Fig. 2.40 Rotational hysteresis loss W_r as a function of flux density B for a typical specimen of 3.13% Si–Fe together with the alternating rotational hysteresis loss W_h (after Brailsford, 1939).

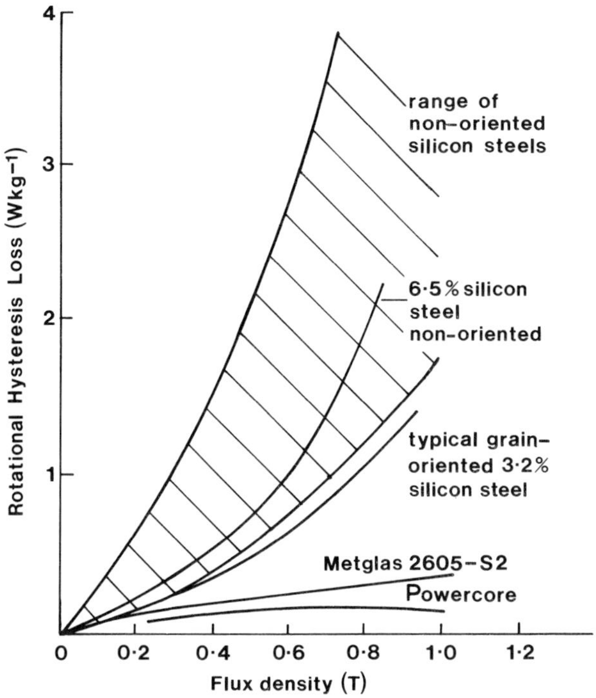

Fig. 2.41 Rotational hysteresis loss as a function of flux density for various materials (after Moses, 1990).

Thompson (1964, 1965), Lommel (1964), Narita and Yamaguchi (1974) and Grimwood *et al.* (1978).

2.16 IRON–ALUMINIUM ALLOYS

The addition of aluminium to iron has similar effects to the addition of silicon, i.e. K_1 decreases, with an accompanying increase in the permeability. The saturation magnetic polarization decreases according to a simple dilution law until the ordered superlattice Fe_3Al (13.9 wt% Al) is formed, for which $J_s = 1.4$ T. Further addition of aluminium causes a much more rapid decrease, and J_s is zero at 18% Al. The Fe–Al alloys have useful magnetic properties up to a maximum aluminium content of 16%. Very high permeabilities are obtained when the 16% Al–Fe alloy is quenched from 600°C so as to suppress the development of short- and long-range order. This alloy, which is often used in magnetic recording heads, is known as Alperm (Japan) or Alfenol (USA), and its properties are summarized in Table 2.2. The addition of aluminium to iron, unlike that of silicon, increases the saturation magnetostriction, and for Fe_3Al (i.e. 13.9 wt% Al) $\lambda_s = 37 \times 10^{-6}$. This material is therefore useful in magnetostriction devices, and is known commercially as Alfer. Since the magnetostriction is also appreciable at 16% Al, this is a disadvantage for a high-permeability material. However, a large magnetostriction is a disadvantage only in the presence of internal stresses, so that the high permeability

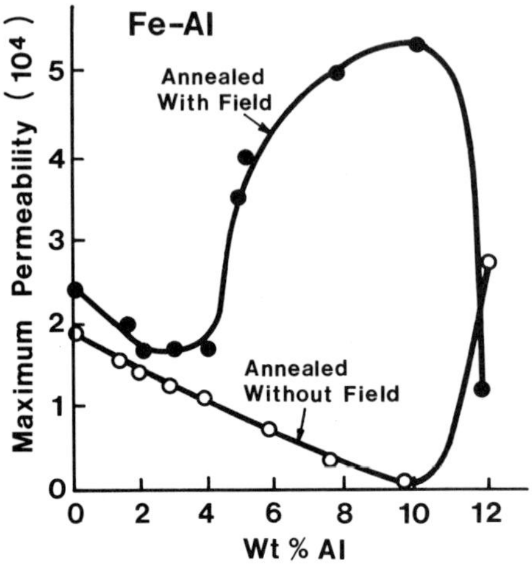

Fig. 2.42 Effect of heat treatment with and without a magnetic field on the maximum permeability of Fe–Al alloys as a function of aluminium content. The heat-treatment field was 140 A m^{-1} (after Sugihara, 1960).

can be obtained if the alloy is prepared in a stress-free form. The exceptionally high resistivity of 16% Al–Fe ($\rho = 1.4 \times 10^{-6}\ \Omega m$) makes it particularly suitable for high-frequency applications.

Sugihara (1960) has shown that the Fe–Al alloys respond to magnetic annealing (Fig. 2.42) and that the effect is most pronounced for the maximum permeability, which may be increased by a factor of five or more. Magnetic and stress annealing in Fe–Al alloys has been investigated by Birkenbeil and Cahn (1962). The magnetic properties of some iron–aluminium alloys are shown in Table 2.9.

Hot and cold rolling to produce the $\{100\}\langle 001\rangle$ cube texture also improves the magnetic properties of the Fe–Al alloys (Al content $<10\%$), but their low ductility and hard oxide scale presents difficulties for large-scale production. Apart from the texture, rolling also produces a slip-induced anisotropy for compositions near Fe_3Al, which can

Table 2.9 Magnetic properties of iron–aluminium alloys (after Adams, 1962).

Al (%)[a]	μ_{20} (μ_0)[d]	μ_{max} (μ_0)	$_BH_c$ (A m^{-1})	B (T)[b]	Core loss at 60 Hz (W kg^{-1} at 1.0 Tesla)
1.0	2041	20 214	26	1.63	1.5
3.0	1923	12 815	36	1.62	1.5
5.0	2381	12 625	26	1.54	1.4
8.0	513	3 461	44	1.34	2.4
8.0[c]	1140	50 900	21	1.28	1.0
13% (Alfer)	700 (μ_i)	3 700	53	1.2 (J_s)	—
16% (Alperm) (Alfenol)	3000 (μ_i)	55 000	3	0.8 (J_s)	—

[a]Annealed at 1000°C in hydrogen.
[b]Measured in a field of 2.4 k A m^{-1}
[c]Annealed in a magnetic field of 136 A m^{-1}.
[d]Measured in a field of 1.6 k A m^{-1}.

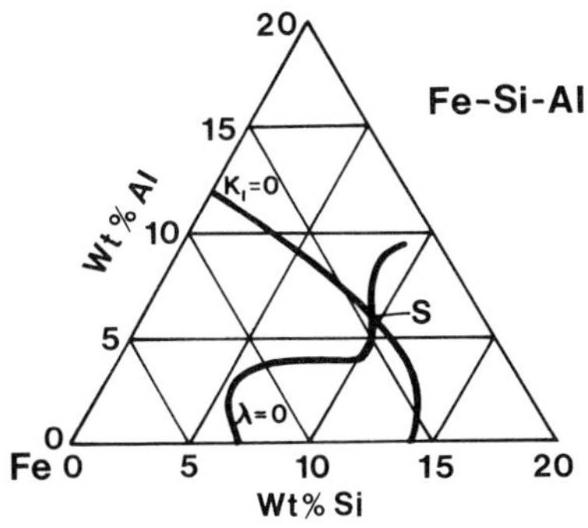

Fig. 2.43 Dependence of the magnetocrystalline anisotropy K_1 and the magnetostriction λ_s of Fe–Si–Al alloys on composition.

be interpreted in a similar way to that observed in rolled Ni–Fe alloys, due in this case to the formation of iron ion pairs.

2.17 IRON–SILICON–ALUMINIUM ALLOYS

The permeabilities of these alloys were first investigated by Masumoto (1936). The alloy containing 10% Si and 5% Al has an initial permeability of about 30 000 and a maximum permeability of about 140 000. These very high values can be attributed to the fact that the cast alloy satisfies the ideal requirement that both $K_1=0$ and $\lambda_s=0$ simultaneously. Indeed, this is the only crystalline material known that does meet this requirement. The variation of K_1 and λ_s with composition is shown in Fig. 2.43. The high permeabilities in the Permalloys are obtained at 'compromise' compositions that make either K_1 or λ_s equal to zero and minimize the other.

Alloys at or close to the composition 85% Fe–10% Si–5% Al also have very low hysteresis loss and high resistivity. Since the alloys are very brittle, they can easily be ground to fine powders with particle diameters of order 10 μm. After annealing to remove residual mechanical damage, they can be pressed to the required shape. These compacted powders have even higher resistivities, so the material, known as Sendust (the alloys were developed by Masumoto in Sendai, Japan), is very suitable for use at high frequencies. Strictly speaking, the name Sendust applies to the alloys 84.9% Fe–9.5% Si–5.6% Al. Adams (1962) has shown that the cast material gives higher permeabilities than sintered powder compacts, although the properties of laminated Sendust compare very favourably with those of Permalloy.

Since cast Sendust is extremely hard and brittle, a very elaborate technique is required to produce the thin sheets necessary for transformer cores.

2.18 NANOCRYSTALLINE HIGH-PERMEABILITY ALLOYS

There is currently considerable interest in using nanocrystalline alloys as soft high permeability materials. The grain size of these alloys is typically about 5–50 nm. Such small grain sizes can be produced by compaction of the powders or by devitrification of the amorphous alloys. Yoshizawa *et al.* (1988) devitrified an alloy 73.5 at% Fe–13.5 at% Si–9 at% B–3 at% Nb–1 at% Cu to produce a nanocrystalline alloy with a grain size of about 10 nm. This technique can also be used to produce nanocrystalline alloys of Co–B–Nn and Fe–B–Zr.

In the nanocrystalline state it is thought that the high magnetic permeabilities are obtained because the magnetocrystalline anisotropy is very much reduced due to local randomization so that there is no distinct easy direction of magnetization. Furthermore the high permeabilities are partly due to the very low magnetostriction. Initial relative permeabilities as high as 100 000 have been observed by Yoshizawa *et al.* (1988) for the above FeSiBNbCu alloy. However, the coercivities and hence the permeabilities are strongly dependent on the grain size. Herzer and Warlimont (1992) have pointed out that even for the same alloy the coercivity may vary from 1.0 A m^{-1} to 5000 A m^{-2} for grain sizes ranging from 10 nm to 40 nm, thus hard and soft magnetic properties can be observed in the same alloy. Schäfer *et al.* (1991) have shown by means of Kerr effect studies on the alloy 73.5 at% Fe–13.5 at% Si–9 at% B–3 at% Nb–1 at% Cu that the domain walls are very wide $\sim 2\ \mu$m indicating that the effective magnetocrystalline anisotropy is low – the domain wall width $\delta = \pi(A/K_1)^{1/2}$ where A is the exchange energy and K_1 is the effect magnetocrystalline anisotropy coefficient.

The high saturation magnetic polarizations $J_s = 1.2–1.3$ T of the new nanocrystalline alloys in combination with initial permeabilities $\sim$100 000 makes them comparable to the nickel–iron Permalloys. Furthermore, like the Permalloys, they respond to heat treatment in a magnetic field so that the shapes of the hysteresis loops can be designed to suit the application requirements. Remanence to saturation ratios ranging from 0.05 to 0.9 may be obtained.

Since the nanocrystalline materials have a relatively high electrical resistivity of about 1.35 $\mu\Omega$m and a ribbon thickness of about 20 μm the eddy current losses are relatively low up to frequencies of about 100 kHz.

There seems to be little doubt that high permeabillity low loss iron-based nanocrystalline alloys will find many applications in the near future, e.g. in inductors, saturable reactors, high frequency transformers and magnetic recording heads. In all of these applications their high saturation magnetic polarization and thermal stability would give them a distinct advantage over many existing materials.

2.19 APPLICATIONS OF HIGH-PERMEABILITY MATERIALS

The above discussion of soft magnetic materials includes only the most important and representative materials in the various groups. For more detailed information and data, including applications to specific devices and machines, the reader is referred to the books by Heck (1974), Chen (1977), Bradley (1971), Tebble and Craik (1969) and Bozorth (1951). For general reviews of the technologically important magnetic materials and their applications, the reader is referred to Chin (1978), Chin and Wernick (1980), Luborsky (1980), Fish (1990) and Moses (1990).

A brief summary of the most important applications of soft magnetic materials, including the high-permeability ferrites, is given in Table 2.10. The range of relative

Table 2.10 Applications of high-permeability materials.

Applications	Requirements
Power applications	
Distribution and power transformers	Low core losses, high permeability, high saturation magnetic polarization
High-quality motors and generators, stators and armatures, switched-mode power supplies	
Instrument transformers	
Audiofrequency transformers	Low core losses, high permeability, high magnetic polarization
Pulse transformers	High permeability
Cores for inductor coils	
Audiofrequency	Low hysteresis, high permeability
Carrier frequency	Very low hysteresis and eddy current loss
Radiofrequency	High permeability at low fields
Miscellaneous	
Relays, switches; Earth leakage circuit	High permeability, low remanence, low coercivity
Magnetic shielding	Low core loss for AC applications
Magnetic recording heads	High initial permeability, low or zero remanence
Magnetic amplifiers; Saturable reactors; Saturable transformers; Transformer cores	Rectangular hysteresis loops, low hysteresis loss
Magnetic shunts for temperature compensation in magnetic circuits	Low Curie temperature, appropriate decrease in permeability with increase in temperature
Electromagnets in indicating instruments, fire detection, quartz watches, electromechanical devices	High permeability, high saturation magnetic polarization
Magnetic yokes in permanent magnet devices, such as lifting and holding magnets, loudspeakers	High permeability, high saturation magnetic polarization

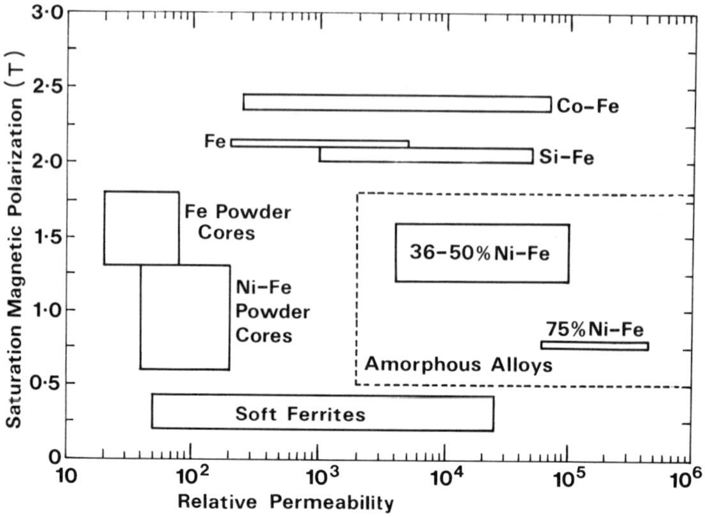

Fig. 2.44 Range of saturation magnetic polarizations and relative maximum permeabilities for various soft magnetic materials.

permeabilities and saturation magnetic polarizations available in soft magnetic materials is shown in Fig. 2.44.

According to Professor H. Warlimont, the world market in 1989 for soft magnetic materials was estimated to be approximately £4500 million pounds sterling, and is shared among the various materials as shown in Table 2.11.

Iron and non-oriented silicon–iron are of course very widely used in motors and magnetic devices in consumer products where high efficiencies are not yet considered of great importance. The other materials are used in applications where high efficiencies are required, and so most current research is devoted to these materials.

Many soft alloys, e.g. 49 Fe–49 Co–2 V Permendur, are now being produced by sintering the powders in controlled atmospheres, but the magnetic properties depend on

Table 2.11 World market for soft magnetic materials in 1989.[a]

Materials	Market share (%)
Fe and non-oriented Si–Fe alloys	54.5
Grain-oriented Si–Fe alloys	30.1
Ferrites	7.6
Ni–Fe, Fe–Co, Ni–Fe–Co alloys	4.5
Powder cores	2.3
Amorphous alloys	1

[a] Magnetic recording media and recording heads have been excluded.

achieving very low porosities. The process is often carried out with a polymeric binder, which is later removed by a special heat treatment.

2.19.1 Special Applications

(*i*) *Transductor core materials*

Transductors are devices that enable the currents and voltages in AC circuits to be controlled by means of the nonlinear properties of magnetic materials. The devices, known as regulators, consist essentially of current coils with ferromagnetic cores. Transductor cores whose inductive reactance can be changed over a wide range by means of a relatively low control current can be used as magnetic amplifiers (Heck, 1974). Transductors can also be used as magnetic switches, voltage and current variation suppressors, and in many other applications.

The requirements for transductor core materials are similar to those for transformer core materials, although ideally the hysteresis loop should be rectangular. The principal materials used are grain-oriented 3.2% Si–Fe, 65% Ni–Fe, 78% Ni–Fe (Permalloy), 50% Ni–Fe (Deltamax) 77% Ni–Fe–5% Cu–2% Cr (Mumetal), 79% Ni–Fe–5% Mo (Supermalloy) and 49% Fe–49% Co–2% V (Supermendur). The principal properties of these alloys have already been given in Table 2.2. The very high permeability (60 000–90 000), saturation magnetic polarization ($J_s = 2.4$ T) and loop rectangularity make Supermendur particularly suitable for transductor cores, but the very high cobalt content makes it very expensive.

(*ii*) *Electromagnetic relay alloys*

An electromagnetic relay is essentially an electromagnet that, by means of a flux change in the core magnetic material, may open or close one or more electrical circuits by attracting an armature that operates the switch. Relays such as those used in telecommunications are operated by very small currents, which may control the switching of much larger currents. The devices are relatively insensitive to temperature variations and to the magnitude of the energizing current, provided that this is above the small threshold value. Relays are generally very simple and reliable devices, and are used in very large numbers. Sometimes the armature may be partially magnetized by a permanent magnet, so that the relay responds to changes in the flux through the armature – these polarized relays are generally much more sensitive than unpolarized relays.

The requirements for the core material are: (a) high saturation magnetic polarization; (b) low coercivity; and (c) high permeability. The remanence need not necessarily be small or zero, because 'freezing' or 'sticking' of the contact in the absence of an energizing current can easily be avoided by interposing a thin non-ferromagnetic spacer material between the contacts. Another requirement of the materials is that their properties should ideally remain constant with time. Repeated operation of the relay in varying temperatures can lead to an increase in the coercivity and a decrease in the permeability. Suitable alloys for relays are listed in Table 2.12.

Table 2.12 Alloys for relays (after Heck, 1974).

Material	$\mu_{max}(\mu_0)$	J_s(T)	H_c (A m^{-1})	B_r(T)
Unalloyed iron:				
Auto-machining iron	4 300–8 000	2.05	60–40	0.8
Open smelted iron	2 200–7 500	2.15	24–120	0.9
Vacuum-smelted iron	—	2.15	16–40	—
Carbonyl iron	30 000	2.15	8–24	0.8
Carbonyl iron strip	40 000	2.15	6–10	0.8
Critically stretched silicon steels:				
Fe–1% Si	4 000–15 000	2.1	30–120	0.9–1.45
Fe–2.5% Si	4 000–12 000	2.0	12–120	0.8–1.2
Fe–4% Si	5 000–20 000	2.0	5–90	0.8–1.2
Nickel steels:				
Fe–36% Ni	6 000–14 000	1.3	8–25	0.8
Fe–50% Ni	15 000–60 000	1.55	5–14	0.8–1.2
Fe–78% Ni	5 000–300 000	0.7	1–8	0.5–0.75

High saturation magnetic polarization and permeabilities are most easily cheaply obtained by using unalloyed iron, which is most suitable for heavy current applications.

(*iii*) *Magnetic recording head materials*

There are four essential requirements for an efficient magnetic recording head material:

(1) high initial permeability;
(2) very low or zero remanence;
(3) good wear resistance;
(4) good corrosion resistance.

Although the 79 Ni–Fe Permalloys have high initial permeability and low remanence, they have poor wear resistance. Miyazaki *et al.* (1972) found that the addition of Nb and Ti to form the alloy Ni–10.5% Fe–3.8% Mo–3% Ti–2.8% Nb resulted in greatly improved wear resistance while maintaining the high initial permeability.

The alloys 16% Al–Fe (Alfenol or Alfer), sintered Fe–9.6% Si–5.4% Al (Sendust) and the soft ferrites Ni–Zn and Mn–Zn are all widely used for recording heads. Although the ferrites have good wear resistance, their initial permeabilities are relatively low. At present, there is no single material in use that combines all of the above material requirements. Currently, magnetic recording heads are usually composites of various materials, e.g. ferrite yokes topped with Sendust poles or ferrite yokes with Sendust film sputtered into the pole gap area.

Amorphous materials such as $Co_{70}Fe_5Si_{15}B_{10}$ and $Co_{90}Zr_{10}$ offer the possibility of combining all the material requirements, including high mechanical strength, and are currently under intensive research. The magnetic properties of various recording head materials are shown in Table 2.13. Another problem associated with recording heads is the 'dead layer' – this is a thin higher-coercivity layer of material produced on the

Table 2.13 Magnetic recording head materials.

Material	J_s (T)	H_c (A m^{-1})	μ (μ_0) at 1 k Hz	T_c (°C)	ρ ($\mu\Omega$ m)	Hardness H_V
77% Ni–17% Fe–4% Mo–2% Cu, Permalloy	0.80	2	20 000	480	1	120
79.9% Ni–10.5% Fe–3.8% Mo–3% Ti–2.8% Nb, Tufperm	0.5	1	40 000	280	0.6	230
84% Fe–16% Al, Alfenol or Alperm	0.9	2	8 000	350	1.5	290
85% Fe–9.6% Si–5.4% Al, Sendust	1.1	2	25 000	500	1	450
Ni–Zn ferrite	0.33	10	2 200	150	10^5	700
Mn–Zn ferrite	0.5	4	5 000	200	10^3	650
$Co_{70}Fe_5Si_{15}B_{10}$ (amorphous)	0.67	—	10 000	—	1.30	1200
$Co_{90}Zr_{10}$ (amorphous)	1.4	4.0	~10 000	—	—	—
$Co_{66}Fe_4(Mo, Si, B)_{30}$ (amorphous)	0.55	—	20 000	—	1.30	900
Co_{69} Mo_6 $(Si,B)_{25}$ (amorphous)	0.8	—	10 000	—	1.30	900

recording head by repeated contact with the tape. This dead layer has a very low permeability, and hence effectively increases the head to tape separation and reduces the efficiency of high-density recording. Although Ni–Zn ferrite heads are more susceptible to this problem than Mn–Zn ferrites, they have higher mechanical hardness. In order to avoid unnecessary wear, the porosity must be kept as low as possible, so the sintering conditions during their manufacture must be very carefully controlled.

Wear and corrosion of the head material in the gap area also have the effect of decreasing recording efficiency by increasing the head–tape separation. Wear of the thin pole tips in composite structures also limits the life of the head. Unfortunately, there is no very clear relationship between hardness and wear rate; for example, some amorphous alloys have twice the hardness of Sendust but wear four times faster. The wear rate will presumably be determined by complex interactions involving friction, elastic and plastic deformation, fracture, corrosion and other chemical properties. Clearly, the design of an efficient magnetic recording head will require a very thorough interdisciplinary approach, since so many material requirements have to be optimized.

A magnetic recording head consists essentially of a coil wound onto a magnetic core with two small gaps, usually filled with non-ferromagnetic materials. This ensures that the flux in the core material and hence in the recording gap is proportional to the signal in the coil. Sometimes the recording gap is filled with a conducting material such as copper. The eddy currents produced in the copper during recording assist the signal flux to bulge into the recording tape. The gap width of the recording head (typically 5–10 μm) must be less than the shortest wavelength to be recorded or read. The quality of the recording depends on the magnetic and mechanical properties of the core material, the gap field and width, the gap–tape separation, the tape material, its thickness, degree of dispersion and alignment, and the tape speed. Thus the recording process is one of great complexity, and many problems remain to be solved. Further details of magnetic recording head materials are given by Heck (1974), Chin and Wernick (1980) and Jones and Mee (1990). For a general discussion of magnetic recording see Middleton (1990), Mee and Daniel (1990) and the book by Mallinson (1987).

(*iv*) *Magnetic shielding*

It is often necessary to shield certain parts of electronic circuits from stray magnetic fields generated by surrounding devices or indeed from the Earth's field. This can be done very effectively by placing the instrument or device inside a cylindrical shell made from magnetic material with a high relative permeability such as Mumetal (77% Ni–16% Fe–5% Cu–2% Cr). The function of the shell is to divert the magnetic flux from the air gap inside the cylinder by conducting it through the high-permeability shell as shown in Fig. 2.45. Since the magnetic potential inside and outside the shell must satisfy the Laplace equation, it can be shown (Page and Adams, 1969) that when the cylindrical shell is placed in a uniform field H_o, the field H_i inside the cylinder is

$$H_i = \frac{4\mu_r H_o}{(\mu_r+1)^2 - (b^2/a^2)(\mu_r-1)^2},$$

where a is the outer radius and b the inner radius of the shell. If $\mu_r \gg 1$, as is usually the case for Ni–Fe and Si–Fe alloys,

$$H_i = \frac{4}{\mu_r}\frac{H_o}{1-b^2/a^2}.$$

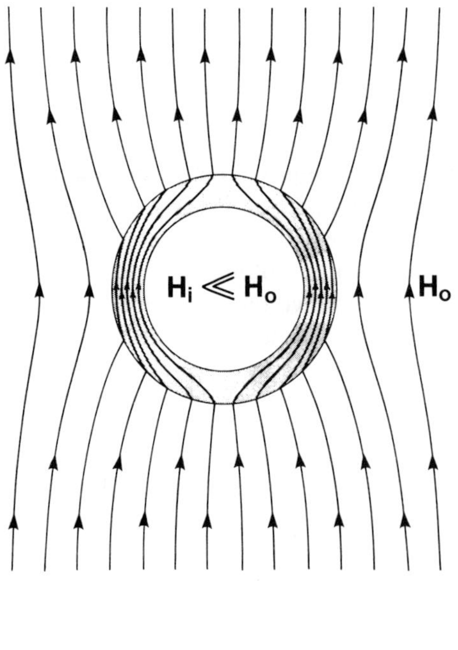

Fig. 2.45 Magnetic shielding by a cylindrical shell of high-permeability material. The field H_i inside the cylinder is very much less than that outside, H_o (after Jackson, 1962).

Thus for effective shielding, i.e. H_i as small as possible, the ratio b/a should be as small as possible. The internal field H_i can of course be reduced to very low values by using several concentric shields.

2.19.2 Low-Curie-Temperature Ni–Fe and Ni–Cu Alloys

There are many applications for which it is important to have a warning of relatively small temperature changes, e.g. in fire alarms or industrial processes. The warning can often be conveniently triggered magnetically by using a ferromagnetic material with a low Curie point, so that when the Curie temperature is reached, the spontaneous magnetization of the material disappears, i.e. the permeability decreases from a large value to a value just greater than one. The Ni–Fe alloys with nickel contents in the range 28–34% have Curie points near or below room temperature, so that for temperature changes in this range they are useful for such applications.

Stäblein (1934) suggested that since the J_s versus T curve is nonlinear near the Curie temperature, alloys with composition in the range 30–33% Ni–Fe be heavily cold-rolled so as to produce some heterogeneity in the composition. Since the Curie temperatures of Ni–Fe alloys in the 28–34% Ni range are very sensitive to composition (Fig. 2.46), the cold rolling produces regions with varying Curie points, which, when superimposed in a single alloy, provide a more accurately linear negative temperature coefficient of saturation magnetic polarization over a range of several tens of degrees Celsius. The precise working range of the alloys can be modified by the addition of small quantities of Cu or Mo, which increase the linearity range.

If a low-Curie-temperature alloy is operated as a transformer with a fixed secondary resistance, the coupling and power dissipation increase as the temperature decreases, so forming an effective de-icing device for power lines.

Nickel–copper alloys with a copper content less than 62 wt% Cu (alloys with Cu content greater than 62% are not ferromagnetic) can also be used in devices requiring

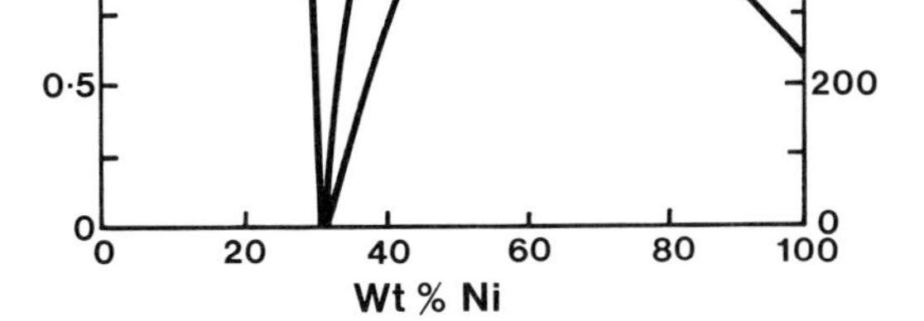

Fig. 2.46 Saturation magnetic polarization J_s and Curie temperature T_c of Ni–Fe alloys as a function of Ni content.

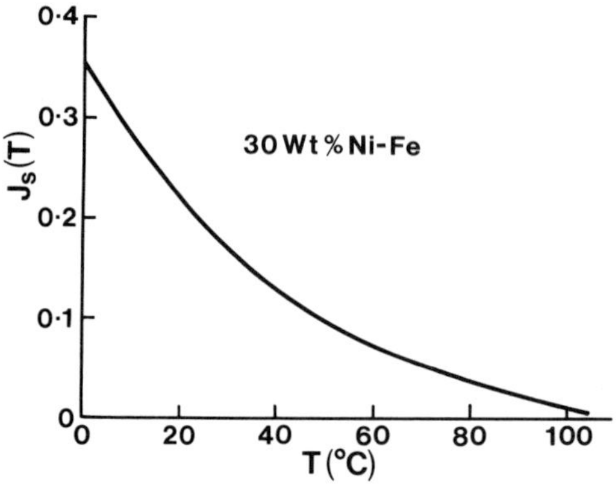

Fig. 2.47 Temperature variation of the saturation magnetic polarization of a 30 wt% Ni–Fe alloy.

a low Curie temperature material. The saturation magnetic polarizations and Curie temperatures decrease approximately linearly with increasing copper content.

2.19.3 Temperature Compensation Alloys

Another important application of Ni–Fe alloys is in magnetic circuits that require compensation for temperature fluctuations. Suppose that in a magnetic circuit a permanent magnet is required to produce a given magnetic flux that must not vary appreciably with temperature, e.g. in a measuring instrument. This can be achieved if the magnetic circuit is shunted with an Ni–Fe alloy whose permeability decreases with increase in temperature. This additional circuit element will of course direct some of the available flux from the main circuit. When there is a rise in temperature, the flux from the permanent magnet falls, but so too does the permeability of the Ni–Fe shunt, so that a larger fraction of the total available flux from the magnet is now available for the main circuit. Since the permeability of this circuit will be chosen to be much less sensitive to temperature fluctuations than the shunt, the available flux in the main circuit will be very nearly constant. The temperature dependence of the saturation magnetic polarization for a 30% Ni–Fe temperature compensation alloy is shown in Fig. 2.47. The field dependences of the permeability are also shown. The 30% Ni–Fe alloys (known as Thermoperm; Stäblein, 1934) can also be used to make temperature-sensitive inductor cores, so that an increase in temperature results in a decrease in the inductance due to the decrease in the permeability of the core. These materials may be used in temperature controllers, temperature–frequency converters and on–off switches.

2.20 NON-FERROMAGNETIC Cu–Ni ALLOYS

In making measurements in the presence of high magnetic fields it is often necessary to use non-ferromagnetic materials in order to avoid problems created by mechanical forces or stray magnetic flux. This is also an important consideration in the design of accurate measuring instruments that depend on magnetic materials. Both of these problems can be avoided by using a completely non-magnetic material, i.e. a material that is not ferromagnetic, paramagnetic or diamagnetic. Such a material has been developed by Pugh (1958), who has shown that an alloy containing 96.3% Cu and 3.7% Ni by weight has a susceptibility of zero at 300 K and a maximum susceptibility of only $+1.1\times10^{-6}$ at 4 K. Throughout the temperature range 300 K to 20 K, the susceptibility lies within the range zero to -0.004×10^{-6}. In order to achieve these exceptionally low values, the alloy must be extremely pure. It should be noted that the diamagnetic susceptibility of pure Cu is -9.6×10^{-6}. The Cu–3.7% Ni alloy is therefore considerably less magnetic than pure Cu.

2.21 NON-FERROMAGNETIC AND VERY-LOW-PERMEABILITY STEELS AND ALLOYS

In some applications it is necessary to avoid the use of ferromagnetic materials, particularly where they are very close to stray fields from permanent magnets, solenoids or other electromagnetic devices. Materials that have a small remanent magnetization may have the undesirable property of attracting unwanted ferromagnetic particles, which can cause contamination or malfunctioning of a particular instrument or device. The simplest way to avoid these problems is to use a diamagnetic material such as copper or a paramagnetic material such as aluminium.

Unfortunately, this is not always practicable or economic, so that materials containing ferromagnetic metals such as Fe, Co or Ni have to be used, either because of their mechanical properties or their corrosion resistance, e.g. the stainless steels. When a non-ferromagnetic material is required, it is usual to specify that its relative permeability should be below a given value. For example, the ferromagnetism of an alloy with a relative permeability $\mu_r = 1.01$ is difficult to detect except by means of a sensitive instrument, so that for most practical purposes it can be regarded as non-ferromagnetic. However, in an exceptionally high magnetic field and field gradient the force on the material may be appreciable. Thus the specification and choice of a non-ferromagnetic material for a particular application must be given careful consideration, particularly if the material is to be used at low temperatures, where it may be strongly paramagnetic. Similar care is necessary if the material is to be cold-worked. Consider, for example, the well-known 18/8 austenitic stainless steel, which contains 18 wt% Cr–8% Ni–Fe. The fully austenitic 18/8 stainless steel is non-ferromagnetic down to about 77 K, owing to the presence of the strongly paramagnetic fcc phase, which is stabilized by the nickel content. However, if the alloy is cold-worked at room temperature, a significant degree of ferromagnetism is induced owing to a transformation from the paramagnetic fcc phase to the ferromagnetic martensitic phase. The alloy remains ferromagnetic until it is heated to at least 600°C, above which it rapidly transforms to the paramagnetic fcc phase. The ferromagnetism of cold-worked 18/8 stainless steel, which can easily be detected by a small magnet, is strongest at highly deformed points such as corners or other areas with a high degree of curvature. Effects due to the increased paramagnetism at low temperatures may also arise, but these may be much less serious than those due to ferromagnetism resulting from cold working.

The magnetic properties of some austenitic stainless steels in the annealed and cold-worked conditions are shown in Tables 2.14 and 2.15.

Table 2.14 Relative permeabilities of some austenitic stainless steels in the annealed and cold-rolled conditions (after Stanley, 1963).

	Composition (wt%) (the balance is Fe)					Magnetic permeability μ_r at 8 kA m^{-1}	Cold reduction (%)			
	C	Mn	Cr	Ni	Others	Annealed	10	20	40	70
(a)	0.11	1.26	17.15	7.41	—	1.010	1.56	6.64	21.3	40.0
(b)	0.09	0.49	18.30	8.34	—	1.007	1.18	3.11	12.3	36.8
(c)	0.06	0.58	18.48	10.18	—	1.004	1.006	1.038	1.36	2.9
(d)	0.06	1.82	17.32	12.54	2.3 Mo	1.003	1.004	1.004	1.005	1.009
(e)	0.13	2.09	27.48	21.48	—	1.005	1.004	1.004	1.004	1.007

Table 2.15 Magnetic properties of austenitic stainless steels after cold reduction (after Stanley, 1963).[a]

Composition (wt%) (the balance is Fe)				Remanence (T) Cold reduction (%)			Coercivity (kA m^{-1}) Cold reduction (%)		
C	Mn	Cr	Ni	70	80	90	70	80	90
0.05	0.63	18.85	8.80	0.385	0.48	0.57	6.4	4.8	3.6
0.05	0.58	18.81	10.59	0.010	0.10	0.22	22.4	18.4	15.2

[a]The maximum applied field was 32 kA m^{-1}.

Table 2.16 Relative permeabilities of non-ferromagnetic iron-based alloys (after Stanley, 1963).

Alloy	Composition (wt%) (the balance is Fe)						Permeability of annealed alloy
	C	Si	Mn	Cr	Ni	Others	
			Wrought alloys				
Manganese steel	1.2	0.6	12	—	—	—	1.03–1.3
Mn+Bi	0.5	—	19	—	2	0.5 Bi	1.003
MnCr 18	0.4	0.5	18	4	—	0.1 N	1.01
MnNiCr 14	0.5	0.5	14	3.5	6	—	1.01
CrNi 12–12	0.1	1.00	2	12	13	—	1.01
Mn–Ni	0.3	—	11	—	8	—	1.003
			Cast alloys				
Mn 12	1.2	0.4	12	—	—	—	1.01
MnCr 18	0.4	0.5	18	3	—	0.1 N	1.01
Cast iron	2.7	2.5	4.2	—	8.5	—	1.05
Cast steel	0.25	0.5	11	—	6	—	1.02

Thus, from Tables 2.14 and 2.15, it can be seen that if the material is to be cold-worked, the stainless steels with compositions near to 18 wt% Cr–8 wt% Ni may not be satisfactory. Note the stability of steels (d) and (e) in Table 2.14, which show very little change in permeability even after a cold reduction of 70%. Apart from the austenitic stainless steels, there are many non-ferromagnetic iron-based alloys, a selection of which is shown in Table 2.16.

2.22 Ni–Fe ALLOYS WITH VERY LOW THERMAL EXPANSION COEFFICIENTS AND OTHER INVAR ALLOYS

Ni–Fe alloys with compositions in the range 35–50% Ni have long been used in applications where very low thermal expansion is required – particularly when it is necessary to match one material to another, such as in glass-to-metal seals. Another application of the alloys is for pipelines to carry liquid gases such as liquid oxygen. The remarkably low thermal expansion anomaly in the 35% Ni–Fe alloy is by no means restricted to this composition nor indeed is it peculiar to Ni–Fe alloys, since it is now known that the effect is exhibited by a large number of alloys, e.g. Ni–Fe with Ni contents between 42 and 50%, as well as Ni–Al alloys.

The thermal expansion coefficient in these alloys is very small, typically $\alpha \approx 50 \times 10^{-6}$, probably because the normal thermal expansion is compensated by a temperature decrease in the volume magnetostriction, although this general conclusion is highly over-simplified. Wohlfarth (1975) has pointed out that the Invar effect is extremely complex, and that the anomalously low thermal expansion coefficient is just one aspect of the behaviour. The alloys are generally known as Invar-type alloys after the original Ni–Fe Invars. Even smaller thermal expansion coefficients have been observed in alloys 31% Ni–Fe–(4–6%)Co, which are known as Superinvars. Stainless Invars have compositions close to Fe–52% Co–10% Cr.

The elastic moduli of many Invar-type alloys are known to be temperature-independent, a property that is also due to compensation via the volume magnetostriction and hence the magnetoelastic coefficients. These alloys are generally known as Elinvars. One such alloy has the composition 35% Ni–12% Cr–Fe.

Other commercially important Elinvars include Fe–40% Ni, Fe–36% Ni–12% Cr, Superelinvar Fe–36% Ni–7% Co–7% Cr, Co-Elinvar Co–30% Fe–10% Cr and the antiferromagnetic Elinvars such as Mn–20% Ni–8% Fe, Cr–14% Fe–1% Mn and Mn–50% Cu. The antiferromagnetic Elinvars are used in applications where the ferromagnetism is a disadvantage, e.g. in hairsprings for watches and delay lines. Small quantities of other elements are added to all the Invar and Elinvar alloys in order to increase their mechanical strength and resistance to ageing effects, but it is essential that the alloys contain at least one 3d transition metal.

The Invar effect has also been observed in amorphous alloys and in 4f-lanthanide (rare earth)–transition metal compounds with Laves structures, e.g. $TbCo_2$ or even in $Nd_2Fe_{14}B$, but there are no purely lanthanide or insulating Invars.

The Invars and Elinvars are used in a very wide range of mechanical and electronic instruments.

Reviews of the theory and properties of various Invar alloys have been given by Wohlfarth (1975), Nakamura (1976) and Wasserman (1990).

2.23 MAGNETOSTRICTION AND MAGNETOELASTICITY

2.23.1 Magnetostriction

When the state of magnetization of a ferromagnetic material is changed by an applied field, there is an accompanying change in its dimensions, provided that part or all of the magnetization change is due to rotation of the magnetization vector, rather than to domain wall displacement.

Although there are several magnetostrictive effects, e.g. volume and torsional changes, we shall consider only the longitudinal effect, i.e. fractional changes in the length. The magnetostriction λ, is defined as

$$\lambda = \frac{\Delta l}{l},$$

where l is the original length of the specimen. Since the effect is strongly dependent on the magnitude of the applied field and is also highly anisotropic, it is essential to state the exact conditions of measurement. When the magnetostriction is that observed when the specimen magnetization is changed from zero to saturation, it is referred to as the saturation magnetostriction and is denoted by λ_s (typically, $\lambda_s \approx 10^{-5}$). The magnetostrictive strain may be positive, negative or, in some special alloys, zero. Since the phenomenon is highly anisotropic, the sign of the magnetostriction may vary for different crystallographic directions, even in the same material. In a single crystal of iron magnetization in the [100] direction produces positive magnetostriction, whereas, with magnetization in the [111] direction, the magnetostriction is negative. In the [110] direction, however, the magnetostriction may be positive, negative or zero, depending on the applied field or the magnetic polarization produced by it. The phenonemon is very complex, and requires many parameters for its complete description (Birss, 1964).

2.23.2 Magnetostrictive Materials and their Applications

Magnetostriction and magnetoelastic effects are fundamental to all ferro- and ferrimagnetic materials. For soft magnetic materials, i.e. those with high permeability, we usually require the magnetostriction to be as small as possible, whereas in hard magnetic materials its presence may be regarded as beneficial, although it is usually of minor importance. However, there are many applications in which the phenomenon can be used directly. If a ferro- or ferrimagnetic material is subjected to an alternating applied field, it will oscillate mechanically. Thus certain magnetostrictive materials can be used as magnetomechanical transducers for vibrators, ultrasonic machines, sonar, mechanical vibration filters and resonators, delay lines, and sound generators. For such applications we require high magnetostriction, i.e. $|\lambda_s| > 20 \times 10^{-6}$, high electrical resistivity to minimize eddy currents, a high magnetomechanical coupling factor, and, for stability, a modulus of elasticity that is not strongly temperature-dependent. The saturation magnetostrictions of various materials at room temperature are given in Table 2.17.

Table 2.17 Saturation magnetostriction λ_s of various materials at room temperature.

Material	Saturation magnetostriction λ_s (10^{-6})
Iron	-7
Fe–3.2% Si	$+9$
Nickel	-33
45 Permalloy, 45% Ni–55% Fe	$+27$
82% Ni–18% Fe Permalloy	0
Permendur, 49% Co–49% Fe–2% V	$+70$
87% Fe–13% Al Alfer	$+30$
Magnetite, Fe_3O_4	$+40$
Cobalt ferrite, $CoFe_2O_4$	-110
Cobalt ferrite, $Co_{1.1}Fe_{1.9}O_4$	-250
$SmFe_2$	-1560
$TbFe_2$	$+1753$
$Tb_{0.3}Dy_{0.7}Fe_{1.93}$ (Terfenol D)	$+2000$
$Fe_{66}Co_{18}B_{15}Si_1$ (amorphous)	$+35$
$Co_{72}Fe_3B_6Al_3$ (amorphous)	0

Note the exceptionally high values of λ_s for $SmFe_2$ and $Tb_{0.3}Dy_{0.7}Fe_{1.93}$. The latter alloy has many applications, and is known commercially as Terfenol D.

According to Clark (1980), these exceptionally high values are due to the highly anisotropic shapes of the Sm (prolate ellipsoidal) and Tb (oblate ellipsoidal) ions and to the large lanthanide–iron exchange energy. Pseudobinaries of $SmFe_2$ and $TbFe_2$ with $DyFe_2$, $CeFe_2$, $PrFe_2$, $YbFe_2$ and $HoFe_2$ also have large magnetostrictions, of about 2000×10^{-6}. All of these materials have the cubic Laves-phase structure, so they can be alloyed very easily. The mixed Laves-phase alloys have the advantage that the magnetocrystalline anisotropy coefficient is lower or even zero, so that large magnetostrictions can be obtained in relatively low applied fields, which is a great advantage in practical applications.

The high magnetostriction in the materials containing cobalt is due to the presence of the cobalt ion Co^{2+}, from which there is a large orbital contribution to the magnetic moment and hence a strong spin–orbit coupling, which increases the magnetostriction. However, an increase in the spin–orbit coupling often leads to a higher magnetocrystalline anisotropy energy, so that materials with lower magnetocrystalline anisotropy energies are preferred, because they can be saturated in lower applied fields. At high frequencies the ferrites have the advantage that the eddy current losses are negligible.

Apart from the very high saturation magnetostriction of $Tb_{0.3}Dy_{0.7}Fe_{1.93}$, the alloy has a very high magnetomechanical coupling coefficient k_{33}, and is also piezomagnetic, i.e. the domain magnetization vectors can be rotated by the application of either a tensile or a compressive stress. The magnetostriction of Terfenol D as a function of the applied field for various values of compressive stress is shown in Fig. 2.48, from which it can be seen that pre-stressing the alloy results in a very significant increase in the

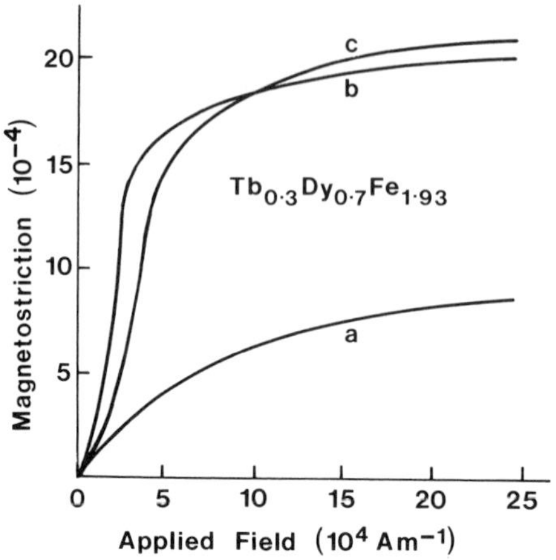

Fig. 2.48 Magnetostriction of Terfenol D ($Tb_{0.3}Dy_{0.7}Fe_{1.93}$) as a function of applied field for three values of applied compressive stress: (a) 0; (b) 6.9 MPa; (c) 13.8 MPa. (Courtesy of Etrema Products, Inc., Ames, USA).

saturation magnetostriction. Unfortunately, there is some hysteresis in the λ_s versus H curves, so careful use of the manufacturer's data must be made when designing practical devices. The performance of $Tb_{0.3}Dy_{0.7}Fe_{1.92}$ for magnetostrictive devices has been investigated by Jenner *et al.* (1990), who give details of the dependence of the magnetomechanical coupling coefficient k_{33}, and the magnetostriction coefficient d_{33} $(= d\lambda/dH)$ as functions of applied compressive stress and applied field.

Currently, Terfenol D is being used in many applications, such as sonar for oceanography, sea farming, oil field mapping and defence systems, micropositioners for fine focusing of laser mirrors, telescopes and electronic microscopes, hydraulics, pumps, high-speed valves and field injection systems, ultrasonic cleaning, emulsification and separation, robotics for the accurate control of joints, electrical power transmission for high contact force in switches and relays and for vibration cancellation.

Clearly, for any particular application, the material must be chosen very carefully, since a number of magnetic, electrical and mechanical requirements must be satisfied simultaneously. For further details of magnetostrictive materials and their applications the reader should consult the book by Heck (1974), and the review by Clark (1980) and the survey of magnetic sensors edited by Boll and Overshott (1989).

2.23.3 Magnetoelastic Energy

When a ferromagnetic material is exposed to a magnetic field or cooled below its Curie temperature, the magnetostriction produces an elastic strain in the material that either increases or decreases the magnetoelastic energy, depending on the conditions. Thus, in general, all ferromagnetic materials possess a small magnetoelastic energy.

The interaction between an applied stress and magnetostriction is an important effect, and may have a marked influence on the field-dependent magnetic properties. The permeability of soft magnetic materials such as the nickel–iron alloys, e.g. Permalloy (68% Ni–32% Fe) is particularly sensitive to stress. A tensile stress of 19.6 MN m^{-2} can increase the permeability from 3000 to 26 000. Thus for materials with positive magnetostriction such as 68% Ni–32% Fe Permalloy a tensile stress assists in the magnetization in the direction of stress, so in general we should expect an increase in permeability. Conversely, if a compressive stress is applied to a material with positive magnetostriction, there is a decrease in permeability. For materials with negative magnetostriction, such as nickel, the permeability is increased by compressive stresses. The effects of stress on polycrystalline iron are very complicated, because the magnetostriction is positive in low fields and negative in high fields. In low fields tensile stresses increase the flux density at a given field strength compared with compressive stresses, but at a critical field strength this situation is reversed, so that it is advantagegous to use compressive stresses.

Although the directions of the domain magnetizations are principally determined by the anisotropy constants, the presence of an applied stress may modify the magnetic properties considerably, especially when the magnetocrystalline anisotropy energy is low, as is the case for soft high-permeability materials such as Ni–Fe Permalloys and Si–Fe alloys. The relative effects of the stress will of course be determined by the magnitude and direction of the stress, and the sign and magnitude of the magnetostriction constants compared with the magnetocrystalline anisotropy energy.

2.24 MAGNETORESISTANCE

The magnetoresistance of a material is the change in electrical resistance (or, strictly speaking, the resistivity) when it is subjected to a magnetic field. The magnetoresistance is generally anisotropic, and its magnitude depends on the angle between the field direction and the direction of current flow. In general, the resistivity increases in the presence of the field, but does so nonlinearly. The effect is greatest when the field is perpendicular to the direction of current flow. In ferromagnetic materials the magnetoresistive change $\Delta\rho/\rho \approx 2$–3%, although a change of about 150% can be observed in diamagnetic bismuth when subjected to a field of order 10 A m^{-1}! Even larger changes can be observed in semiconductors, but these are very temperature-sensitive.

Magnetoresistors, provided they are calibrated, are very useful for the measurement and detection of magnetic fields, and can be used to read magnetic tickets, detect the presence of magnetic bubbles in memories, and to read magnetically recorded tapes and disks.

Magnetoresistors for reading magnetically recorded information are commonly made from thin films of Ni–Fe (often 80% Ni–Fe), Ni–Co and Co–Fe, for which the magnetoresistive change is quite large (Hunt, 1971). A more nearly linear response can be obtained by biasing the magnetoresistor by means of a thin-film permanent magnet such as Fe_3O_4, CoPt (Hill *et al.*, 1989) or TbCo. For further details on magnetoresistivity the reader is referred to Thompson *et al.* (1975) and especially to the review by McGuire and Potter (1975), who give tables of the properties of magnetoresistive materials. Negative magnetoresistivity of about 2.6% has been observed in Co–3% Ir and many other related alloys by McGuire *et al.* (1984). A review of magnetoresistance and its applications has been published by Dibbern (1989).

REFERENCES

Adams, E. (1962). *J. Appl. Phys.* **33**, 1214.
Archenhold, W. F., Sandham, H. F. and Thompson, J. E. (1960). *Br. J. Appl. Phys.* **11**, 46.
Assmus, F., Detent, K. and Ibe, G. (1957). *Z. Metallk.* **48**, 341.
Beckley, P., Snell, D. and Lockhart, C. (1985). *J. Appl. Phys.* **57**, 4212.
Benford, J. G. (1974). *IEEE Trans. Magn.* **MAG-10**, 1052.
Berkowitz, A. E. and Kneller, E. (eds) (1969). *Magnetism and Metallurgy*, Vols 1 and 2 (Academic Press, New York).
Birkenbeil, H. J. and Cahn, R. W. (1962). *Proc. Rhys. Soc. Lond.* **79**, 831.
Birss, R. R. (1964). *Magnetism and Symmetry* (North-Holland, Amsterdam).
Bishop, J. E. L. (1976a). *IEEE Trans. Magn.* **MAG-12**, 21.
Bishop, J. E. L. (1976b). *IEEE Trans. Magn.* **MAG-12**, 248.
Bishop, J. E. L. (1978). *J. Magn. Magn. Mater.* **10**, 75.
Bishop, J. E. L. and Lee, E. W. (1963). *Proc. R. Soc. Lond.* **A276**, 96.
Boll, R. (ed.) (1979). *Soft Magnetic Materials: Fundamentals, Alloys, Properties, Products, Applications* (Siemens AG/Heydon & Sons Ltd, London).
Boll, R. and Overshott, K. J. (eds) (1989). *Magnetic Sensors* (Vol. 5 of *Sensors: A Comprehensive Survey*, ed. W. Göpel, J. Hesse and J. N. Zemel) (VCH, Weinheim).
Boon, C. R. and Robey, J. A. (1969). *Phys. Stat. Sol.* **33**, 617.
Boon, C. R. and Thompson, J. E. (1964). *Proc. IEE* **111**, 605.
Boon, C. R. and Thompson, J. E. (1965). *Proc. IEE* **112**, 2147.
Bozorth, R. M. (1951). *Ferromagnetism* (Van Nostrand, New York).
Bozorth, R. M. (1953). *Rev. Mod. Phys.* **25**, 42.
Bradley, F. N. (1971). *Materials for Magnetic Functions* (Hayden, New York).
Brailsford, F. (1938). *J. IEE* **83**, 566.
Braislford, F. (1939). *J. IEE* **84**, 399.
Brailsford, F. (1966). *Physical Principles of Magnetism* (Van Nostrand, London).
Carr, W. J. (1959). *J. Appl. Phys.* **30** (Suppl.), 90S.
Carr, W. J. (1976) *J. Appl. Phys.* **47**, 4176.
Chen, C. W. (1961). *J. Appl. Phys.* **32** (Suppl.), 348S.
Chen, C. W. (1977). *Magnetism and Metallurgy of Soft Magnetic Materials* (North-Holland, Amsterdam).
Chen, C. W. (1978). *J. Magn. Magn. Mater.* **7**, 308.
Chikazumi, S. (1952). *Phys. Rev.* **85**, 918.
Chikazumi, S. (1958). *J. Appl. Phys.* **29**, 346.
Chikazumi, S. and Graham, C. D. (1969). *Magnetism and Metallurgy*, ed. A. E. Berkowitz and E. Kneller (Academic Press, New York).
Chikazumi, S. and Oomura, T. (1955). *J. Phys. Soc. Japan* **10**, 842.
Chikazumi, S., Suzuki, K. and Iwata, H. (1957). *J. Phys. Soc. Japan* **12**, 1259.
Chin, G. Y. (1978). *J. Magn. Magn. Mater.* **9**, 283.
Chin, G. Y. and Wernick, J. H. (1980). *Ferromagnetic Materials: A Handbook on the Properties of Magnetically Ordered Substances*, Vol. 2, ed. E. P. Wohlfarth (North-Holland, Amsterdam), p. 55.
Cioffi, P. P. (1934). *Phys. Rev.* **45**, 742.
Cioffi, P. P., Williams, H. J. and Bozorth, R. M. (1937). *Phys. Rev.* **51**, 1009.
Clark, A. E. (1980). *Ferromagnetic Materials: A Handbook on the Properties of Magnetically Ordered Substances*, Vol. 1, ed. E. P. Wohlfarth (North-Holland, Amsterdam), p. 531.
Crangle, J. (1954). *Br. J. Appl. Phys.* **5**, 151.
Cullity, B. D. (1972). *Introduction to Magnetic Materials* (Addison-Wesley, Reading, Mass.).

Dibbern, U. (1989). in *Sensors: A Comprehensive Survey*, ed. W. Göpel, J. Hesse and J. N. Zemel, Vol. 5; *Magnetic Sensors*, ed. R. Boll and K. J. Overshott (VCH, Weinheim).
Dragoshansky, Y. N., Zyakova, V. A., Khan, Y. B. and Veksler, A. Z. (1972). *Phys. Met. Metallov.* **34**, 82.
Dunn, C. G. and Walter, J. L. (1959). *J. Appl. Phys.* **30**, 1067.
Enoch, R. D. and Fudge, A. D. (1966). *Br. J. Appl. Phys.* **17**, 623.
Fiedler, H. C. (1977). *IEEE Trans. Magn.* **MAG-13**, 1433.
Fish, G. E. (1990). *Proc. IEEE* **78**, 947.
Foster, K. and Littmann, M. F. (1985). *J. Appl. Phys.* **57**, 4203.
Fukawa, K. and Yamamoto, T. (1982). *IEEE Trans. Magn.* **MAG-18**, 963.
Goss, N. P. (1935). *Trans. Am. Soc. Metals* **23**, 511.
Goto, I., Matoba, I., Imanaka, T., Gotoh, T. and Kan, T. (1975). *Proc. EPS Conf. Soft Magnetic Materials*, Part 2 (University College, Cardiff), p. 262.
Gould, J. E. (1971). *Cobalt Alloy Permanent Magnets* (Centre d'Information du Cobalt, Brussels).
Graham, C. D. (1969). *Magnetism and Metallurgy*, Part 2, ed. A. E. Berkowitz and E. Kneller (Academic Press, New York), p. 723.
Grimwood, G. C., Campbell, A. M. and Evetts, J. E. (1978). *IEEE Trans. Magn.* **MAG-14**, 359.
Heck, C. (1974). *Magnetic Materials and their Applications* (Butterworth, London).
Herzer, G. and Warlimont, H. (1992). *Concise Encyclopedia of Magnetic and Superconducting Materials*, ed. J. Evetts (Pergamon, Oxford).
Hill, E. W., Halter, D. S. and Birtwhistle, J. K. (1989). *IEEE Trans. Magn.* **MAG-25**, 3836.
Hill, S. and Overshott, K. J. (1978). *IEEE Trans. Magn.* **MAGN-14**, 773.
Hunt, R. (1971). *IEEE Trans. Magn.* **MAG-7**, 150.
Ichijima, I., Nakamura, M., Nozawa, T. and Nakata, T. (1984). *IEEE Trans. Magn.* **MAG-20**, 1557.
Iuchi, T., Yamaguchi, S., Ichiyama, T., Nakamura, M., Ishimoto, T. and Kuroki, K. (1982). *J. Appl. Phys.* **53**, 2410.
Jackson, J. D. (1962). *Classical Electrodynamics* (Wiley, New York) p. 164.
Jenner, A. G., Greenough, R. D., Wilkinson, A. J. and Parvinmher, A. (1990). *IEEE Trans. Magn.* **MAG-26**, 2589.
Jones, Jr, R. E. and Mee, C. D. (1990). *Magnetic Recording Handbook*, ed. C. D. Mee and E. D. Daniel (McGraw-Hill, London).
Joslin, P. C., Moses, A. J. and Thompson, J. E. (1970). *Proc. IEE* **119**, 709.
Kornetzki, M. and Lucas, I. (1955). *Z. Phys.* **142**, 70.
Lee, E. W. (1958). *Proc. IEE* **C105**, 337.
Lee, E. W. (1960). *Proc. IEE* **C107**, 257.
Lee, E. W. and Lynch, A. C. (1959) *Adv. Phys.* **8**, 292.
Littmann, M. F. (1967). *J. Appl. Phys.* **38**, 1104.
Littmann, M. E. (1971). *IEEE Trans. Magn.* **MAG-7**, 48.
Lommel, J. M. (1964). *J. Appl. Phys.* **35**, 970.
Luborsky, F. E. (1980). *Ferrogmagnetic Materials: A Handbook on the Properties of Magnetically Ordered Substances*, Vol. 1, ed. E. P. Wohlfarth (North-Holland, Amsterdam), p. 451.
McCurrie, R. A., Martin, P. J. and Ashurst, S. J. (1975). *J. Mater. Sci.* **10**, 1788.
McGuire, T. R. and Potter, R. I. (1975). *IEEE Trans. Magn.* **MAG-11**, 1018.
McGuire, T. R., Aboaf, J. A. and Klokholm, E. (1984). *IEEE Trans. Magn.* **MAG-20**, 972.
Major, R. V. and Samadian, V. (1989). *J. Mater. Eng.* **11**, 27.
Mallinson, J. C. (1987). *The Foundations of Magnetic Recording* (Academic Press, London).
Masumoto, H. (1936). *Sci. Rep. Tohoku Imp. Univ.* **25**, 388.
Middleton, B. K. (1990). *Magnetic and Recording Handbook*, ed. C. D. Mee and E. D. Daniel (McGraw-Hill, London).

Miyazaki, T., Sawada, R. and Ishijima, Y. (1972). *IEEE Trans. Magn.* **MAG-8**, 501.
Moses, A. (1974). *J. Mater. Sci.* **9**, 217.
Moses, A. J. (1990). *Proc. IEE* **A137**, 233.
Moses, A. J. and Thomas, B. (1973a). *IEEE Trans. Magn.* **MAG-9**, 651.
Moses, A. J. and Thomas, B. (1973b). *IEEE Trans. Magn.* **MAG-9**, 655.
Moses, A. J., Thomas, B. and Thompson, J. E. (1972). *IEEE Trans. Magn.* **MAG-8**, 785.
Nakamura, Y. (1976). *IEEE Trans. Magn.* **MAG-12**, 278.
Nakamura, Y., Hirose, K., Nozawa, T. and Matsuo, M. (1987). *IEEE Trans. Magn.* **MAG-23**, 3074.
Narita, K. and Yamaguchi, T. (1974). *IEEE Trans. Magn.* **MAG-10**, 165.
Narita, K. and Yamishiro, Y. (1980). *IEEE Trans. Magn.* **MAG-16**, 728.
Néel, L. (1954). *J. Phys. Radium* **15**, 225.
Nozawa, T., Yamamoto, T., Matsuo, Y. and Ohya, Y. (1977). *Proc. 3rd Conf. Soft Magnetic Materials.*
Nozawa, T., Yamamoto, T., Matsuo, Y. and Ohya, Y. (1978). *IEEE Trans. Magn.* **MAG-14**, 252.
Nozawa, T., Yamamoto, T., Matsuo, Y. and Ohya, Y. (1979). *IEEE Trans. Magn.* **MAG-15**, 972.
O'Kelly, D. (1975). *IEEE Trans. Magn.* **MAG-11**, 55.
Overshott, K. J. (1976). *IEEE Trans. Magn.* **MAG-12**, 840.
Overshott, K. J. and Thompson, J. E. (1970). *Proc. IEE* **117**, 865.
Overshott, K. J., Preece, I. and Thompson, J. E. (1966). *Proc. IEE* **115**, 1840.
Page, L. and Adams, N. I. (1969). *Principles of Electricity*, 4th Edn (Van Nostrand, Princeton).
Paxton, W. S. and Nilan, T. G. (1955). *J. Appl. Phys.* **26**, 994.
Pfutzner, H. and Zehetbauer, M. (1982). *Japan. J. Appl. Phys.* **21**, L580.
Polivanov, K. M. (1952). *Izv. Akad. Nauk SSSR, Ser. Fiz.* **16**, 449.
Pry, R. H. and Bean, C. P. (1958). *J. Appl. Phys.* **29**, 532.
Pugh, E. W. (1958). *Rev. Sci. Instr.* **29**, 1118.
Rassmann, G. and Hofmann, U. (1968). *J. Appl. Phys.* **39**, 603.
Sato, T., Kuroki, K. and Tanaka, O. (1978). *IEEE Trans. Magn.* **MAG-14**, 350.
Schafer, R., Hubert, A. and Herzer, G. (1991). *J. Appl. Phys.* **69**, 5325.
Shilling, J. W. and Houze, J. L. (1974). *IEEE Trans. Magn.* **MAG-10**, 195.
Shilling, J. W., Morris, W. G., Osborn, M. L. and Rao, P. (1978). *IEEE Trans. Magn.* **MAB-14**, 104.
Shimanaka, H., Matoba, I., Ichida, T., Koyabashi, S. and Funahashi, T. (1975). *Proc. EPS Conf. on Soft Magnetic Materials*, Part 2 (University College, Cardiff), p. 269.
Stäblein, F. (1934). *Tech. Mitt. Krupp.* **2**, 127.
Stanley, K. J. (1963). *Electrical and Magnetic Properties of Metals* (American Society for Metals, Metals Park, Ohio).
Sugihara, M. (1960). *J. Phys. Soc. Japan* **15**, 1456.
Swift, W. M. (1973). *IEEE Trans. Magn.* **MAG-9**, 46.
Swift, W. M., Reynolds, W. T. and Thornburg, D. R. (1975). *IEEE Trans. Magn.* **MAG-11**, 1649.
Taguchi, S. and Sakakura, A. (1969). *J. Appl. Phys.* **40**, 1539.
Taguchi, S., Yamamoto, T. and Sakakura, A. (1974). *IEEE Trans. Magn.* **MAG-10**, 123.
Taguchi, S., Sakakura, A., Matsumoto, F., Takashima, K. and Kuroki, K. (1976). *J. Magn. Magn. Mater.* **2**, 121.
Taniguchi, S. (1955). *Sci. Rep. Tohoku Univ.* **A7**, 269.
Taniguchi, S. and Yamamoto, M. (1954). *Sci. Rep. Tohoku Univ.* **A6**, 330
Tebble, R. S. and Craik, D. J. (1969). *Magnetic Materials* (Wiley-Interscience, London).
Thompson, D. A., Romanikiw, L. T. and Mayadas, A. F. (1975). *IEEE Trans. Magn.* **MAG-11**, 1039.
Wassermann, E. F. (1990). *Ferromagnetic Materials: A Handbook on the Properties of Magnetically Ordered Substances*, Vol. 5, ed. K. H. J. Buschow and E. P. Wohlfarth (Elsevier, Amsterdam), p. 237.

Westerstrand, B., Nordblad, P. and Nordberg, L. (1975). *Physica Scripta* **11**, 383.
Williams, H. J. (1937). *Phys. Rev.* **52**, 747.
Williams, H. J. and Goertz, M. (1952). *J. Appl. Phys.* **23**, 316.
Williams, H. J., Sherwood, R. C., Goertz, M. and Schnettler, F. J. (1953). *AIEE Comm. Elec.* **72** (Pt I), 531.
Wohlfarth, E. P. (1975). *IEEE Trans. Magn.* **MAG-11**, 1638.
Yamamoto, T. and Nozawa, T. (1970). *J. Appl. Phys.* **41**, 2981.
Yamamoto, T., Taguchi, S., Sakakura, A. and Nozawa, T. (1972). *IEEE Trans. Magn.* **MAG-8**, 677.
Yoshizawa, Y., Oguma, S. and Yamauchi, K. (1988) *J. Appl. Phys.* **64**, 6044.

Chapter 3

Amorphous Ferromagnetic Alloys and Ferrofluids

3.1 INTRODUCTION

Although ferromagnetism is generally associated with the highly ordered crystalline state, Gubanov (1960) predicted from theoretical arguments that ferromagnetism should be possible in an amorphous material, provided of course that the Heisenberg exchange integral is positive. In contrast to the constant exchange coupling in crystalline ferromagnetics, where the lattice parameter does not vary appreciably, the exchange coupling in an amorphous material varies about the mean value of the interatomic distance, and therefore depends on the radial distribution function. This means that, although the material may have all the fundamental properties associated with ferromagnetism, the individual magnetic moments may not be excactly parallel to each other.

One of the most important series of amorphous alloys comprises those based on the composition $T_{80}M_{20}$, where T is one or more of the transition metals Fe, Co, Ni, Mo, Cr, Mn, Pd etc., and M is one or more of the metalloid or glass-forming elements P, B, Si or C. These alloys have attracted a great deal of research effort, since many of them are potentially useful as high-permeability alloys with comparable or superior properties to Si–Fe and Ni–Fe alloys. Amorphous ferromagnetic alloys can also be prepared from the transition and lanthanide metals, e.g. Gd–Co, Tb–Fe, Dy–Fe etc.

3.2 PREPARATION OF AMORPHOUS FERROMAGNETIC ALLOYS

Unlike silicate glasses, amorphous metallic materials can be produced only if the molten material is splat-cooled with a cooling rate typically of order 10^6 °C s^{-1}. The stringency of the latter requirement can be reduced if glass-forming elements are added to the alloy, which usually contains one or more of the ferromagnetic transition metals Fe, Co or Ni. The glass-forming elements which have so far proved to be effective are P, B, Si and C. The function of these elements is to increase the viscosity and its temperature dependence of the liquid alloy, so that if the molten alloy is cooled sufficiently rapidly, there is a large increase in the viscosity, which impedes the rearrangement of the atoms on the equilibrium crystallographic sites. This results in the formation of an amorphous material that is in a metastable energy state higher than that associated with the fully crystalline state. At low temperatures or ambient temperatures, where the atomic diffusion rates are low, the material may exist in the amorphous state almost indefinitely.

Amorphous alloys can be produced as follows. A jet of the molten alloy is ejected under pressure from a small orifice into the interior of a cold copper drum, which is rotating at high speed so as to cool the alloy by conduction. As the jet strikes the surface of the drum, it spreads into a flat ribbon before becoming rigid. The width and thickness of the ribbon can be controlled (within limits) by the dimensions of the orifice, the drum velocity and the ejection pressure. Since the whole volume of the alloy must be cooled very quickly, the upper limit to the thickness of the ribbons is about 50 μm, with a width of 30 mm. The technology of the preparation of amorphous alloys is currently a subject of considerable interest, since if the alloys are to be suitable for commercial applications, it is desirable to increase the thickness and size of the resulting ribbons. The maximum thickness is of course limited by the thermal conductivity of the alloy itself and the substrate onto which it is quenched.

Amorphous metallic alloys in thin film form can also be prepared by electrodeposition, chemical or electrodeless deposition or by sputtering.

3.3 STRUCTURE OF AMORPHOUS FERROMAGNETICS

An amorphous or glassy material is one in which the ions are in non-periodic arrangements, with the mean distances between particular pairs of ions approximately equal to those in the crystalline material. The magnetic structure of an amorphous ferromagnetic material is shown in Fig. 3.1(a). For simplicity, it has been assumed that there is only one atomic species with a magnetic moment and that the magnetic moments are parallel. The small angular deviation between the magnetic moments, which may vary from point to point, gives rise to a small magnetic susceptibility in very high fields. The most useful parameter for describing the structure of an amorphous alloy is the radial distribution function. This gives the average number of ions in a spherical shell at distances between r and $r+dr$ from an arbitrarily chosen origin ion.

From X-ray diffraction measurements of the spacing between Gd and Co ions in amorphous and crystalline $Gd_{36}Fe_{64}$, Cargill (1974) has concluded that the interatomic distances in the amorphous and crystalline material are very similar, a result that is also indicated by measurements of their densities. The close similarity of the ion coordination numbers in the amorphous material confirms that the structure is chemically disordered, as we should expect. From neutron scattering experiments on amorphous and crystalline $Tb_{33}Fe_{67}$, Rhyne *et al.* (1974) reached similar conclusions.

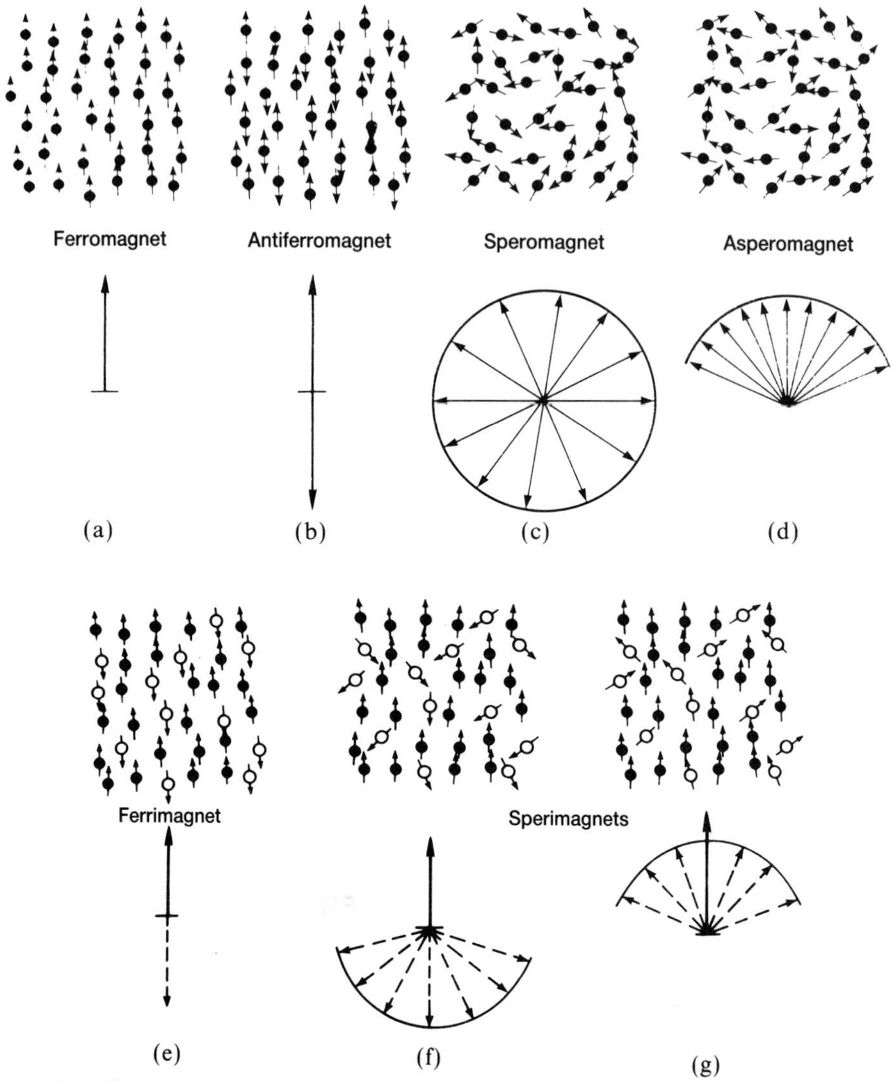

Fig. 3.1 (a)–(d) Possible single-network magnetic moment structures in amorphous alloys, showing the spatial distributions of the resulting magnetic moments. (e)–(g) Possible two-subnetwork magnetic moment structures in amorphous alloys, showing the spatial distributions of the resulting magnetic moments. Solid and dashed lines represent the magnetic moments of the two chemical species (after Moorjani and Coey, 1984).

3.3.1 Magnetic Structure

The ferromagnetic structure shown in Fig. 1(a) has been observed in a large number of amorphous alloys, e.g. Fe–Ni–P–B, Fe–Ni–B–Si, Fe–B and Fe–B–Si–C, in which there are positive exchange interactions between neighbouring ions. However, in some amorphous alloys the exchange interactions may be negative, in which case the alloys are antiferromagnetic (Fig. 3.1b). If there are superexchange interactions, the alloy may be ferrimagnetic (Fig. 1e). Examples of possible magnetic structures are shown in Figs 3.1(a)–(g). The ferrimagnetic structure shown in Fig. 1(e) is that suggested by Rhyne *et al.* (1974) for amorphous $TbFe_2$, in which the Fe–Fe and Tb–Tb interactions are positive while those for Tb–Fe are negative, i.e. antiferromagnetic. The saturation magnetization of $TbFe_2$ at 0 K corresponds to $4.2\mu_B$ per formula unit, which is significantly lower than the value $5.4\mu_B$ for the corresponding crystalline material, i.e. the Laves-phase intermetallic compound. If we assume that the magnetic moments of the Tb and Fe ions are respectively $9\mu_B$ and $2.2\mu_B$, the expected magnetic moment for the ferrimagnetic structure is

$$\mu = (9 - 2 \times 2.2)\mu_B = 4.6\mu_B$$

per formula unit, which is about 10% larger than the observed value of $4.2\mu_B$. This may be attributed to slight deviations from colinear alignment of the Tb and Fe magnetic moments due to local variations in the local anisotropy and/or to variations in the exchange interactions. It is also possible that some of the differences in the magnetic moments of the amorphous and crystalline forms of a material could be due to electron transfer from one ion to another or to changes in the local 'band' structure. However, these mechanisms will generally be very complex, and as yet no detailed theory is available.

Similar arguments can be applied to the observed magnetic moments in amorphous materials containing transition metals and glass-formers such as P, B, C or Si in which the saturation magnetization is in general lower than that for the crystalline form. Amorphous $GdCo_3$ is also known to have the ferrimagnetic structure shown in Fig. 3.1(e), in which the ferromagnetic Gd and Co subnetworks are antiferromagnetically coupled, with a net theoretical magnetic moment

$$\mu = (7 - 3 \times 1.7)\mu_B = 1.9\mu_B$$

per formula unit.

If for a particular material the two subnetwork magnetic moments cancel then the material is an antiferromagnetic, with no net magnetization (Fig. 3.1b).

It has been suggested that the lack of any net magnetization in the amorphous alloy $DyCo_3$ can be explained by the sperimagnetic structure shown in Fig. 3.1(f). The exchange interactions are similar to those in $GdCo_3$ and $TbFe_2$, except that in this case the local crystal fields of the cobalt ions define locally preferred directions for the dysprosium magnetic moments, which, in combination with the exchange interaction, gives rise to the sperimagnetic structure with no net magnetic moment.

Note that the dysprosium subnetwork has a net magnetic moment antiparallel to the cobalt sublattice, but since the magnetic moment of dysprosium is $10\mu_B$ while that of cobalt is only $1.7\mu_B$, the net magnetic moment is

$$\mu = (10\langle\cos\theta\rangle_{av} - 3 \times 1.7)\mu_B = 0,$$

where θ is the angle between a dysprosium moment and the moment of the cobalt sublattice; hence

$$\langle\cos\theta\rangle_{av} \approx 0.5,$$

implying that the angular deviations θ of the dysprosium ion magnetic moments are quite large, about 60°.

In the structure shown in Fig. 1(c), which is referred to as speromagnetic, the magnetic moments have a large number of orientations, so that the net magnetic moment is zero. There may of course be some degree of localized magnetic order – either antiferromagnetic or ferromagnetic – but when this is averaged over the whole material, the net result is the speromagnetic structure shown. Such a structure may exist in YFe_2, which shows no evidence for long-range magnetic order above about 40 K. However, when a speromagnetic material is subjected to an applied field, it may become quite strongly magnetic, as indeed YFe_2 does. At very low temperatures <22 K, YFe_2 shows a small magnetic remanence, suggesting that there is some genuine long-range order at these temperatures (Rhyne *et al.*, 1974).

If in a speromagnetic material the exchange interaction is anisotropic, the material may show a spontaneous magnetization, as can be seen from Fig. 1(d). Such materials are known as asperomagnetics.

For further details of the magnetic structures of amorphous ferromagnetic materials the reader is referred to Coey (1978) and Moorjani and Coey (1984). A review of the atomic structures of metallic alloy glasses has been given by Cargill (1975).

3.3.2 Magnetic Properties

According to Gubanov (1960), the fundamental exchange interactions are similar to those in crystalline materials, but modified by the disorder and range of interaction distances of nearest neighbours. Furthermore, there is some distortion of the ionic electron orbits, and hence a possibility of electron transfer from one ion to another, with a consequent change in the effective magnetic moments. If we assume that the ferromagnetism is non-localized but due to an itinerant electron interaction then the electron transfer may modify the local 'band' structure, so that the magnetic behaviour of amorphous ferromagnets will be extremely complex. However, from the Heisenberg model, the exchange interaction E_{ex} depends on the ionic separation r_{ij}, the angle θ_{ij} between the effective localized magnetic moments, and of course the sign and magnitude of the exchange integral, i.e.

$$E_{ex} = -2J(r_{ij})S^2 \cos \theta_{ij}.$$

Thus we may expect the exchange energy to be weaker in amorphous materials, and this can indeed be inferred from the observation that the Curie temperatures of amorphous materials are in general lower than those for the corresponding crystalline materials. Furthermore, we might expect the saturation magnetizations and hence the effective ionic magnetic moments to be lower, and indeed this is generally the case. The variation of the saturation magnetization with temperature has the same general form as that given by the Weiss–Brillouin theory. This seems to be the case for many of the amorphous alloys in the system $(Fe, Ni)_{80}(P, B)_{20}$. From ferromagnetic resonance experiments on amorphous $Fe_{40}Ni_{40}P_{14}B_6$, Hasegawa (1976) has concluded that the g factor for this alloy is 2.05, which suggests that the ferromagnetic moment is due almost entirely to electron spin, again confirming that the fundamental mechanisms in amorphous ferromagnetics are similar to those in crystalline ferromagnetics. The fundamental aspects of ferromagnetism in amorphous alloys have been discussed by Moorjani and Coey (1984).

3.3.3 Magnetic Anisotropy

Since there are no crystallographic directions in amorphous alloys, their physical properties are expected to be isotropic. However, since a particular magnetic ion interacts with its nearest neighbours and perhaps next-nearest neighbours in a way that depends on their separation and relative orientations, we may assume that there will at least be some kind of local 'magnetocrystalline' anisotropy due to distortions in the electron orbitals. This anisotropy will be very short range, so that the average macroscopic anisotropy is very close to or equal to zero (in a thin ribbon we may expect surface effects to be important). This propspect of a zero or near-zero isotropic 'magnetocrystalline' anisotropy has stimulated much interest in amorphous alloys such as $(FeNi)_8$, $(B, Si, C)_{20}$, $Fe_{80}(B, Si, C)_{20}$ and $(FeCo)_{80}(B, Si, C)_{20}$, which have very high permeabilities and are therefore of potential commercial importance as soft magnetic materials. In these and similar materials the absence of grain boundaries and a significant 'magnetocrystalline' anisotropy means that there is very little resistance to domain wall propagation even in applied fields as small as 5 A m^{-1}. The observed permeabilities are usually in the range 2000–800 000. However, the upper limit to the permeability is set by the unfortunate fact that there is a local ionic anisotropy, so that the effective macroscopic anisotropy is not exactly zero. Thus finite applied fields are required to saturate the materials, typically in the range 1–1000 A m^{-1}.

The induced anisotropy is a consequence of the method of preparation. The very rapid cooling rates that are necessary to retain the amorphous state produce thermally induced stresses and hence associated magnetoelastic interactions, which manifest themselves as a magnetic anisotropy. Additional uniaxial anisotropy can be induced by heat treatment in a magnetic field, i.e. by thermomagnetic treatment.

As the ribbons cool on the drum, there is clearly an anisotropy in the cooling rates, which differ for various points on the ribbon. The jet stream also tends to induce a preferred axis or easy direction of magnetization parallel to the axis of the ribbon, so that the magnetization curves depend on the direction of measurement relative to the ribbon axis as shown in Fig. 3.2.

The anisotropy energy K_u associated with magnetization in the easy direction can be calculated from the area enclosed by the magnetization curve and the magnetization axis, i.e.

$$K_u = \mu_0 \int_0^{M_s} \boldsymbol{H} \cdot \mathrm{d}\boldsymbol{M},$$

and similarly for magnetization at other angles to the easy direction. Alternatively, the anisotropy energy can be estimated by measuring the field required to saturate the material in a given direction, so that the induced anisotropy energy constant K_u can be estimated from the anisotropy field H_A, i.e.

$$H_A = \frac{2K_u}{\mu_0 M_s},$$

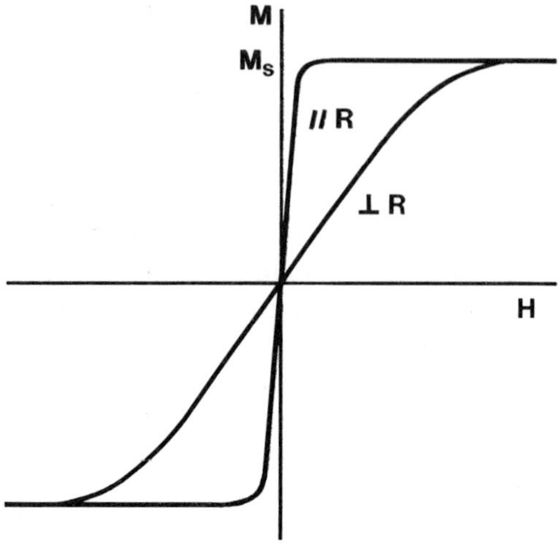

Fig. 3.2 Magnetization curves parallel and perpendicular to the rolling direction for a typical soft amorphous alloy. Note the very low coercivities. The remanence may sometimes be a substantial fraction of the saturation magnetization.

assuming that magnetization involves rotational processes, even if only those of the magnetic spins on the domain walls. The individual magnetic spins have to overcome small energy barriers whose height, width and distribution are determined by the local values of the magnetic anisotropy, due to atomic ordering or single-ion effects in combination with the stress-induced anisotropy. These anisotropies are strongly dependent on the alloy composition and the previous mechanical, thermal or thermomagnetic treatment. Detailed measurements by Graham *et al.* (1976) have shown that if the alloys $Fe_{40}Ni_{40}P_{14}B_6$ and $Fe_{29}Ni_{49}P_{14}B_6Si_2$ are annealed at temperatures in the range 200–240°C, the anisotropy is substantially reduced after about 30 min. Values of the induced anisotropy range, for example, from 80 J m^{-3} in a magnetically annealed $Fe_{40}Ni_{40}P_{14}B_6$ alloy (Luborsky *et al.*, 1975) to 5000 J m^{-3} in a quenched $Fe_{80}P_{13}C_7$ (Fujimori *et al.*, 1974). Luborsky *et al.* (1975) have also shown that when $Fe_{40}Ni_{40}P_{14}B_6$ is annealed in a magnetic field of 1680 A m^{-1}, the coercivity is reduced by a factor of 10–50. The lowest coercivity obtained was 0.24 A m^{-1}, which is comparable to the lowest values for magnetically annealed Ni–Fe crystalline alloys, i.e. the Permalloys. It is now well established that magnetic annealing of amorphous magnetic alloys parallel to the ribbon axis decreases the coercivity and increases the permeability very substantially.

Luborsky *et al.* (1975) have shown that the values of K_u can also be conveniently determined from measurements of the initial susceptibility parallel and perpendicular to the specific or intended easy direction of magnetization, which may be induced by an applied stress, by magnetic annealing or during the preparation of the material. Since the material is usually in the form of a thin ribbon a few millimetres wide, it is essential to correct for the effects of demagnetizing fields.

The induced uniaxial anisotropy that results from heat treatment in a magnetic field is thought to be due to a mechanism similar to the atomic pair ordering proposed for the induced anisotropy and improved magnetic properties of high-permeability Ni–Fe alloys (see Chapter 2). The uniaxial induced anisotropy produced by thermomagnetic treatment of amorphous alloys in the system $(Fe, Ni)_{80}(P, B)_{20}$ has been investigated by Luborsky and Walter (1977a, b). Although the results were interpreted in terms of the atomic pair-ordering theory, Luborsky and Walter (1977b) found that the theory could not account for all their results, and suggested that other ordering effects and the presence of the metalloid elements P and B were also important.

3.3.4 Magnetic Domain Structures

The domain structures in amorphous alloys are usually very complex, and vary from point to point in the material. This arises because the splat cooling process leaves residual stresses in the material and regions in which the anisotropy may be effectively uniaxial either in the plane of the ribbon or perpendicular to it. Becker (1976) observed that the regions with uniaxial anisotropy often occurred in small patches distributed over the surface of the ribbon. The very fine cellular domain structure of these patches has a cell size is considerably smaller than the ribbon thickness. By studying the behavior of these patches of domains in various applied fields, Becker (1976) confirmed that they represent regions of the material in which there is a component of magnetization perpendicular to the plane of the film, and suggested that these patch domains impede magnetization processes. If the ribbons are annealed at about 200–300°C, the patch domains disappear and the field required to saturate the material along the ribbon axis decreases from about 1600–8000 A m^{-1} to about 80 A m^{-1}. Since the local uniaxial anisotropy is due to local stress (magnetostriction), it may be assumed that the local stresses are removed by annealing. Similarly, Becker (1976) found that if a stress of

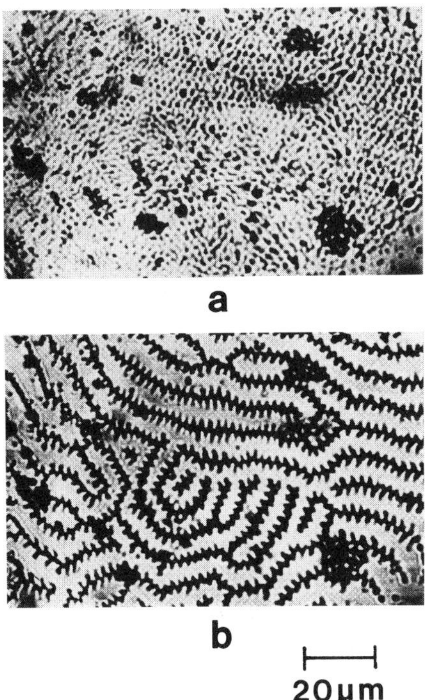

Fig. 3.3 Maze domain structure of amorphous $Fe_{80}P_{13}C_7$ (a) without external field; (b) with an applied field of 8 kA m^{-1} perpendicular to the ribbon surface (after Obi *et al.*, 1976).

about 100 MN m^{-2} was applied together with a field of about 80 A m^{-1}, the magnetization was also increased to near saturation. Since the magnetostriction λ in this alloy is about 10^{-5}, a tensile stress of 100 MN m^{-2} produces an anisotropy field

$$H_A = \frac{3\lambda\sigma}{\mu_0 M_s} = 5000 \text{ A m}^{-1},$$

which is of the same order of magnitude as the field required to saturate the alloy. From the above discussion, we may conclude that the magnetic properties of amorphous magnetic alloys are very strongly dependent on the local anisotropy and magnetoelastic properties of the ribbons. Domains in amorphous $Fe_{80}P_{13}C_7$ are shown in Fig. 3.3.

Gyorgy *et al.* (1976) investigated the domain structure of the amorphous alloy $Fe_{80}P_{13}C_7$, which has zero magnetostriction, $\lambda_s = 0$. From the domain structure of this alloy, there is clear evidence that the domain magnetization vectors tend to follow the flow lines produced during the splat cooling, and it can be assumed that these are parallel to the easy directions of magnetization. Near the edges of the ribbon, the easy direction of magnetization is almost perpendicular to the axis of the ribbon. After magnetic annealing parallel and perpendicular to the rolling direction (i.e. the axis of the ribbon), the domain structure was modified, which clearly showed that the easy direction of magnetization is determined by, and is parallel to, the direction of the applied field during the anneal.

3.3.5 Coercivity of Amorphous Alloys

The coercivity of amorphous alloys is strongly affected by all of the parameters that are relevant to those of crystalline materials, so that the coercivity is determined by domain wall pinning or by obstacles to domain wall motion. This implies that the most important general parameter is the variation of the domain wall energy as a function of position within the material. The origin of coercivity and hysteresis due to this has already been discussed earlier in chapter 1.

According to Kronmüller (1981a, b), there are at least five domain wall pinning effects that may contribute to the coercive field. For soft magnetic materials these are as follows, in decreasing order of importance.

1. In alloys that have a significant magnetostriction there can be volume pinning of domain walls by defect structures ($H_c \approx 1$–10 A m^{-1}). The coercivity due to pinning arises because the defects in the alloy exert elastic stresses. This idea is supported by the fact that H_c generally increases with increase in magnetostriction. Quasi-dislocation dipoles are the main sources of elastic stresses, and result in the strongest domain wall pinning.
2. Local structural rearrangements can lead to relaxation effects ($H_c \approx 10^{-2}$–1 A m^{-1}). The domain wall pinning or impedance to domain wall motion arises from the fact that a rearrangement of interstitial impurities can result in an induced local anisotropy. This magnetic after-effect has been shown to occur in Fe–B and Fe–B–P alloys. Since the rearrangements of the ions are thermally activated, the domain wall mobility is time- and temperature-dependent.

3. The production of amorphous ribbons by quenching the molten alloy onto a rotating wheel results in the inner surface of the ribbon, which is in contact with the wheel, having a surface that replicates the surface of the wheel plus any other variables such as air pockets and other effects due to the casting conditions. The outer surface of the ribbon, on the other hand, is smooth, with a superimposed relatively long-wavelength undulation. In general, the coercivity increases as the surface roughness ratio $\Delta T/T$ increases, where ΔT is the amplitude of the local surface irregularity and T is the average thickness of the ribbon. The contribution to the coercivity is typically $<0.5\ \mathrm{A\ m^{-1}}$.
4. A contribution to the coercivity may also arise due to clusters of short-range chemically ordered regions with $H_c < 0.1\ \mathrm{A\ m^{-1}}$. The increase in domain wall pinning is due to an induced local anisotropy and a change in the exchange energy, thus giving rise to fluctuations in the domain wall energy as a function of position. Ordered clusters of 12 ions are expected to produce a measurable effect.
5. A very small contribution to the coercivity ($H_c \approx 10^{-4}$–$10^{-1}\ \mathrm{A\ m^{-1}}$) results from intrinsic fluctuations in the local exchange and anisotropy energies. The exchange energy fluctuations are expected to make the larger contribution to the coercivity, although both effects are very small.

The coercivity is in general proportional to the rate of change of the domain wall energy with position, and depends on the nature, number per unit volume and strength of the pinning sites.

In order to produce high-permeability amorphous alloys with the lowest possible coercivities, all of the above contributions must be minimized. Ideally, the material should be homogeneous, and free from compositional variations, structural defects and residual stresses, i.e. the domain wall energy should be independent of position x, so that the graph of γ versus x is a straight line parallel to the x axis.

Unfortunately, defect structures, compositional fluctuations and stresses are inevitably produced during the rapid quenching, and cannot be completely removed by annealing. Thus the domain wall pinning that arises via magnetostriction and the associated elastic stresses can be reduced only by using alloys with zero magnetostriction. This can be done by careful choice of the composition. The domain wall pinning due to short-range order and relaxation effects, which now become relatively more significant, can be reduced by annealing. If these conditions are achieved, the minimum intrinsic coercivity can be observed only in very smooth ribbons with $\lambda = 0$ ideally.

For detailed discussions of the theories of coercivity in amorphous alloys the reader is referred to the papers by Kronmüller (1981a, b) and the short review by Luborsky (1983).

3.4 APPLICATION OF AMORPHOUS ALLOYS AS SOFT MAGNETIC MATERIALS

It has already been mentioned that the relatively high saturation magnetization, low coercivity and exceptionally low anisotropy of amorphous ferromagnetic alloys make them particularly suitable for use as soft magnetic materials. The electrical resistivities of high-permeability amorphous alloys are higher than those of crystalline high-permeability alloys, so that in alternating current applications such as transformer cores the energy losses due to eddy currents are expected to be much lower.

Assuming that the magnetization changes occur by coherent rotation of the magnetization vector, against anisotropy forces, the lower bound to the relative permeability is given by

$$\mu_r = \frac{\mu_0 M_s^2}{2K_u} + 1.$$

Ideally, therefore, we require a material with M_s as large as possible and K_u as small as possible. Both these requirements are satisfied by a large number of amorphous alloys based on the composition $T_{80}M_{20}$, where T is one or more of the transition metals Fe, Ni or Co, and M is one or more of the metalloid or glass-forming elements P, B, Si or C. If we take $M_s \approx 6 \times 10^5$ A m^{-1} and $K_u \approx 100$ J m^{-3} as representative values for the alloy $Fe_{40}Ni_{40}P_{14}B_6$, we find, after substitution of these in the above expression, that the lower bound to the relative permeability $\mu_r \approx 25\,000$, which is certainly of the correct order of magnitude. Since changes in the magnetization in amorphous alloys will occur predominantly by domain wall movement – a process requiring much lower fields than coherent rotation – the experimental values of μ_r should be very much higher than this, as is the case. It is now well established that the permeability of many alloys in the system $T_{80}M_{20}$ can be increased very substantially by heat treatment in a magnetic field parallel to the required direction of magnetization. The effect of magnetic annealing is illustrated by Fig. 3.4, which shows the hysteresis loops for as-cast and magnetically annealed $Fe_{40}Ni_{40}P_{14}B_6$. The heat treatment reduces or removes the residual stresses and hence the associated anisotropy, and at the same time induces a small uniaxial anisotropy parallel to the field, which increases the permeability in this direction at the expense of that perpendicular to it. Direct evidence for the induced uniaxial anisotropy

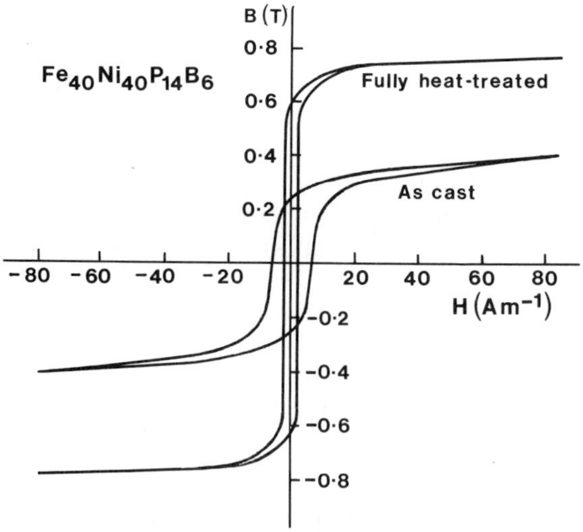

Fig. 3.4 *B–H* hysteresis loops for amorphous $Fe_{40}Ni_{40}P_{14}B_6$ (Metglas 2826) (after Mendelsohn *et al.*, 1976).

has been obtained by Gyorgy *et al.* (1976), who showed that the field direction was also the preferred direction from the domain magnetization vectors.

Since the magnetostriction in amorphous transition metal alloys is positive, the permeability in the preferred direction can be increased still further by a tensile stress in this direction. Becker (1976) has also shown that the B–H loops show increased squareness in the presence of stresses in the range 15–70 MN m^{-2} and a reduction in the coercivity by a factor of about 2 (similar behaviour is observed for stressed 68 Permalloy i.e. 68% Ni–32% Fe). Twisting the ribbons also increased the B–H loop squareness, but this was accompanied by an increase in the coercivity. The above and similar experiments lead to the important conclusion that, as a consequence of the exceptionally low values of the induced anisotropy, the magnetic behaviour of the amorphous transition metal alloys is strongly dependent on the residual stresses and the associated magnetoelastic interactions. Thus, in order to obtain high permeabilities, amorphous alloys must be annealed, preferably in a magnetic field parallel to the required direction of magnetization.

In view of the sensitivity of amorphous alloys to stress and the associated magnetoelastic interactions, this can be minimized by using materials with very low or zero magnetostriction. Chen *et al.* (1975) have shown that the amorphous alloy $Fe_3Co_{72}P_{16}B_6Al_3$ has zero magnetostriction. By means of a magnetic anneal with the field parallel to the rolling direction of the ribbon, they have shown that a very high permeability can be obtained, namely 500 000 compared with the already high value of 200 000 in the as-cast alloy. The B–H loops are shown in Fig. 3.5.

The magnetic properties of various amorphous alloys are given in Table 3.1. The electrical resistivities of the alloys are in the range 1.3–1.8 $\mu\Omega$ m, so they are expected to show very low eddy current losses as well as very low hysteresis losses. The total energy losses for the amorphous alloy $Fe_{80}B_{20}$ are much lower than the best-quality

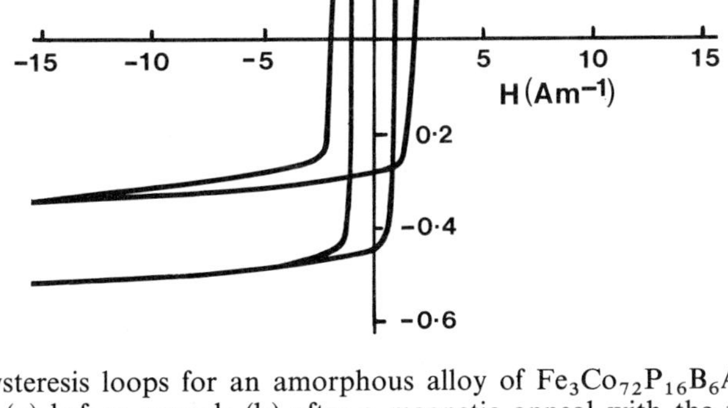

Fig. 3.5 B–H hysteresis loops for an amorphous alloy of $Fe_3Co_{72}P_{16}B_6Al_3$, which has zero magnetostriction: (a) before anneal; (b) after a magnetic anneal with the field parallel to the rolling direction (after Chen *et al.*, 1975).

Table 3.1 Magnetic properties of high-permeability, low-loss amorphous alloys under DC conditions.

Material	J_s (T)	H_c (A m^{-1})	T_c (°C)	λ (10^{-6})	μ_{max} [a]
$Fe_{80}B_{20}$ [b]	1.6	8	375	30	80 000*
$Fe_{80}B_{20}$ [b]	1.6	3	375	—	300 000
$Fe_{40}Ni_{40}P_{14}B_6$ [c]	0.78	—	250	—	58 000*
$Fe_{40}Ni_{40}P_{14}B_6$ [c]	0.78	5	250	—	275 000
$Fe_{78}B_{13}Si_9$ [d]	1.56	—	415	27	600 000
$Fe_{81}B_{13.5}Si_{3.5}C_2$ [e]	1.61	6	370	30	300 000
$Fe_{66}Co_{18}B_{15}Si_1$ [f]	1.8	—	415	35	400 000
$Fe_{40}Ni_{38}B_{18}Mo_4$ [g]	0.88	8	350	—	12 000*
$Fe_{40}Ni_{38}B_{18}Mo_4$ [g]	0.88	0.6	350	8	800 000
$(Co, Fe)_{70}MoSiB)_{30}$ [h]	0.55	0.4	250	0.3	600 000
$(Co, Mo, Fe, Mn)_{77}(Si, B)_{23}$ [i]	0.8	1.0	350	0.3	300 000
$Co_{58}Ni_{10}Fe_5(B, Si)_{27}$	0.55	0.24	270	0	600 000
$Co_{72}Fe_3P_{16}B_6Al_3$	0.63	1.2	260	0	—

Resistivities are in the range 1.2–1.8 $\mu\Omega$ m.
[a] All alloys have been annealed in a magnetic field except those marked *.
[b] Metglas 2605; [c] Metglas 2826; [d] Metglas 2605S-2; [e] Metglas 2605SC; [f] Metglas 2605CO; [g] Metglas 2826MB; [h] Vitrovac 6025; [i] Vitrovac 6030.

grain-oriented Si–Fe alloys, though not quite so low as those for the Ni–Fe Permalloys. A comparison of the core losses as a function of the flux density and frequency in amorphous crystalline materials has been made by Luborsky (1978), and it appears that they have essentially the same flux density and frequency dependence. A relative permeability of 880 000 has been obtained for amorphous $Fe_{40}Ni_{40}P_{14}B_6$. This exceptionally high value was obtained by annealing the alloy in a magnetic field of 800 A m^{-1} for 2 h at 325°C, after which treatment the coercivity was only 0.5 A m^{-1}.

For all the potentially useful amorphous alloys the best properties are obtained after minimization of the magnetoelastic anisotropy by a stress-relief anneal, preferably in a magnetic field. Complete stress relief can be achieved in most alloys before the start of crystallization, but the rate at which the stress is relieved depends on the original melting and quenching conditions. Typical *B–H* hysteresis loops for the alloy $Fe_{66}Co_{18}B_{15}Si_1$ (Metglas 2605CO) are shown in Fig. 3.6.

Amorphous alloys such as $Fe_{81}B_{13.5}Si_{3.5}C_2$ are especially useful as transformer cores for shipborne and airborne power transformers, which operate at 400 Hz, where losses are still relatively low. The *B–H* loop for this alloy at 50 Hz is compared with that for 3% Si–Fe in Fig. 3.7.

Very efficient magnetic shielding of cables and other devices is achievable by using amorphous alloys with low or zero magnetostriction such as $Co_{66}Fe_4(Mo, Si, B)_{30}$, which also have a very high initial permeability. Since the magnetostriction is very low or zero, little change in the permeability is produced by mechanical bending. The amorphous ribbons can easily be woven into meshes, so that very large devices can be magnetically screened.

The mechanical strength and flexibility of high-permeability amorphous ribbons is useful in various types of security device; thus library books and expensive consumer goods can be monitored by including a very small thin strip (about 20 μm thick) of an amorphous alloy, which can be detected by high-frequency electromagnetic waves.

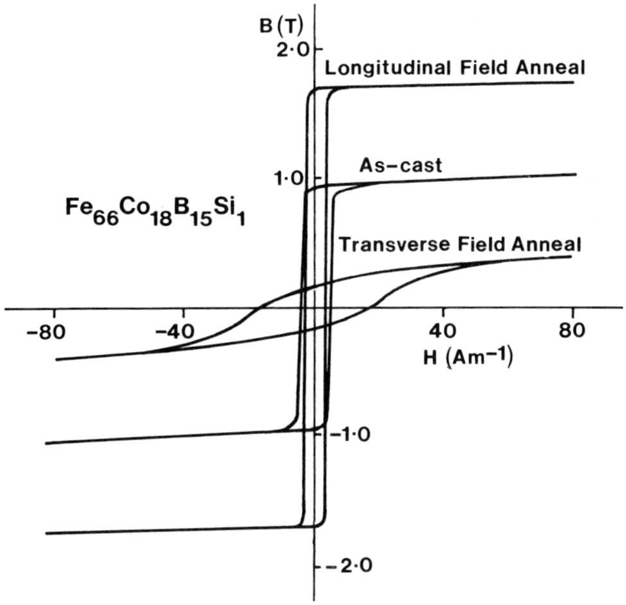

Fig. 3.6 B–H loops for amorphous $Fe_{66}Co_{18}B_{15}Si_1$ (Metglas alloy 2605CO) after various heat treatments, after Allied Signal, USA).

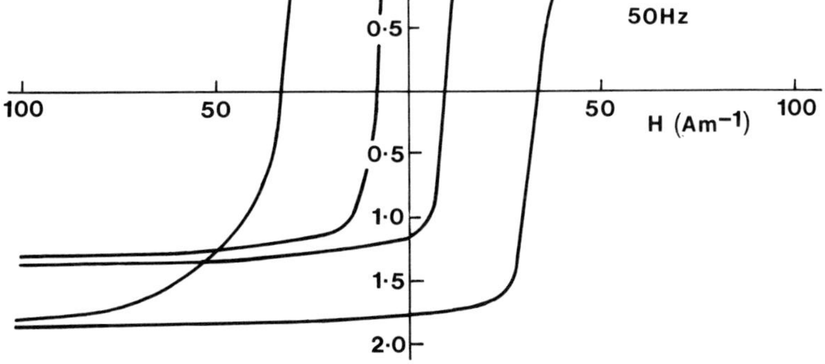

Fig. 3.7 Comparison of B–H loops for amorphous $Fe_{81}B_{13.5}Si_{3.5}C_2$ with crystalline 3% Si–Fe (after Boll *et al.*, 1983).

The very high permeabilities of amorphous alloys combined with their high mechanical strength, hardness, low or zero magnetostriction, wear and corrosion resistance makes them very suitable for application in magnetic recording heads. A comparison of some of the properties for crystalline and amorphous alloys is shown in Table 3.2.

Table 3.2 Magnetic recording head materials.

Material group	Composition	Initial permeability (μ_i)	Saturation magnetic polarization (T)	Electrical resistivity ($\mu\Omega$ cm)	Vickers hardness H_V
Crystalline	77 Ni–CuMoFe (Permalloy)	20 000	0.8	60	120
	77 Ni–CuMoFe–5 (Nb, Ti)	40 000	0.5	90	240
	85 Fe–9.6 Si–5.4 Al (Sendust)	25 000	1.1	88	450
	84 Fe–16 Al	8 000	0.9	145	280
Amorphous	$Co_{66}Fe_4(Mo, Si, B)_{30}$	20 000	0.55	130	900
	$Co_{66}Mn_6(Si, B)_{25}$	10 000	0.8	130	900

Table 3.3 Applications of high-permeability amorphous alloys (Allied Signal, USA).

Alloys	Saturation magnetic polarization	Core losses	Typical applications
Iron-based:			
2605S-2	High	Low	Inductors Power transformers Distribution transformers Ground fault interrupters Motors
2605SC 2605CO	High	Low to medium	Pulse transformers Magnetic switches Magnetoelastic transducers
2605S-3A 2605SM	High	Very low	High-frequency transformers Current transformers High-sensitivity ground fault interrupter cores
Nickel–iron-based:			
2826MB	Medium	Medium	Shielding Magnetic field sensors Recording heads Earth leakage cores Sensors
Cobalt-based:			
2705M 2714A 2705MN	Low to medium	Lowest	High-frequency inverter transformers High-frequency magnetic amplifiers Recording heads Sensors Flexible shielding

The potential applications of high-permeability amorphous materials can be divided into two main groups:

(1) cores for power transformers, inductor cores and armatures, and pole pieces for electric motors for which the minimization of losses is important;
(2) electronic applications, e.g. low-power high-frequency transformer cores, pulse transformers, magnetic amplifiers, transducers, sensors and electronically controllable acoustic delay lines.

Many applications of amorphous magnetic alloys are summarized in Table 3.3, which includes the commercial Metglas alloys produced by Allied Signal (USA) and Vitrovac produced by Vacuumschmelze (Germany).

The properties of amorphous and crystalline alloys are compared in Fig. 3.44.

The properties and applications of amorphous magnetic alloys have been discussed in the books and reviews by Luborsky (1978, 1980, 1983). Luborsky *et al.* (1983), Chen (1980), Gilman and Leamy (1978), Kobe and Ferchmin (1977), Levy and Hasegawa (1977), Ferchmin and Kobe (1983), Hasegawa (1983), Moorjani and Coey (1984), Egami (1984), Warlimont (1988), Fish (1989, 1990) and Hilzinger (1990).

3.5 ELASTOMAGNETIC PROPERTIES AND APPLICATIONS

It has already been mentioned that oriented domains are produced by magnetic annealing of amorphous ribbons parallel to the ribbon axis. This decreases the magnetostriction, since there is little or no domain magnetization vector rotation during magnetization. Apart from the resulting high permeability, the alloys are also acoustically quiet due to the absence of magnetostriction. If, however, the alloys are magnetically annealed perpendicular to the ribbon axis, the magnetostriction can be maximized because magnetization parallel to the ribbon axis will require a large rotation of the domain magnetization vectors. It has been shown that this perpendicular annealing can result in magnetomechanical coupling coefficients k as large as 0.95, implying an efficiency of $k^2=0.9$ (i.e. 90%) for the conversion of magnetic energy into mechanical energy and vice versa, e.g. for amorphous $Fe_{81.5}B_{13}Si_{3.5}C_2$. Even in cast alloys such as $Fe_{66}Co_{18}B_{15}Si_1$, $k=0.4$. Thus the very high magnetostriction and magnetomechanical coupling coefficients combined with high mechanical strength make amorphous alloys very suitable for application as elastomagnetic transducers or sensors. Various configurations can be used for measurement of tension, bending or pressure. The dependence of the magnetization curve on applied stress in an amorphous $Co_{75}Si_{15}B_{10}$ alloy is shown in Fig. 3.8. Further details of elastomagnetic effects and devices are given by O'Handley and Chou (1978), Boll *et al.* (1983) and Gibbs and Squire (1989).

Alloys such as $Fe_{80}P_{13}C_7$ and $Fe_{78}Si_{16}B_{12}$ are particularly suitable for application in electronically controllable surface acoustic wave delay lines since they show a strong dependence of Young's modulus on the applied field via the ΔE effect, and the velocity of sound is proportional to the square root of Young's modulus. The time delay is therefore controllable by the application of a very small magnetic field. The absence of grain boundaries in amorphous alloys has the advantage that they produce very little acoustic attenuation.

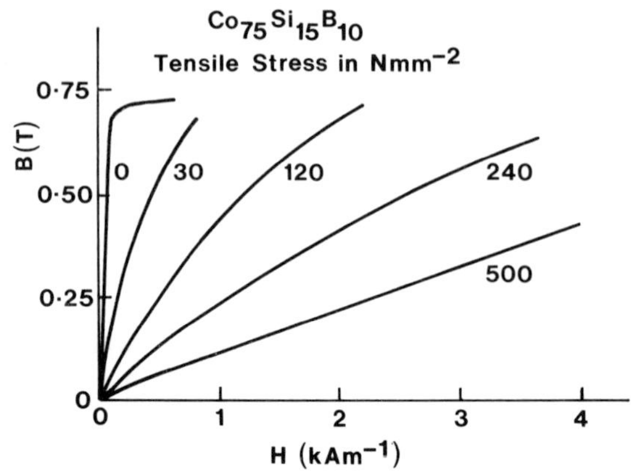

Fig. 3.8 Effects of applied stress on the B–H curves of amorphous $Co_{75}Si_{15}B_{10}$ (after Boll, 1980).

3.6 MAGNETIC PROPERTIES OF HARD AMORPHOUS ALLOYS

In contrast to the extremely soft magnetic properties of the amorphous transition metal–metalloid alloys described above, Rhyne *et al.* (1974) and Savage *et al.* (1974) have reported that amorphous $TbFe_2$ has a maximum coercivity of about 2.4×10^6 A m^{-1} at 4.2 K, which decreases rapidly as the temperature is increased. This suggests that in amorphous $TbFe_2$ there is exceptionally strong domain wall pinning or that domain wall nucleation after saturation is exceptionally difficult. Since the hysteresis loop is not rectangular, domains are still present in positive fields, so that the mechanism responsible for the coercivity is probably domain wall pinning. This will be most effective when the domain walls are very narrow and therefore interact strongly with the local atomic structure and the associated magnetic anisotropy. In this context, the magnetic anisotropy is due to single-ion or dipolar ionic effects, namely that the directions of the lanthanide magnetic moments are strongly affected by 'crystal' fields originating from the asymmetry of their local environments. The magnetic anisotropy associated with lanthanide ions that are not in S states is very large. The local anisotropy for amorphous materials should also be large, but since the anisotropies are randomly orientated, their effects may be expected to cancel. However, detailed analysis has shown that this does not usually occur, because there are additional effects due to the exchange forces, which for amorphous lanthanide–transition metal alloys lead to canted ferrimagnetic structures as shown in Fig. 3.9. The temperature variation of the saturation magnetic polarization for crystalline and amorphous $TbFe_2$ is shown in Fig. 3.10. Note that the values of both $J_s(0)$ and T_c for the amorphous material are lower than those for the crystalline material.

Savage *et al.* (1974) have suggested that the rapid decrease in the coercivity of amorphous $TbFe_2$ is due to a thermal activation process, since they observed that at 4.2 K the remanence has an infinite relaxation time, which becomes time-dependent at higher temperatures and can easily be observed to decay at 81 K. This temperature- and time-dependent decay of the remanence accounts for the decrease in coercivity. It is interesting to note that the energy product of $TbFe_2$ at 4.2 K is about 230 kJ m^{-3}.

Clark (1973) has suggested the possibility of producing a high-coercivity form of

Fig. 3.9 Canted ferrimagnetic structure of a lanthanide–transition metal amorphous alloy. The canting of the magnetic moments is due to the combined effects of the 'crystal' fields and exchange forces from the ions in the immediate neighbourhood of a particular ion.

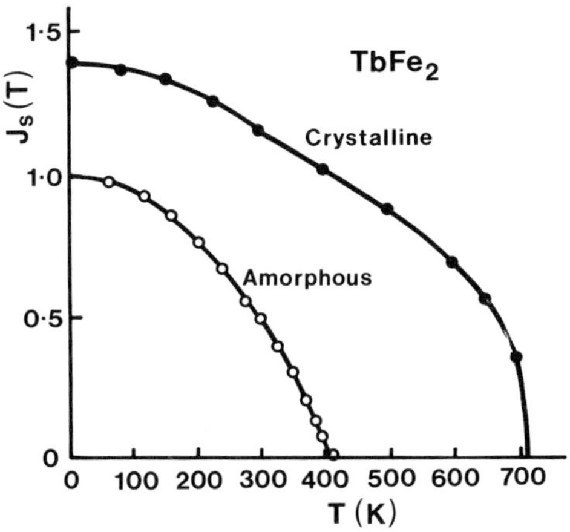

Fig. 3.10 Temperature dependence of the saturation magnetic polarization of amorphous and crystalline $TbFe_2$ (after Rhyne *et al.*, 1974).

$TbFe_2$ by controlled crystallization. If the precipitated crystals are sufficiently small, they could act as very strong domain wall pinning sites, and hence increase the coercivity. Experiments have been carried out by Savage *et al.* (1974), who found a very steep rise in the coercivity (measured at 20°C) at the onset of crystallization (400°C), as shown in Fig. 3.11. The hysteresis loop with the maximum coercivity of 3×10^5 A m^{-1} is shown

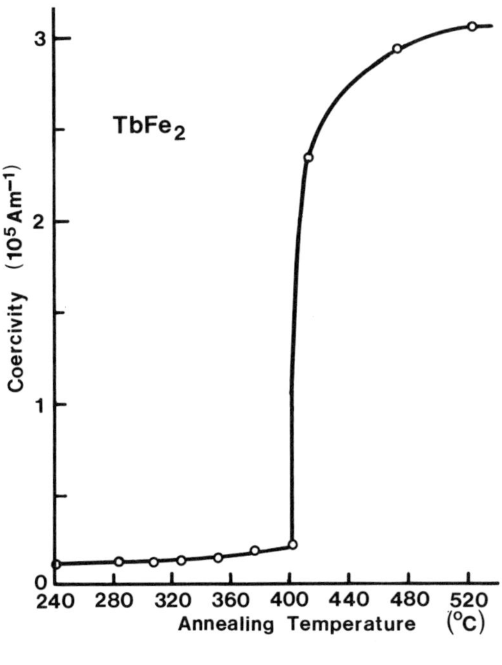

Fig. 3.11 Coercivity at 20°C of amorphous $TbFe_2$ as a function of the annealing temperature. The rapid rise at 400°C is due to the onset of crystallization. The annealing times were 2 h each, except for the two highest temperatures, which were for 17 h (after Savage *et al.*, 1974).

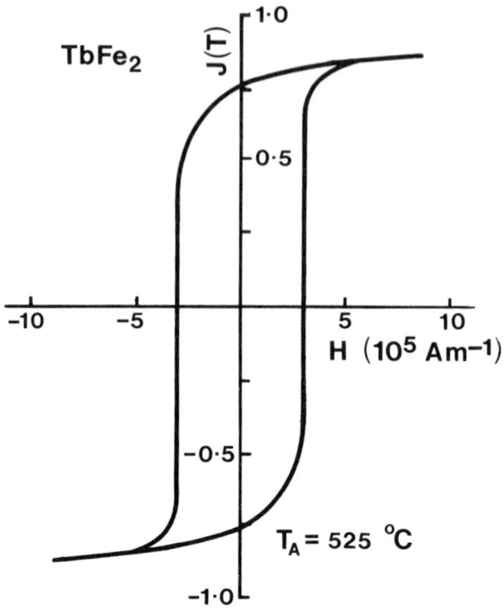

Fig. 3.12 Hysteresis loop at 20°C with maximum coercivity of 3×10^5 A m^{-1} for $TbFe_2$ annealed for 17 h at 525°C (after Savage *et al.*, 1974).

in Fig. 3.12. The maximum energy product is about 70 kJ m^{-3}, which shows that the material could be used as a permament magnet.

Very high coercivities have also been observed in melt-spun $Pr_{40}Fe_{60}$ and $Nd_{40}Fe_{60}$ and Sm–Fe alloys, namely 224, 624 and 1920 kA m^{-1} respectively. High coercivities have also been obtained by controlled crystallization of $Fe_{82}B_{18}$ containing 5% Tb or La. It is possible that controlled crystallization of amorphous alloys may result in the formation of new low-temperature phases that would otherwise be inaccessible, thus generating a wide range of new materials with unusual properties (see Moorjani and Coey, 1984).

Selective annealing in $Pr_{27}Co_{23}$ by laser beams has been used to create high-coercivity regions within the magnetically soft amorphous host alloy. Such alloys could be used for magnetically coded security keys with permanent read-only memories.

3.7 AMORPHOUS ALLOYS FOR ERASABLE MAGNETO-OPTICAL RECORDING AND CURIE-POINT WRITING

Many amorphous alloys can be prepared in the form of thin films in which the preferred or easy direction of magnetization is perpendicular to the plane of the film. In this case, the alloys can be used for the digital storage of information in the form of reverse bubble domains in a similar way to that used in the ferrimagnetic garnet crystals as described in Chapter 4. The digits are 'written' by means of a focused laser beam, which locally raises the temperature of the film above the Curie or compensation temperature, and then, after passage of the laser beam, the local region cools below T_c or T_{comp} and becomes a fully saturated reverse 'bubble' domain. The reversal in the direction of magnetization occurs in order to reduce the magnetostatic energy of the perpendicularly magnetized film. The recorded information, i.e. the presence or absence of a reverse bubble, is read by means of a scanning plane-polarized laser beam, which experiences a clockwise or anticlockwise polar Kerr-effect rotation, depending on the direction of magnetization in the field, i.e. $+\theta$ over a bubble or $-\theta$ elsewhere.

The principal requirement for such a recording medium is that the uniaxial anisotropy energy coefficient K_u should be greater than the magnetostatic energy associated with magnetization perpendicular to the plane of the film, i.e.

$$K_u > \tfrac{1}{2}\mu_0 M_s^2,$$

assuming that the demagnetization factor perpendicular to the plane of the film is $D_z = 1$. If the above condition is satisfied then the film can be fully saturated perpendicular to its plane. The reverse 'bubble' domains can be detected optically by the polar Kerr effect, provided that the latter is sufficiently large – typically, values of $\theta_K \approx 0.25°$ are required. Furthermore, for stable high-density storage the film should have a coercivity in the range 50–500 kA m^{-1}.

These requirements can be satisfied by many amorphous ferrimagnetic alloys such as Gd–Co, GdTb–FeCo, Tb–FeCo and DyTb–FeCo. Compensation-point writing was first demonstrated in amorphous $Gd_{22}Co_{78}$ by Chaudhari *et al.* (1973), who used magneto-optical readout. A disk-reading and playback system using $Gd_{27}Co_{73}$ has been described by Togami *et al.* (1982). Bits as small as 1 μm in diameter were recorded with a laser beam.

The saturation magnetization, which should not be too high, can be adjusted by a suitable choice of composition, while the value of the induced anisotropy K_u is determined both by the composition and the method of deposition of the film onto the disk. The Curie and compensation temperatures, which should be at least 100 K above room temperature, are determined by the temperature dependences of the two sublattice magnetizations of the lanthanide (rare earth) and transition metal ions. Although the above conditions can also be satisfied by crystalline materials, amorphous alloys have the advantage that the signal-to-noise ratio is larger because there are no grain boundaries. Furthermore, the above amorphous alloys can be deposited onto inexpensive polymer or glass substrates at relatively low temperatures. Crystalline CoPt and CrCoPt/Cr multilayers are also being considered for magneto-optical recording.

Amorphous alloys of Tb–Fe and Dy–Fe have low Curie temperatures and other magnetic properties that make them suitable for Curie-point thermomagnetic writing. For a thorough review of erasable magneto-optical recording materials the reader is referred to Hansen and Heitmann (1989) and Bloomberg and Connell (1990).

3.8 FERROFLUIDS

Ferromagnetic liquids can be made by suspending very fine or colloidal iron, γ-Fe_2O_3, Fe_3O_4 or Co particles in water, oil or a mixture of oils. The preparation, magnetic properties and stability of ferromagnetic liquids has been reviewed by Charles and Popplewell (1980). The particles are single domains, and are usually sufficiently fine to be superparamagnetic. They are coupled to the liquid by a stabilizing agent such as oleic acid. The function of the latter is to coat the particles by attaching its polar carboxylic end-group to the particle surfaces. The other end of the molecule is compatible

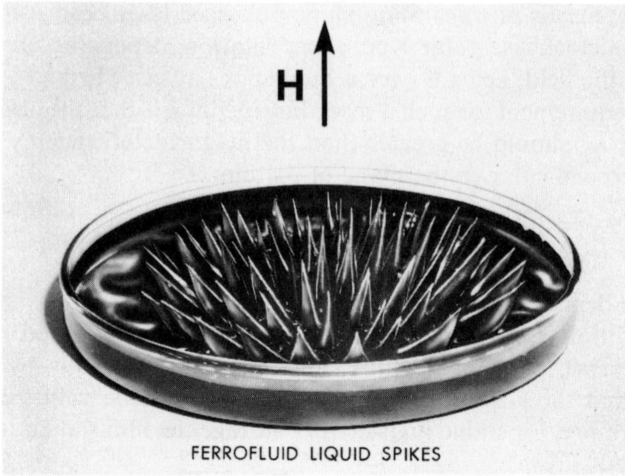

Fig. 3.13 Surface structure of a ferrofluid in the presence of a magnetic field H; approximately full size (courtesy of Ferrofluidics Corporation, Nashua, NH, USA).

with the solvent, which is usually water, a hydrocarbon oil or a fluorocarbon. The surfactant produces a short-range repulsion between the particles, and therefore prevents agglomeration. The resulting liquids, usually known as ferrofluids, are very stable and have many unusual properties when they are subjected to a magnetic field. These are due to the complex interactions of the magnetic forces, viscosity, surface free energy and gravity. For example, if a small permanent magnet is held above a ferrofluid so as to subject it to a critical field strength, the ferrofluid is drawn up into a regular arrangement of peaks as shown in Fig. 3.13. When the magnetic field is removed, the ferrofluid simply falls under gravity. Since the ferrofluid is under constant agitation, there is no remanent magnetization and therefore no hysteresis.

Although ferrofluids are usually oil- or water-based and contain Fe, Co, γ-Fe_2O_3 or Fe_3O_4 particles, ferrofluids containing Ni–Fe, Gd and Fe particles dispersed in mercury have also been prepared (Charles and Popplewell, 1980).

3.8.1 Magnetization of Ferrofluids

Since ferrofluids contain single-domain particles with diameters less than about 15 nm, they are generally superparamagnetic, so that the magnetization is spontaneously reversed by the thermal agitation energy kT. Furthermore, since the particles are freely suspended in a liquid, they are also subjected to Brownian agitation and are therefore in constant vibration and rotation. According to Kaiser and Miskolczy (1970), the magnetization at temperature T in a field H is given by the superparamagnetic Langevin function. A wide range of ferrofluids are now available, with saturation magnetic polarizations up to a maximum of about 0.1 T. Ferrofluids can therefore be suspended in space by a relatively small applied field, and if the saturation magnetic polarization is not too low, this can be done by a small permanent magnet.

3.8.2 Viscosity of Ferrofluids

In common with all colloidal suspensions, the viscosity of ferrofluids increases with the volumetric concentration of the solvated particles and the ratio of the thickness of the stabilizing layer and the particle diameter δ/D. The viscosity also depends on the net magnetization of the liquid (Rosensweig, 1985), and hence on the magnitude and direction of the applied field. Since the colloidal particles are single domains that are usually small enough to be superparamagnetic, the magnetization of ferrofluids is also very strongly dependent on the temperature.

A theory of magnetoviscosity of magnetic suspensions has been presented by Hall and Busenberg (1969). The increase in the viscosity of a ferrofluid in the presence of an applied field can be understood qualitatively as follows. Since the viscosity of a liquid can be considered to be due to the friction between adjacent concentric cylindrical shells of liquid, the suspended particles experience a rotation due to the difference in the velocities in adjacent layers. However, in the presence of a magnetic field, their easy axes of magnetization tend to align parallel to the field, so that any rotation about an axis perpendicular to the field direction is hindered by a restoring torque $\mu_0 \boldsymbol{M} \times \boldsymbol{H}$. This resistance to the rotation of the particles introduces an additional frictional coupling between the fluid layers, and hence results in an increase in the viscosity.

3.8.3 Magnetic Levitation

Rosensweig (1966a, b; 1985) has shown that in the presence of a magnetic field a ferrofluid has an additional pressure P_m determined by the area under the magnetization curve, i.e. between the curve and the applied field axis:

$$P_\mathrm{m} = \mu_0 \int_0^H \boldsymbol{M} \cdot \mathrm{d}\boldsymbol{H}.$$

Bernouilli's equation, which can be used to describe the behaviour of fluids in general, must be modified to take account of the magnetic pressure P_m, so that the augmented Bernouilli equation is

$$p + \tfrac{1}{2}\rho v^2 + h\rho g - \mu_0 \int_0^H \boldsymbol{M} \cdot \mathrm{d}\boldsymbol{H} = \mathrm{const},$$

where the symbols have their usual meanings. Detailed discussions of magnetic pressure effects and the ferrohydrodynamics of ferrofluids have been given by Rosensweig (1966a, b; 1985). One very interesting consequence of the augmented or magnetic Bernouilli equation is that in the presence of a magnetic field the pressure inside the ferrofluid may be so high that an immersed ferromagnetic or non-ferromagnetic body with a density higher than that of the ferrofluid may float. Since the magnetic levitation is controlled by the magnitude of the applied field, the body may be made to sink or float as required, so that the application of a field changes the apparent specific gravity of the ferrofluid. Tungsten, which has a specific gravity of 19.3, has been levitated in this way, even though the specific gravity of the ferrofluid was only 1.3. This property has already been applied to the rapid measurement of the specific gravities of various materials (Kaiser and Miskolczy, 1970). Another potentially important application of this effect is in the separation of minerals or scrap metals. Rosensweig (1966b) has also shown that if a permanent magnet is immersed in a ferrofluid, it suspends itself free from contact with either the bottom or the walls of the container; this effect is due to the pressure exerted on the magnet by its own magnetic field. If the magnet is displaced from its equilibrium position, the surrounding magnetic fields between the magnet and the container are compressed so that the magnet is acted upon by restoring forces, which return it to its equilibrium position. A theoretical investigation of this phenomenon has been carried out by Byrne (1978).

3.8.4 Applications of Ferrofluids

Oil-based ferrofluids may be used as lubricants in vertically mounted bearings, where the ferrofluid can be held in position by a small permanent magnet ring. Ferrofluids can also be used to seal low-friction gas bearings. The latter may prove to be among the most important applications of ferrofluids, since they could be used to reduce the starting torque of electric motors and any other kinds of rotating machines such as record decks, where the elimination of rumble is particularly desirable.

Romanikiw *et al.* (1975) have shown that if a ferrofluid is sandwiched between two glass plates and is subjected to a magnetic field, it divides into a domain structure. Such a ferrofluid sandwich, provided that it is sufficiently thin, can also be used for the direct observation of magnetic field configurations, e.g. on magnetic recording tapes, since the light transmission of the ferrofluid layer depends on its thickness, which in turn depends on the strength of the magnetizing field at each point.

Johnson (1968) has devised and patented an oscillographic writing system in which a jet of ferrofluid ink is directed at the writing surface and deflected by electrically controlled magnetic field gradients.

Since ferrofluids have relative permeabilities in the range 1–5, they can be used in the magnet and coil systems of loudspeakers in order to increase the magnetic flux density between the magnet and the coil, thereby increasing the efficiency of the loudspeaker.

Ferrofluids have proved to be very useful in the study of the static and dynamic configurations of electromagnetic fields, particularly those produced by linear motors.

The use of water-based ferrofluids containing Fe_3O_4 (i.e. the well-known 'magnetic colloid') for observing magnetic domains has already been mentioned in Chapter 1, although Puchalska *et al.* (1978) have shown that when the domains are observed by reflected polarized light, the birefringence of the ferrofluid introduces new and informative features into the appearance of the domain structure. The change in the refractive index for ferrofluids containing Fe_3O_4 is proportional to the flux density squared, i.e. $\Delta n \propto B^2$.

For a discussion of the potential applications of ferrofluids, the reader is referred to Kaiser and Miskolczy (1970), Charles and Popplewell (1980) and Raj and Moskowitz (1990). A list of references and a bibliography of ferrofluids has been compiled by Zahn and Shenton (1980). A recent review by Blums *et al.* (1990) contains about 870 references.

REFERENCES

Becker, J. J. (1976). *Proc. Conf. Magnetism and Magnetic Materials* (AIP Conf. Proc. No. 29), p. 204.

Bloomberg, D. S. and Connell, G. A. N. (1990). *Magnetic Recording Handbook*, ed. C. D. Mee and E. D. Daniel (McGraw-Hill, London).

Blums, E., Gogosov, V. V., Charles, S. W. and Rosensweig, R. E. (1990). *J. Magn. Magn. Mater.* **85**, 1.

Boll, R. (1980). *Umsch. Wiss. Tech.* **80**, 403.

Boll, R., Hilzinger, H.-R. and Warlimont, H. (1983). *Glassy Metals: Magnetic, Chemical and Structural Properties* (CRC Press Inc., Boca Raton, Florida).

Buschow, K. H. J. (1988). *Ferromagnetic Materials: A Handbook on the Properties of Magnetically Ordered Substances*, Vol. 4, ed. E. P. Wohlfarth and K. H. J. Buschow (Elsevier, Amsterdam).

Byrne, J. V. (1978). *IEEE Trans. Magn.* **MAG-14**, 560.

Cargill, G. S. (1974). *Proc. Conf. Magnetic Materials* (AIP Conf. Proc. No. 18), p. 631.

Cargill, G. S. (1975), *Solid State Phys.* **30**, 227.

Charles, S. W. and Popplewell, J. (1980). *Ferromagnetic Materials: A Handbook on the Properties of Magnetically Ordered Substances*, Vol. 2, ed. E. P. Wohlfarth (North-Holland, Amsterdam).

Chaudhari, P., Cuomo, J. J. and Gambino, R. J. (1973). *Appl. Phys. Lett.* **22**, 337.

Chen, H. S. (1980). *Rep. Prog. Phys.* **43**, 353.

Chen, H. S., Ferris, S. D., Georgy, E. M., Leamy, H. J. and Sherwood, R. C. (1975). *Appl. Phys. Lett.* **26**, 405.

Clark, A. E. (1973). *Appl. Phys. Lett.* **23**, 642.

Coey, J. M. D. (1978). *J. Appl. Phys.* **49**, 1646.

Egami, T. (1984). *Rep. Prog. Phys.* **47**, 1601.

Ferchmin, A. R. and Kobe, S. (1983). *Amorphous Magnetism and Metallic Magnetic Materials Digest* (North-Holland, Amsterdam).

Fish, G. E. (1989). *Mater. Sci. Engng* **B3**, 457.

Fish, G. E. (1990). *Proc. IEEE* **78**, 947.

Fujimori, H., Masumoto, T., Obi, Y. and Kikuchi, M. (1974). *Japan. J. Appl. Phys.* **13**, 1889.

Gibbs, M. R. J. and Squire, P. T. (1989). *Magnetic Sensors*, ed. R. Boll and K. J. Overshott (Vol. 5 of *Sensors: A Comprehensive Survey*, ed. W. Göpel, J. Hesse and J. N. Zemel) (VCH, Weinheim).

Gilman, J. J. and Leamy, H. J. (eds) (1978). *Metallic Glasses* (ASM, Metals Park, Ohio).

Graham, Jr, C. D., Egami, T., Williams, R. S. and Takei, Y. (1976). *Proc. Conf. Magnetic Materials* (AIP Conf. Proc. No. 29), p. 218.

Gubanov, A. I. (1960). *Sov. Phys. Sol. Stat. Phys.* **2**, 468.

Gyorgy, E. M., Leamy, H. J., Sherwood, R. C. and Chen, H. S. (1976). *Proc. Conf. Magnetism and Magnetic Materials* (AIP Conf. Proc. No. 29), p. 198.

Hall, W. F. and Busenberg, S. N. (1969). *J. Chem. Phys.* **51**, 137.

Hansen, P. and Heitmann, H. (1989). *IEEE Trans. Magn.* **MAG-25**, 4390.

Hasegawa, R. (1976). *Proc. Conf. Magnetism and Magnetic Materials* (AIP Conf. Proc. No. 29), p. 216.

Hasegawa, R. (ed.) (1983). *Glassy Metals: Magnetic, Chemical and Structural Properties* (CRC Press, Boca Raton, Florida), p. 163.

Hilzinger, H. R. (1990). *J. Magn. Magn. Mater.* **83**, 370.

Johnson, C. E. (1968). US Patent 3,510,878, 2 April 1968.

Kaiser, R. and Miskolczy. (1970). *J. Appl. Phys.* **41**, 1064.

Kobe, S. and Ferchmin, A. R. (1977). *J. Mater. Sci.* **12**, 1713.

Kronmüller, H. (1981a). *J. Magn. Magn. Mater.* **24**, 159.

Kronmüller, H. (1981b). *J. Appl. Phys.* **52**, 1859.
Levy, R. A. and Hasegawa, R. (eds) (1977). *Amorphous Magnetism*, Vol. 2 (Plenum Press, New
Luborsky, F. E. (1978). *J. Magn. Magn. Mater.* **7**, 143.
Luborsky, F. E. (1980) *Ferromagnetic Materials: A Handbook on the Properties of Magnetically Ordered Substances*, Vol. 2, ed. E. P. Wohlfarth (North-Holland, Amsterdam), p. 451.
Luborsky, F. E. (ed.) (1983). *Amorphous Metallic Alloys* (Butterworths, London), p. 360.
Luborsky, F. E. and Walter, J. L. (1977a). *IEEE Trans. Magn.* **MAG-13**, 1635.
Luborsky, F. E. and Walter, J. L. (1977b). *IEEE Trans. Magn.* **MAG-13**, 953.
Luborsky, F. E., Becker, J. J. and McCary, R. O. (1975). *IEEE Trans. Magn.* **MAG-11**, 1644.
Luborsky, F. E., Livingston, J. D. and Chin, G. Y. (1983). *Physical Metallurgy*, 3rd edn, Vol. 2, ed. R. W. Chan and P. Haasen (North-Holland, Amsterdam), p. 1673.
Mendelsohn, L. I., Nesbitt, E. A. and Bretts, G. R. (1976). *IEEE Trans. Magn.* **MAG-12**, 924.
Moorjani, K. and Coey, J. M. D. (1984). *Magnetic Glasses* (Elsevier, Amsterdam).
Neuringer, J. L. and Rosensweig, R. E. (1964). *Phys. Fluids* **7**, 1927.
Obi, Y., Fujimori, H. and Saito, H. (1976). *Japan J. Appl. Phys.* **15**, 611.
O'Handley, R. C. and Chou, C. P. (1978). *J. Appl. Phys.* **49**, 1659.
Puchalska, I. B., Jones, G. A. and Jouve, H. (1978). *J. Phys. D: Appl. Phys.* **11**, L175.
Raj, K. and Moskowitz, R. (1990). *J. Magn. Magn. Mater.* **85**, 233.
Rhyne, J. J., Schelling, J. H. and Koon, N. C. (1974). *Phys. Rev.* **B10**, 4672.
Romanikiw, L. T., Slusarczuk, M. M. G. and Thompson, D. A. (1975). *IEEE Trans Magn.* **MAG-11**, 25.
Rosensweig, R. E. (1966a). *Nature* **210**, 613.
Rosensweig, R. E. (1966b). *AIAA J.* **4**, 1751.
Rosensweig, R. E. (1985). *Ferrohydrodynamics* (Cambridge University Press).
Savage, H. T., Clark, A. E., Pickart, S. J., Rhyne, J. J. and Alperin, H. A. (1974). *IEEE Trans. Magn.* **MAG-10**, 807.
Togami, Y., Kobayashi, K., Kajiura, M., Sato, K. and Teranishi, T. (1982). *J. Appl. Phys.* **53**, 2335.
Warlimont, H. (1988). *Mater. Sci. Engng* **99**, 1.
Zahn, M. and Shenton, K. E. (1980). *IEEE Trans. Magn.* **MAG-16**, 387.

Chapter 4

Ferrimagnetic Materials

4.1 INTRODUCTION

Ferrimagnetic materials form a very large and important group that, like ferromagnetic materials, possess a spontaneous magnetization and most of the other properties of ferromagnetics. An important difference, however, is that the spontaneous magnetization is the resultant of two oppositely magnetized sublattices, each of which is ferromagnetically ordered. Indeed, the first permanent magnetic material to be discovered was the mineral magnetite or lodestone, which is essentially ferrous ferrite, i.e. $FeFe_2O_4$. This material has the spinel structure and is ferrimagnetic rather than ferromagnetic. Magnetite ($Fe_3O_4 \equiv FeFe_2O_4$) has been used as a compass for many centuries – hence its popular name lodestone, i.e. leading stone.

Most ferrimagnetics are oxide magnetic materials, so that they are usually insulators or semiconductors, a property that is of great importance in practical applications, where induced eddy currents must be minimized.

4.2 FERRIMAGNETIC MATERIALS WITH THE SPINEL STRUCTURE

The general chemical formula for the ferrites that have the spinel structure (i.e. the same structure as the mineral spinel $MgAl_2O_4$) is MFe_2O_4, where M is a divalent metal ion, usually with an ionic radius in the range 0.06–0.1 nm. For the most important ferrites M is usually a transition metal such as Mn, Fe, Co, Ni or Cu, or Zn, Mg or Cd. The properties of the more important ferrimagnetic spinels are shown in Table 4.1. It is of course possible to make ferrimagnetic spinels that are solid solutions of two or more ferrites, e.g. Mn–Zn and Ni–Zn ferrite find wide commercial application. Furthermore, it is sometimes possible for M to be a mixture of ions that have an average valence of 2, e.g. Li^+ and Fe^{3+} in $Li_{0.5}Fe_{2.5}O_4$. Another important example of this is the ferrite γ-Fe_2O_3 (the recording material on most magnetic tapes), which has the spinel structure

Table 4.1 Properties of ferrimagnetic spinels at 20°C.

Ferrite	$J_s = \mu_0 M_s$	Ion moment (μ_B)	Curie temperature T_c (°C)	Lattice parameter (nm)	Density (10^3 kg m^{-3})
$MnFe_2O_4$	0.50	5	300	0.850	5.00
$FeFe_2O_4$	0.60	4.1	585	0.839	5.24
$CoFe_2O_4$	0.53	3.7	520	0.838	5.29
$NiFe_2O_4$	0.34	2.3	585	0.834	5.38
$CuFe_2O_4$[a]	0.17	1.3	455	0.8445	5.42
$MgFe_2O_4$	0.14	1.1	440	0.836	4.52
$Li_{0.5}Fe_{2.5}O_4$	0.39	2.6	670	0.833	4.75
γ-Fe_2O_3	0.52	2.3	575	0.834	4.89
$ZnFe_2O_4$	0	0	—	0.84	5.4
$CdFe_2O_4$	0	0	—	0.873	5.76

[a] $CuFe_2O_4$ may also have a tetragonal structure with $a = 0.822$ nm and $c = 0.870$.

and a chemical formula that can be written in the form

$$\square_{1/3}Fe_{8/3}O_4 \quad \text{or} \quad (\square_{1/3}Fe^{3+}_{2/3})Fe^{3+}_2O_4$$

where the symbol $\square$ represents a vacant site. Note that the iron:oxygen ratio is still 2:3. The trivalent Fe^{3+} ions can also be wholly or partially replaced by Cr^{3+} and Al^{3+} to give pure or mixed ferrites, many of which are ferrimagnetic at room temperature. Thus a whole series of ferrimagnetic spinels can be prepared.

The spinel structure is shown in Fig. 4.1. The relatively large oxygen ions form an fcc lattice in which there are 32 octahedral sites (i.e. sites surrounded by 6 oxygen ions) and 64 tetrahedral sites (surrounded by 4 oxygen ions). However, in the spinel structure only 8 of the tetrahedral sites and only 16 of the octahedral sites are occupied by metal ions. These tetrahedral and octahedral sites are referred to as A and B sites respectively. (Note that a useful mnemonic is that the letter 'A' is similar to a tetrahedron, while the letter 'B' is rather like the digit 8, and hence octahedral.) The atomic arrangement in the spinel structure is best understood by dividing the unit cell into eight cubes. If we now consider two octants of the unit cell in more detail, we can see that the metal ion

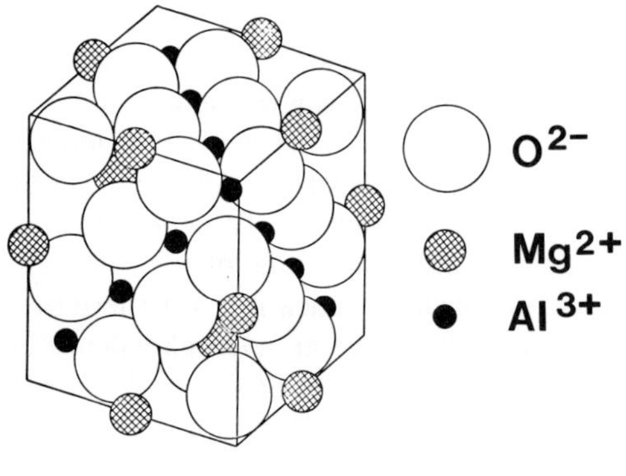

Fig. 4.1 The spinel structure (from Kittel, 1956).

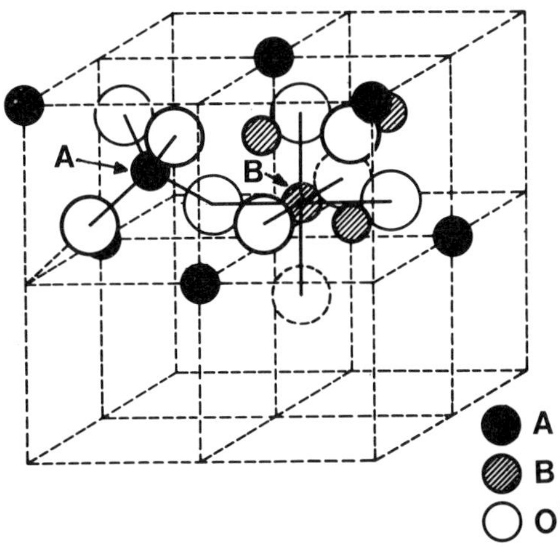

Fig. 4.2 Arrangements of metal ions in the two octants A and B, showing tetrahedrally (A) and octahedrally (B) coordinated sites.

A in Fig. 4.2 is tetrahedrally coordinated, while that at B is octahedrally coordinated. Note that many of the ions have been omitted for clarity. It should be understood that the structure shown in Fig. 4.2 is known as the inverse spinel structure because eight of the octahedral B sites are occupied by the divalent ions M^{2+}, whereas in the mineral spinel the divalent Mg^{2+} ions occupy tetrahedral A sites only. The ferrites $MnFe_2O_4$, $ZnFe_2O_4$ and $CdFe_2O_4$ are known as normal spinels because the divalent ions Mn^{2+}, Zn^{2+} and Cd^{2+} ions occupy tetrahedral A sites. This distinction between normal and inverse spinels is not rigid, since there are many examples where ions of a particular species can be found on both A and B sites in the same material, where the relative occupation depends on composition, preparation conditions and heat treatment. The magnetic structures of normal and inverse spinels are shown in Fig. 4.3. Since the magnetic moments of the Mn^{2+} and Fe^{3+} ions are both $5\mu_B$, we can see that the net magnetic moment of normal $MnFe_2O_4$ is due solely to the eight uncompensated Fe^{3+} ions on the octahedral sites, i.e.

$$\text{magnetic moment/unit cell} = 8 \times 5\mu_B = 40\mu_B.$$

Since the volume of the unit cell is simply equal to the cube of the lattice parameter, we find that the expected saturation magnetic polarization per unit volume at 0 K is

$$J_s(0) = \frac{\mu_0 40\mu_B}{a^3} = 0.6 \text{ T}.$$

This is larger than the experimental value of 0.5 T shown in Table 4.1, but it is sufficiently accurate to illustrate and confirm the basic theory of ferrimagnetism. It is known from neutron diffraction experiments that $MnFe_2O_4$ is a mixed normal and inverse spinel, so that some of the Mn^{2+} ions do occupy octahedral sites. However, this does not alter the expected magnetic moment per unit cell, because the magnetic moments of the Mn^{2+} and Fe^{3+} ions are both $5\mu_B$. The discrepancy between the theoretical and experimental magnetic moment of $MnFe_2O_4$ is due to changes in the valences of the ions, e.g. Mn^{2+}

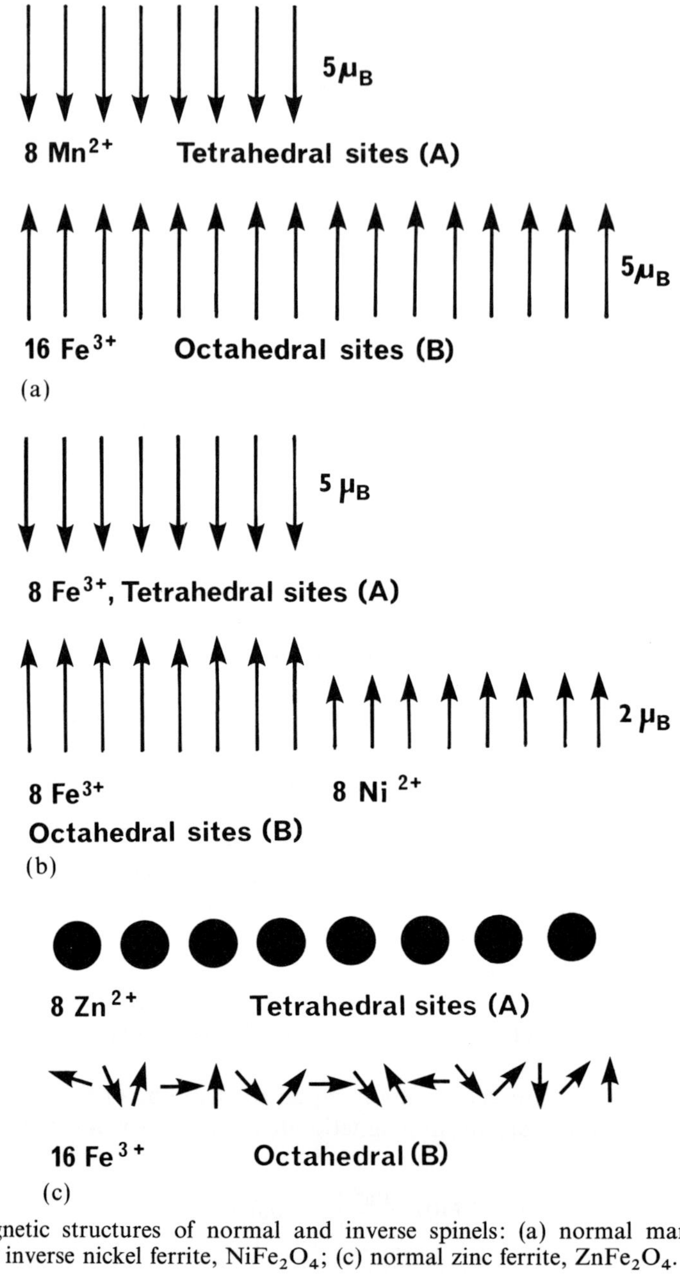

Fig. 4.3 Magnetic structures of normal and inverse spinels: (a) normal manganese ferrite, $MnFe_2O_4$; (b) inverse nickel ferrite, $NiFe_2O_4$; (c) normal zinc ferrite, $ZnFe_2O_4$.

may be oxidized to Mn^{3+} or even Mn^{4+}, depending on the conditions of preparation – similarly for the Fe^{3+} ions, which may be reduced to Fe^{2+}. Incomplete quenching of the orbital magnetic moments or more complex superexchange interactions between the ions via the O^{2-} ions may also contribute to the discrepancy.

Similarly, the magnetic moment of nickel ferrite $NiFe_2O_4$, which has the inverse spinel structure, is due entirely to the uncompensated magnetic moments of the eight Ni^{2+}

ions on the octahedral sites (Fig. 4.3b), i.e.

$$\text{magnetic moment/unit cell} = 8 \times 2\mu_B = 16\mu_B,$$

and the saturation magnetic polarization at 0 K is

$$J_s(0) = \mu_0 \frac{16\mu_B}{a^3} = 0.32 \text{ T}.$$

The discrepancy between the theoretical and experimental magnetic moment given in Table 4.1 can largely be attributed to incomplete quenching of the orbital magnetic moment and to more complex superexchange interactions between the Ni^{2+} ions via the O^{2-} ions.

Normal zinc ferrite, $ZnFe_2O_4$, for which the magnetic moment of the Zn^{2+} is zero, shows rather different behaviour. From Fig. 4.3(c), we can see that it is paramagnetic because, unlike normal $MnFe_2O_4$, the magnetic moments of the ions on the octahedral sites are not antiferromagnetically ordered as we should expect but assume random orientations with respect to each other due to thermal agitation. Thus, in the absence of an applied field, the net magnetic moment is zero, so that the pure material is not ferrimagnetic. However, we shall see later that $ZnFe_2O_4$ still has very important magnetic effects when it is mixed with ferrites that have the inverse or normal spinel structures. Similarly, $CdFe_2O_4$ is paramagnetic.

A comparison of the theoretical and observed magnetic moments of various ferrites is shown in Fig. 4.4, from which it can be seen that there is very good agreement between theoretical and experimental values. The unusual behaviour of $CuFe_2O_4$ is discussed later.

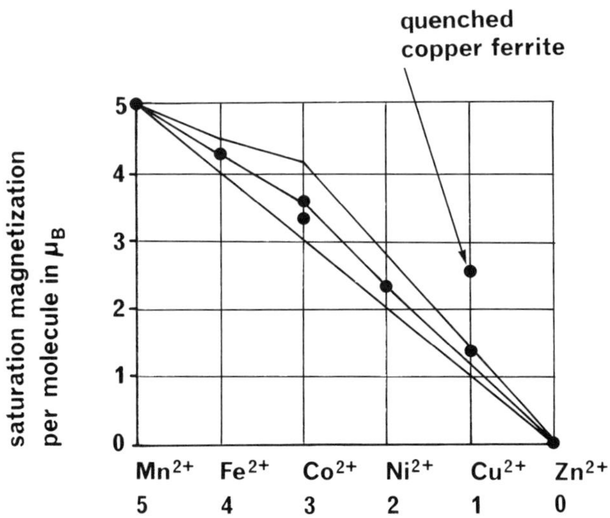

Fig. 4.4 Comparison of the theoretical and observed magnetic moments of ferrites. The behaviour of $CuFe_2O_4$ is discussed later.

4.3 NÉEL TWO-SUBLATTICE THEORY OF FERRIMAGNETISM

We can understand how the phenomenon of ferrimagnetism arises in the following general way using an argument suggested by Kittel (1986), which contains the essential features of Néel's (1948) detailed theory. If we assume that the Weiss internal field theory can be applied to ferrimagnetic materials then, since there are two magnetic sublattices, there are two internal Weiss fields, which are proportional to the sublattice magnetizations. We also assume that the sublattice magnetizations M_A and M_B are due to the magnetic moments of ions in tetrahedral and octahedral sites respectively. The internal fields and hence the magnetizations are due to exchange interactions between A–A ions, B–B ions and A–B ions. Thus the internal fields acting on the A and B sites are

$$\boldsymbol{B}_A = -\lambda \boldsymbol{M}_A - \nu \boldsymbol{M}_B,$$

$$\boldsymbol{B}_B = -\mu \boldsymbol{M}_B - \nu \boldsymbol{M}_A.$$

The internal fields and the magnetizations must be specified as vectors, since we do not yet know what their relative orientations are; λ, μ and ν are all positive constants incorporating the magnetic constant μ_0. In the presence of these fields, the total internal magnetic energy E_T is given by the sum of the interactions of the individual sublattice magnetizations with their effective field acting on them, i.e. the sum of the scalar products:

$$\begin{aligned} E_T &= -\tfrac{1}{2}\boldsymbol{M}_A \cdot \boldsymbol{B}_A - \tfrac{1}{2}\boldsymbol{M}_B \cdot \boldsymbol{B}_A, \\ &= \tfrac{1}{2}\boldsymbol{M}_A \cdot (\lambda \boldsymbol{M}_A + \nu \boldsymbol{M}_B) + \tfrac{1}{2}\boldsymbol{M}_B \cdot (\mu \boldsymbol{M}_B + \nu \boldsymbol{M}_A), \\ &= \tfrac{1}{2}\lambda |\boldsymbol{M}_A|^2 + \tfrac{1}{2}\nu \boldsymbol{M}_A \cdot \boldsymbol{M}_B + \tfrac{1}{2}\mu |\boldsymbol{M}_B|^2 + \tfrac{1}{2}\nu \boldsymbol{M}_A \cdot \boldsymbol{M}_B \end{aligned}$$

Now

$$|\boldsymbol{M}_A|^2 = M_A^2, \qquad |\boldsymbol{M}_B|^2 = M_B^2,$$

and hence

$$E_T = \tfrac{1}{2}\lambda M_A^2 + \tfrac{1}{2}\mu M_B^2 + \nu \boldsymbol{M}_A \cdot \boldsymbol{M}_B.$$

We should expect the most stable magnetic structure to be that which minimizes the total magnetic energy E_T. The first two terms in this expression are always positive because λ, μ and the square of any real quantity are positive. Thus the total energy of the crystal is determined by the magnitude and sign of the last term. If we expand the scalar product, we obtain

$$\nu \boldsymbol{M}_A \cdot \boldsymbol{M}_B = \nu |\boldsymbol{M}_A|\,|\boldsymbol{M}_B| \cos\theta,$$

where θ is the angle between the magnetization vectors $\boldsymbol{M}_A$ and $\boldsymbol{M}_B$. If $\theta = 90°$, the above term is zero, so that some reduction in the energy is achieved in this case. However, the greatest reduction occurs when the expression is negative, and the largest possible negative value is obtained when $\theta = 180°$, since $\cos 180° = -1$. This corresponds to an antiferrimagnetic alignment of the sublattice magnetizations M_A and M_B. The total energy of the crystal is then

$$E_T = \tfrac{1}{2}\lambda M_A^2 + \tfrac{1}{2}\mu M_B^2 - \nu M_A M_B.$$

If we take our 'zero' of energy to be that when the two sublattice magnetizations are zero, i.e. $M_A = M_B = 0$, then an even lower energy (i.e. a negative energy) can be obtained if the last term in the equation is greater than the sum of the first two, i.e.

$$\nu M_A M_B > \tfrac{1}{2}(\lambda M_A^2 + \mu M_B^2).$$

This expression gives a criterion for ferrimagnetism and a phenomenological explanation for its occurrence, i.e. if the two sublattices are antiferromagnetically aligned with respect to each other then there is a consequent lowering of the total magnetic energy of the crystal. The antiferromagnetic alignment of the two sublattice magnetizations can be achieved only if each sublattice is ferromagnetically aligned. The magnetic structures of various ferrimagnetic materials are shown in Fig. 4.3. The net magnetization, i.e. the net spontaneous magnetization at 0 K, is simply the numerical difference between the sublattice magnetizations:

$$M_s = |M_A - M_B| \quad \text{or} \quad J_s = |J_A - J_B|.$$

Since ferrimagnetism depends on magnetic interactions of the type A–A, B–B and A–B, it is, like ferromagnetism, a cooperative phenomenon. As in the Weiss–Brillouin theory of ferromagnetism, all the magnetic spins experience constant thermal agitation, so that at non-zero temperatures we should expect that many individual spins will effectively be reversed, thus reducing the internal field, which in turn reduces the alignment, and so on. It is clear that in ferrimagnetic materials this cooperative interaction will be even more complex than that in ferromagnetics because of the presence of two magnetic sublattices and at least three types of magnetic interactions. A further complication arises because the ionic magnetic moments on the sublattices are not identical – usually the octahedral sites contain $8Fe^{3+}$ ions with $5\mu_B$ each and $8M^{2+}$ ions with magnetic moments ranging from 1 to $5\mu_B$ each. It has also been shown that the octahedral Fe^{3+} and M^{2+} may not be completely ordered, depending on composition and heat treatment.

The temperature dependence of the magnetization of ferrites and many ferrimagnetic oxide materials is therefore rather more complex than that for the ferromagnetic metals such as Fe, Co, Ni and their alloys. This important property is discussed in Section 4.4. The fundamental properties of oxide spinels have been reviewed by Krupička and Novák (1982).

4.4 TEMPERATURE DEPENDENCE OF THE SPONTANEOUS MAGNETIZATION OF FERRIMAGNETIC MATERIALS

In a classic paper Néel (1948) suggested that there are at least six basic types of $M_s(T)$ versus T curves for ferrimagnetics with two sublattice magnetizations. These are shown in Fig. 4.5. The curves M, P, Q and R all show a single Curie temperature, which

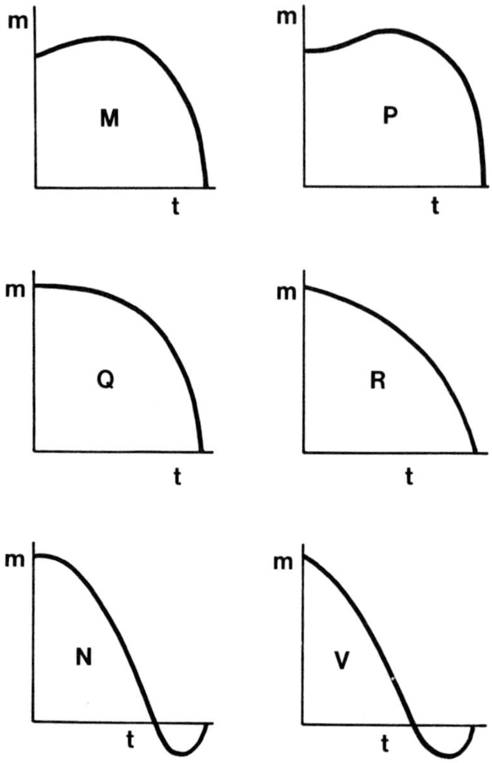

Fig. 4.5 Six basic types of $M_s(T)$ versus T curves for ferrimagnetic materials with two sublattice magnetizations, plotted in reduced units $m = M_s(T)/M_s(0)$ and $t = T/T_c$.

suggests that the ferromagnetic alignment on both the A and B sublattices is broken down at the same temperature, although at temperatures less than T_c the thermal energy may produce different rates of change in the magnetizations. The important distinguishing features of the curves M, P, Q and R are the rates of change in the magnetization at 0 K and T_c. The slopes at 0 K are finite for curves M and R, while they are zero for curves P and Q. Finite slopes at 0 K suggest that one of the sublattices is not saturated: a situation in conflict with the third law of thermodynamics. However, Néel's theory oversimplifies the physical situation because he assumes that the spins are either parallel or antiparallel throughout the whole temperature range. Yafet and Kittel (1952) have shown that a lower magnetic energy can be obtained if it is assumed that the B sublattice divides into two equivalent magnetic substructures B_1 and B_2, with magnetizations $\boldsymbol{M}_{B_1}$ and $\boldsymbol{M}_{B_2}$ that are not parallel, thus giving rise to what are now known as triangular spin structures. The Yafet–Kittel theory removes the difficulty of incomplete saturation at 0 K and hence the conflict with the third law of thermodynamics.

Thus the only types of $M_s(T)$ versus T curve that are physically possible for ferrimagnetic materials are the Q, P and N types; the Q-type behaviour is similar to that for ferromagnetic materials. The temperature variation of $M_s(T)$ for a number of spinel ferrites is shown in Fig. 4.6. This type of behaviour can be understood by considering the individual temperature variation of each sublattice. The observed magnetization is simply the numerical difference between the sublattice magnetizations.

P-type behaviour is observed in $NiFe_{1.37}Al_{0.63}O_4$ (Maxwell and Pickart, 1953), and its schematic interpretation is shown in Fig. 4.7. A very important type of ferrimagnetic

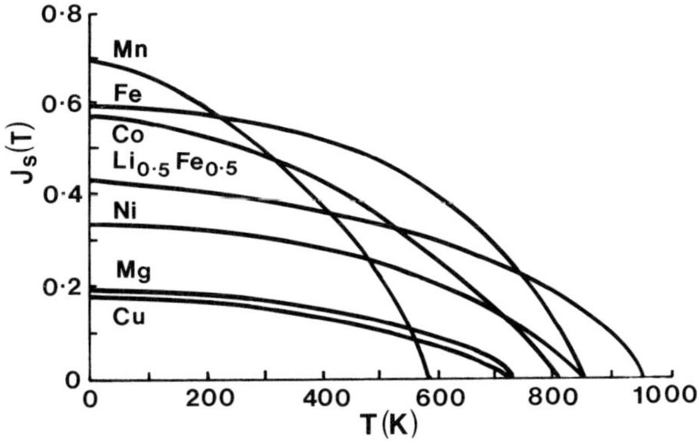

Fig. 4.6 Temperature variation of the saturation magnetic polarization for various spinel ferrites (after Smit and Wijn, 1959).

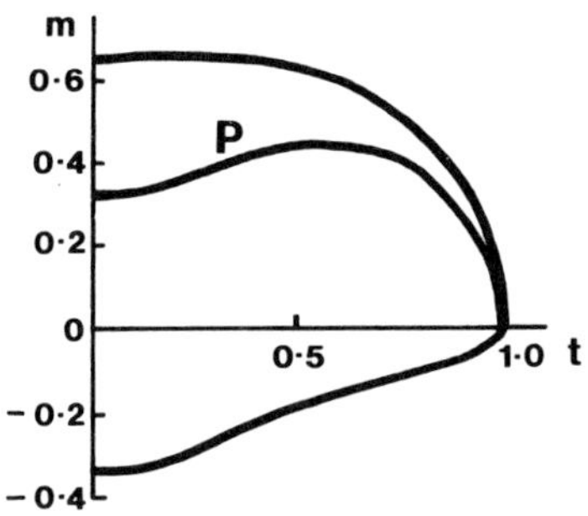

Fig. 4.7 Schematic interpretation of P-type behaviour in terms of the sublattice magnetizations.

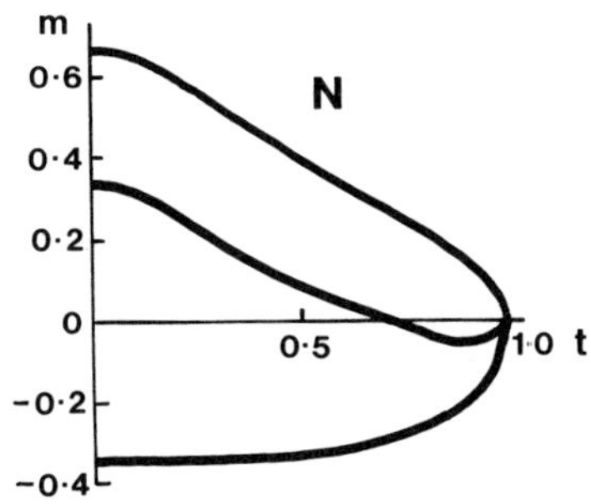

Fig. 4.8 The sublattice magnetizations for an N-type material

behaviour is that of type N. The sublattice magnetizations are shown in Fig. 4.8, from which we can see that the net magnetization is zero before either of the sublattice magnetizations is zero. However, as the temperature is increased further beyond the compensation temperature, the net magnetization increases, although its sign is reversed with respect to that of the original magnetization direction. This behaviour, first predicted by Néel, has been observed for inverse spinel ferrites with compositions in the range $Li_{0.5}Fe_{2.5-\delta}Cr_\delta O_4$ ($1<\delta<1.6$) (Gorter and Schulkes, 1953) is very important, since it provides direct and convincing experimental evidence for the general validity of Néel's two-sublattice theory of ferrimagnetism. The lanthanide garnets also show similar behaviour, and although their crystallographic and magnetic structures are different, it can be interpreted in a similar way.

Above the Néel temperature T_N, the paramagnetic susceptibilities χ of many ferrimagnetic materials obey the relation

$$\frac{1}{\chi}=\frac{1}{C}-\frac{1}{\chi_0}-\frac{C'}{T-T_N},$$

where C, C' and χ_0 are constants characteristic of the material.

4.5 MAGNETIC STRUCTURES OF FERRITES

Since the magnetic properties of the ferrimagnetic spinels depend on the interaction between the ions on the tetrahedral (A) and octahedral (B) sites, the distribution of the ions on these sites is of critical importance. In some ferrites there is considerable variation in site preference for particular ions, which is also strongly dependent on the temperature of preparation, rate of cooling and subsequent heat treatment. The cation occupation of A and B sites in the most important spinel ferrites is shown in Table 4.2.

The magnetic structures can be conveniently written as follows, so that the magnetic moment directions of the ions are clearly shown:

$$FeFe_2O_4 \quad \overrightarrow{[Fe^{3+}]} \quad [\overleftarrow{Fe^{2+}}\ \overleftarrow{Fe^{3+}}] \quad O_4$$

$$CoFe_2O_4 \quad \overrightarrow{[Fe^{3+}]} \quad [\overleftarrow{Co^{2+}}\ \overleftarrow{Fe^{3+}}] \quad O_4$$

$$NiFe_2O_4 \quad \overrightarrow{[Fe^{3+}]} \quad [\overleftarrow{Ni^{2+}}\ \overleftarrow{Fe^{3+}}] \quad O_4$$

$$CuFe_2O_4 \quad \overrightarrow{[Fe^{3+}]} \quad [\overleftarrow{Cu^{2+}}\ \overleftarrow{Fe^{3+}}] \quad O_4$$

The magnetic structures of these ferrites are shown in Fig. 4.3(b), so that the magnetic moment is due to the divalent ions only; for example, the magnetic moment of the Fe^{2+} ion is $4\mu_B$, so we should expect a magnetic moment of $8 \times 4\mu_B$ per unit cell, and hence a magnetic polarization per unit volume at OK of

$$J_s = \mu_0 M_s = \frac{32\mu_B\mu_0}{a^3} = \frac{32 \times 9.27 \times 10^{-24} \times 4\pi \times 10^{-7}}{(0.839 \times 10^{-9})^3} = 0.63\ \mathrm{T},$$

which is close to the experimental value of 0.6 T. The large experimental values of the magnetic moments for $CoFe_2O_4$ and $NiFe_2O_4$, which are known to have the completely inverse spinel structure, are attributed to a contribution from the orbital magnetic moment that is incompletely quenched by the crystal field.

Table 4.2 Cation occupation of A and B sites in ferrites.

Ferrite	Tetrahedral A sites	Octahedral B sites	A (μ_B)	B (μ_B)	$\lvert A+B \rvert$ (μ_B)	Exptl (μ_B)
$MnFe_2O_4$	$Fe^{3+}_{0.2}Mn^{2+}_{0.8}$	$Mn^{2+}_{0.2}Fe^{3+}_{0.8}$	5	5+5	5	4.6
$FeFe_2O_4$	Fe^{3+}	$Fe^{2+}Fe^{3+}$	5	4+5	4	4.1
γ-Fe_2O_3	Fe^{3+}	$_{1/3}Fe^{3+}_{5/3}$	5	0+8.3	3.3	3.2
$CoFe_2O_4$	Fe^{3+}	$Co^{2+}Fe^{3+}$	5	3+5	3	3.7
$NiFe_2O_4$	Fe^{3+}	$Ni^{2+}Fe^{3+}$	5	2+5	2	2.3
$CuFe_2O_4$	Fe^{3+}	$Cu^{2+}Fe^{3+}$	5	1+5	1.0	1.3
$MgFe_2O_4$	$Mg^{2+}_{0.1}Fe^{3+}_{0.9}$	$Mg^{2+}_{0.9}Fe^{3+}_{1.1}$	0+4.5	0+5.5	1.0	1.0
$Li_{0.5}Fe_{2.5}O_4$	Fe^{3+}	$Li^{+}_{0.5}Fe^{3+}_{1.5}$	5	0+7.5	2.5	2.6
$ZnFe_2O_4$	Zn^{2+}	$Fe^{3+}Fe^{3+}$	5	5−5	Random	Paramagnetic

4.5.1 The Magnetic Structure of $MnFe_2O_4$

From Table 4.2 we can see that the ions Mn^{2+} and Fe^{3+} have no preference for either octahedral or tetrahedral sites if it is assumed that the crystal field stabilization energy is the only factor determining the site occupation. In this case the distribution must be determined by other factors, such as ionic size, polarization and Madelung energy effects. It is now well established that $MnFe_2O_4$ may have both the normal and inverse spinel structures (Fig 4.3a and b), although the former is preferred. From neutron diffraction experiments, it has been shown that the site occupation is

$$\overset{A}{[\overset{\longrightarrow}{Fe^{3+}_{0.2}} + \overset{\longrightarrow}{Mn^{2+}_{0.8}}]} \quad \overset{B}{[\overset{\longleftarrow}{Mn^{2+}_{0.2}} + \overset{\longleftarrow}{Fe^{3+}_{1.8}}]} \quad O_4$$

Since the magnetic moments of the Mn^{2+} and Fe^{3+} ions are both $5\mu_B$, we should still expect a magnetic moment of $8 \times 5\mu_B$ per unit cell. However, from experimental measurements on stoichiometric $MnFe_2O_4$, the magnetic moment per unit cell is only $8 \times 4.6\mu_B$, but this can be explained by the simultaneous occurrence of Mn^{3+} and Fe^{2+} ions, each of which has a magnetic moment of $4\mu_B$. Thus the observed magnetic moment of $4.6\mu_B$ is a weighted mean of the theoretical limits $4\mu_B$ and $5\mu_B$.

4.5.2 The Magnetic Structure of γ-Fe_2O_3

The oxide γ-Fe_2O_3 (widely used for magnetic recording) has the inverse spinel structure, with the Fe^{3+} ions unequally divided on octahedral and tetrahedral sites as shown in Fig. 4.9, from which we can see that some of the B sites are vacant, and that the magnetic moment is due to the excess of Fe^{3+} ions on B sites over those on A sites. It is convenient to write the formula for γ-Fe_2O_3 in the form

$$\tfrac{4}{3}[Fe_2O_3] \equiv Fe_{8/3}O_4 \equiv FeFe_{5/3}O_4$$

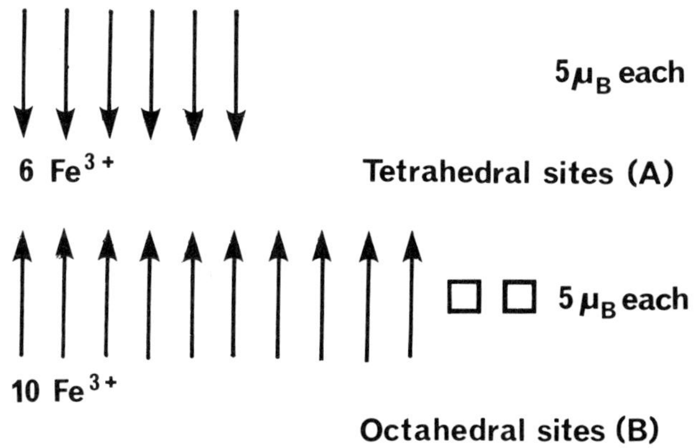

Fig. 4.9 Magnetic structure of the inverse spinel γ-Fe_2O_3, showing that some of the B-sites are vacant (□).

and its site occupation as

$$\begin{array}{cc} \overset{\text{A}}{\longrightarrow} & \overset{\text{B}}{\longleftarrow} \\ [Fe^{3+}] & [\square_{1/3}Fe^{3+}_{5/3}]\ O_4 \end{array}$$

where the symbol $\square$ represents an octahedral site vacancy. The magnetic moment per formula unit is therefore

$$\mu = (\tfrac{5}{3} - 1) \times 5\mu_B = \tfrac{10}{3}\mu_B = 3.33\mu_B,$$

or $8 \times 3.33\mu_B$ per unit cell, giving an expected saturation magnetic polarization $J_s = 0.535$ T.

4.5.3 Magnetic Structure of $CuFe_2O_4$

It has been shown that $CuFe_2O_4$ has a partly normal spinel structure and that the larger experimental magnetic moment given in Table 4.2 can be attributed to the presence of Cu^{2+} on tetrahedral A sites. The magnetic moment of copper ferrite can be increased still further if it is quenched from a high temperature, and it appears that there is a relatively small energy difference for the Cu^{2+} ion in octahedral and tetrahedral sites. This, coupled with relatively high diffusion rates at about 1000°C, explains why the effective magnetic moment can be increased to as much as $2\mu_B$ per formula unit. The magnetic structure of copper ferrite in the ideal inverse spinel state is shown in Fig. 4.10(a). If we replace one of the Fe^{3+} ions on the tetrahedral sites with a Cu^{2+} ion, the structure is now as shown in Fig. 4.10(b) – the displaced Fe^{3+} now appears on the octahedral sites. With this structure, the magnetic moment is

$$\mu = 7 \times 1\mu_B + 2 \times 5\mu_B - 1\mu_B = 16\mu_B.$$

Thus we can see that just one Cu^{2+} ion per unit cell on a tetrahedral site doubles the magnetic moment. Since the magnetic moment of $CuFe_2O_4$ cannot be increased much beyond $16\mu_B$ per unit cell, it is clear that the transfer of a second Cu^{2+} to an A site is difficult. The transfer process is thermally activated, and will therefore depend on the usual Boltzmann factor $\exp(-E_A/kT)$, where E_A is the activation energy of the process. We may assume that the presence of one Cu^{2+} per unit cell on an A site increases the activation energy E_A, thus making the transfer of a second Cu^{2+} ion very unlikely.

4.5.4 Magnetic Structure of $MgFe_2O_4$

If $MgFe_2O_4$ had a completely inverse spinel structure, the Fe^{3+} would be equally divided among the octahedral and tetrahedral sites, and the Mg^{2+} would occupy the remaining octahedral sites. The magnetic moment of the Mg^{2+} ion (electronic structure $1s^2 2s^2 2p^6 3s^2$) is zero, since all its electrons are spin-paired, so that inverse $MgFe_2O_4$ is not expected to be ferrimagnetic. However, from neutron diffraction experiments, it has been found that there is an unequal distribution of Fe^{3+} ions over the A and B sites. The observed magnetic moment is therefore due to the difference in this occupation, i.e.

$$\mu = (1.1 - 0.9)5\mu_B = 1.0\mu_B,$$

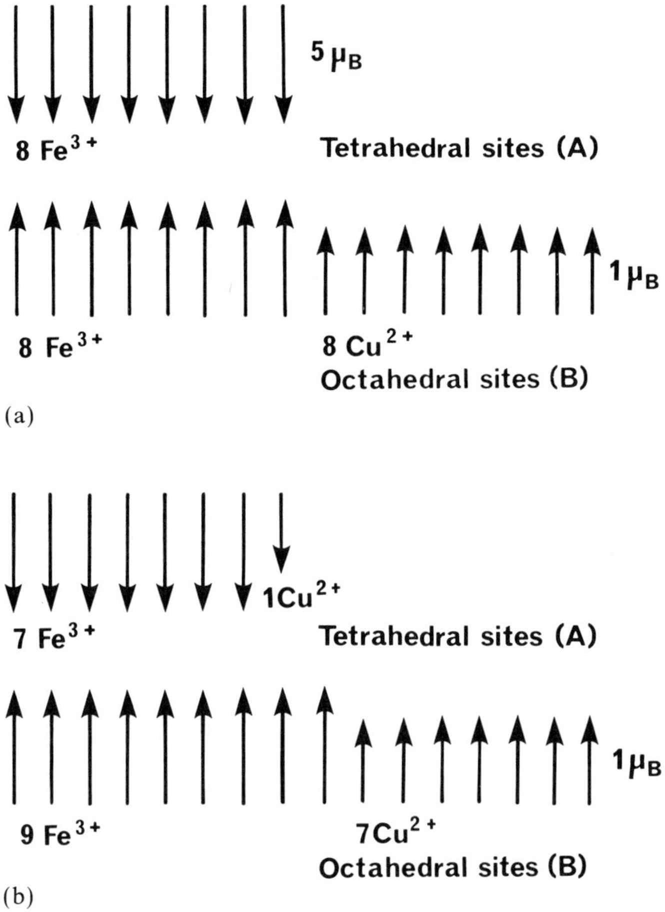

Fig. 4.10 (a) Magnetic structure of copper ferrite in the ideal inverse spinel state. (b) Magnetic structure of copper ferrite when one of the Fe^{3+} ions on the tetrahedral sites is replaced by a Cu^{2+} ion. This structure can be obtained by quenching the copper ferrite from high temperatures.

or $8 \times 1\mu_B$ per unit cell. The equilibrium cation distribution in $MgFe_2O_4$ is given by the formula

$$\overset{\xrightarrow{\mathrm{A}}}{[Fe^{3+}_{0.9}+Mg^{2+}_{0.1}]}\quad \overset{\xleftarrow{\mathrm{B}}}{[Mg^{2+}_{0.9}+Fe^{3+}_{1.1}]}\quad O_4$$

Like $CuFe_2O_4$, the site occupation depends markedly on heat treatment. If the cation distribution is described by a parameter x then mixed normal and inverse spinels such as $CuFe_2O_4$ and $MgFe_2O_4$ can be presented by the formula

$$\overset{\mathrm{A}}{[\overrightarrow{Fe^{3+}_{1-x}}\ \overrightarrow{M^{2+}_{x}}]}\quad \overset{\mathrm{B}}{[\overleftarrow{M^{2+}_{1-x}}\ \overleftarrow{Fe^{3+}_{1+x}}]}\quad O_4$$

so that the magnetic moment per formula unit, assuming antiferromagnetic alignment of the moments on A and B sites, is

$$\begin{aligned}\mu &= [(1+x)-(1-x)]5\mu_B + (1-x-x)n\mu_B \\ &= [10x + (1-2x)n]\mu_B,\end{aligned}$$

where n is the number of unpaired spins on the M^{2+} ion, i.e. Cu^{2+} or Mg^{2+}. The saturation magnetizations of $CuFe_2O_4$ and $MgFe_2O_4$ have been measured after quenching from a range of temperatures, and it has been found that the parameter x could be represented reasonably accurately by the Boltzmann distribution formula:

$$\frac{x(1+x)}{(1-x)^2} = \exp\left(\frac{-E}{kT}\right),$$

where E is the energy required to move an M^{2+} ion from a B site onto an A site and to return an Fe^{3+} ion to a B site. For annealed and quenched samples of $MgFe_2O_4$ values of x in the range 0.11–0.14 have been observed, and these orders of magnitude have been corroborated by neutron diffraction measurements.

4.5.5 Magnetic Structure of $ZnFe_2O_4$

Zinc ferrite, $ZnFe_2O_4$, unlike all the other ferrites shown in Table 4.4, has a completely normal spinel structure with all the Fe^{3+} ions on the octahedral sites and the Zn^{2+} ions on the tetrahedral sites. Unfortunately, $ZnFe_2O_4$ does not have a magnetic moment of $16 \times 5\mu_B$ per unit due to the Fe^{3+} ions on octahedral sites. Since the magnetic moment of the Zn^{2+} ion is zero (Zn^{2+} has a full $3d^{10}$ shell containing spin-paired electrons), the normal antiferromagnetic coupling between the ions on A and B sites is replaced by the antiferromagnetic coupling on octahedral sites, and we should expect the $ZnFe_2O_4$ to be antiferromagnetic with no net spontaneous magnetization. However, careful experiments have failed to reveal any ordered antiferromagnetic coupling of the Fe^{3+} ions on the octahedral sites, so that $ZnFe_2O_4$ is paramagnetic. The magnetic structure of $ZnFe_2O_4$ is shown in Fig. 4.3(c). It would appear from this that $ZnFe_2O_4$ is of little interest as a magnetic material, but, as we shall see in Section 4.7, this is far from the case.

4.6 MAGNETIC STRUCTURE AND PROPERTIES OF MIXED INVERSE SPINEL FERRITES

The inverse spinel ferrites form continuous solid solutions with each other, enabling a whole range of magnetic properties to be obtained. Fortunately, the magnetic moments can still be interpreted in terms of Néel's two-sublattice theory, and, broadly speaking, the expected magnetic moments are in good agreement with the observed values. We can represent a mixed ferrite containing two divalent ions by the formula

$$\overset{\text{A}}{\overrightarrow{[Fe^{3+}]}}\quad \overset{\text{B}}{[\overleftarrow{(M_1^{2+})_{1-x}(M_2^{2+})_x}\,\overleftarrow{Fe^{3+}}]}\quad O_4$$

The net magnetic moment, assuming antiferromagnetic coupling between A and B sites, is

$$\mu = [(1-x)n_1 + xn_2]\mu_B,$$

where n_1 and n_2 are the ionic magnetic moments of the M_1^{2+} and M_2^{2+} ions respectively. Thus μ is simply a linear function of x. The variation with x of other properties such as the magnetocrystalline anisotropy energy and magnetostriction is usually rather more complicated. However, this effect can be used to great advantage, since by suitable choice of composition, we can prepare ferrites with specific properties for particular applications.

4.7 MAGNETIC STRUCTURE AND PROPERTIES OF MIXED NORMAL AND INVERSE SPINEL FERRITES

If we add increasing proportions of normal $ZnFe_2O_4$, which has zero magnetic moment, to any of the inverse spinel ferrites, we find rather surprisingly that the saturation magnetization increases in every case and reaches a maximum when about 50 mol% of $ZnFe_2O_4$ has been added. With further addition of $ZnFe_2O_4$, the magnetic moment decreases and is zero again at 100 mol% $ZnFe_2O_4$. The saturation magnetizations at 0 K (by extrapolation) for various mixed ferrites are shown in Fig. 4.11. A range of values for the saturation magnetization of Mg–Zn ferrites is shown because the site occupation is very sensitive to heat treatment. The mixed Mn–Zn, Ni–Zn and Mg–Zn ferrites are among the most important magnetic materials, and are used in large quantities for a very wide range of commercial applications. Their relatively large saturation magnetizations compared with the single pure ferrites can readily be understood in terms of the Néel two-sublattice theory of ferrimagnetism and our knowledge of tetrahedral and octahedral site occupation. Consider the effect of replacing one formula unit of $NiFe_2O_4$ with one formula unit of $ZnFe_2O_4$. Since there are eight formula units in each unit cell, the mixed ferrite has a composition $(NiFe_2O_4)_7(ZnFe_2O_4)_1$. The magnetic structure of this ferrite is shown in Fig. 4.12(a), and it is important to note that the Zn^{2+} occupies a tetrahedral site because Zn^{2+} ions have a preference for tetrahedral rather than octahedral sites. Since the magnetic moment of the Zn^{2+} is zero,

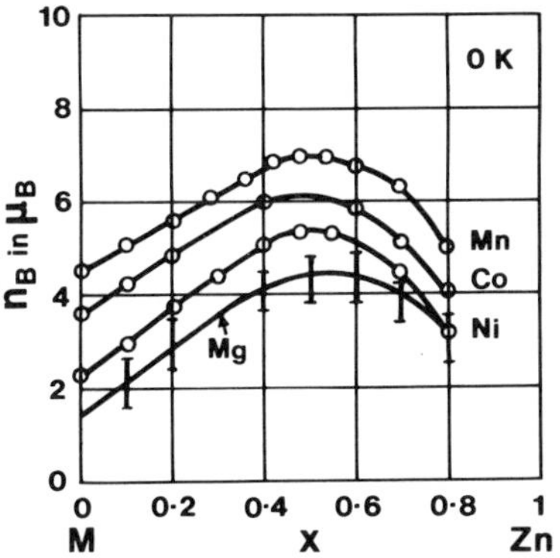

Fig. 4.11 Saturation magnetization at 0 K expressed as the number n_B of μ_B per molecule for various mixed ferrites, $M_{1-x}Zn_xFe_2O_4$ (after Guillaud, 1951).

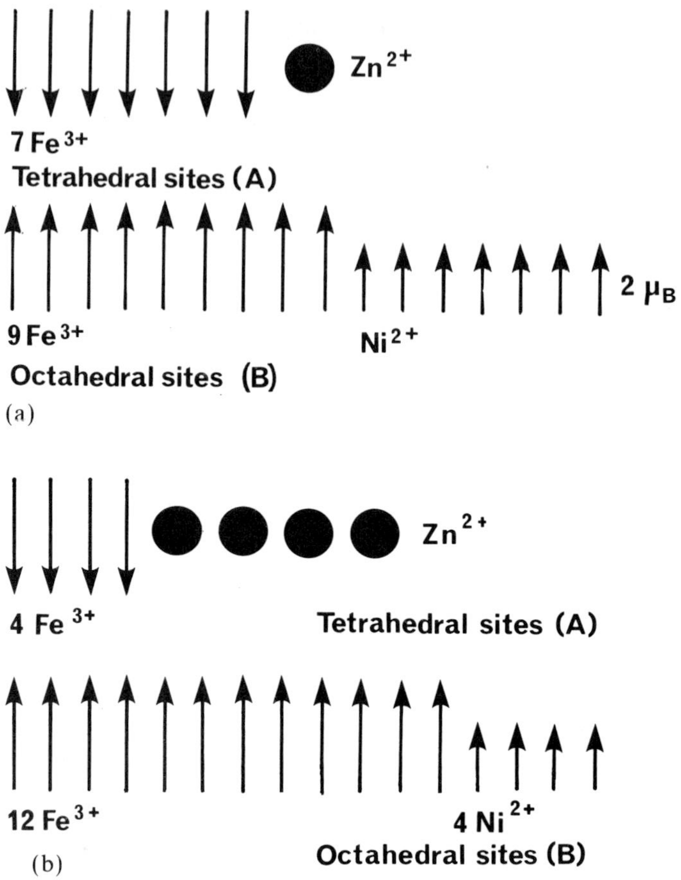

Fig. 4.12 (a) Magnetic structure of the ferrite $(NiFe_2O_4)_7$ $(ZnFe_2O_4)_1$, showing the effect of mixing normal spinel $ZnFe_2O_4$ with the inverse spinel $NiFe_2O_4$. (b) Effect of adding more $ZnFe_2O_4$ to $NiFe_2O_4$. The addition of even more $ZnFe_2O_4$ results in a decrease in the saturation magnetization due to antiferromagnetic alignment of the magnetic moments on the B sites.

we can now see that the magnetic moment of the unit cell has been increased from $16\mu_B$ $(8\times 2\mu_B)$ in the pure $NiFe_2O_4$ to $7\times 2\mu_B + 2\times 5\mu_B$, i.e. $24\mu_B$, in the mixed ferrite.

Note that the Fe^{3+} on the tetrahedral site, which was replaced by Zn^{2+}, now occupies an octahedral site. If we add another formula unit of $ZnFe_2O_4$ to the unit cell, the magnetic moment is increased still further to $6\times 2\mu_B + 4\times 5\mu_B$, i.e. $32\mu_B$. It might appear that we could continue this process indefinitely. However, as more and more $ZnFe_2O_4$ is added, the B–B antiferromagnetic exchange interaction increases relative to the A–B interaction, so that the addition of more than about four $ZnFe_2O_4$ molecules results in the gradual appearance of antiferromagnetically aligned ion spins on the B sites, and the net magnetic moment decreases to zero for pure $ZnFe_2O_4$, which is paramagnetic. The maximum effect is illustrated in Fig. 4.12(b). If more $ZnFe_2O_4$ molecules are added, the ions on the B sites are nearly all antiferromagnetically aligned, until, finally, in pure $ZnFe_2O$ even the antiferromagnetic alignment disappears and the material becomes paramagnetic as shown in Fig. 4.3(c). For a mixed ferrite in which a fraction x of the

M^{2+} ions are replaced by Zn^{2+} ions the formula and magnetic structure can be written

$$\overset{\xrightarrow{\quad A \quad}}{[Fe^{3+}_{1-x}\ Zn^{2+}_{x}]}\ \overset{\xleftarrow{\quad B \quad}\ \overline{\qquad}}{[Fe^{3+}_{1+x}\ M^{2+}_{1-x}]}$$

so that the magnetic moment per molecule is

$$\begin{aligned} \mu &= [5(1+x)+n(1-x)-5(1-x)]\mu_B \\ &= [(10-n)x+n]\mu_B, \qquad \text{provided that } x<0.35, \end{aligned}$$

where n is the number of unpaired spins on the M^{2+} ions.

4.8 MAGNETIC STRUCTURE OF THE MIXED NORMAL SPINEL $MnFe_2O_4$ WITH NORMAL SPINEL $ZnFe_2O_4$

Although $MnFe_2O_4$ has an 80% preference for the normal spinel structure, we find that when it is mixed with $ZnFe_2O_4$, the Zn^{2+} ions preferentially occupy the tetrahedral sites. However, since the magnetic moment of the Fe^{3+} ion is $5\mu_B$ and the Zn^{2+} has zero magnetic moment, the replacement of each Mn^{2+} ion by a Zn^{2+} ion increases the magnetic moment of the unit cell by $5\mu_B$ for each added Zn^{2+} ion. For a mixed ferrite in which a fraction x of the Mn^{2+} ions is replaced by Zn^{2+} ions, the formula and magnetic structure can be written as

$$\overset{\xrightarrow{\quad A \quad}}{[Mn^{2+}_{1-x}\ Zn^{2+}_{x}]}\ \overset{\xleftarrow{\quad B \quad}}{[Fe^{3+}_{2}]}\ O_4$$

and the magnetic moment per formula unit is

$$\begin{aligned} \mu &= [2\times 5-5(1-x)]\mu_B \\ &= 5(1+x)\mu_B, \qquad \text{provided that } x<0.4. \end{aligned}$$

If x is greater than about 0.5, the Fe^{3+} ions on the B sites gradually become antiferromagnetically aligned relative to each other, so that the magnetic moment decreases with increasing x. For values of x close to or equal to one, the antiferromagnetic order on the B sites disappears and the material becomes paramagnetic.

The effects of mixing inverse spinel $MnFe_2O_4$ with normal spinel $ZnFe_2O_4$ are readily understood in terms of their preferred magnetic structures. Since the magnetic moments of the Mn^{2+} and Fe^{3+} ions are both $5\mu_B$, the net magnetic moment that is observed when normal $MnFe_2O_4$ is mixed with normal $ZnFe_2O_4$ is the same as that observed when inverse $MnFe_2O_4$ is mixed with normal $ZnFe_2O_4$.

4.9 PREPARATION OF FERRITES

The magnetic properties of the ferrites, particularly their permeabilities, depend markedly on the composition and microstructure, so that the conditions of preparation are of great importance. The principal parameters, which require careful control, are grain size, porosity and oxidation state. In view of their commercial importance, the preparation of ferrites has been studied in considerable detail, but in the following only a brief summary of the most important effects will be given. The preparation and properties of cubic spinel ferrites have been reviewed by Slick (1980). The books by Smit and Wijn (1959), Tebble and Craik (1969), Standley (1972), Heck (1974) and Snelling (1988) give detailed information on the preparation and properties of ferrites.

4.9.1 The Sintering Process

The spinel ferrites are usually prepared from powdered carbonates or oxides, which are milled to very fine powders, compacted into the required shapes and sintered at about 1000°C. According to Chol (1971), the particle size range in the pre-sintered powder rather than the average particle size has the greatest effect on the microstructure and the electrical properties. In the case of magnetic ceramics such as the ferrites it is not possible to lower the sintering temperature by the addition of a glassy phase, because this would have very undesirable magnetic effects. For example, the presence of this non-magnetic phase would introduce internal demagnetizing fields and domain wall barriers, thus raising the coercivity and lowering the permeability.

The driving force for the sintering process depends on the difference between the surface free energy of the particle compact and that of the ideal single-crystal state. The energy available for sintering is known as the sinter reactivity.

$$E_s = 6\gamma_s/D$$

where γ_s is the surface free energy, which for oxides is of order $1\ \mathrm{N\,m^{-1}}$, and D is the diameter of the particles (assumed to be spherical for simplicity). Thus for efficient sintering we require very fine powders and compacts with a high degree of particle contact so as to provide a large number of diffusion paths. It is now well established that in ferrites and garnets the main mechanism of densification is by volume diffusion, although surface diffusion is important in the initial stages in order to form larger contact areas between the particles. The mechanism of material transport is thought to be that in which the pore surfaces act as sources of vacancies, which diffuse through the particles to the grain boundaries with a corresponding counterflow of ions from the grain boundaries to the pores. The sintering process is of course thermally activated, so that as time proceeds, the particles grow together, gradually filling the interparticle pores with material. The latter process leads of course to a shrinkage in the original compact volume. If the sintering is continued, a further reduction in the interfacial energy can be obtained due to an increase in the grain size, so reducing the intergranular area still further. Theoretical calculations show that the rate of grain growth should be proportional to $t^{1/2}$, but the presence of impurities at the grain boundaries lowers this to $t^{1/3}$. Since commercial-grade materials invariably contain impurities, the control of the magnetic properties is rather complex, and requires constant monitoring of the manufacturing process.

Sintering in ferrites can be assisted by the addition of borates and alkali fluorides, which produce a liquid phase at the sintering temperature. Additions of SiO_2 and CaO appear as second phases at the grain boundaries, and increase the bulk resistivities of

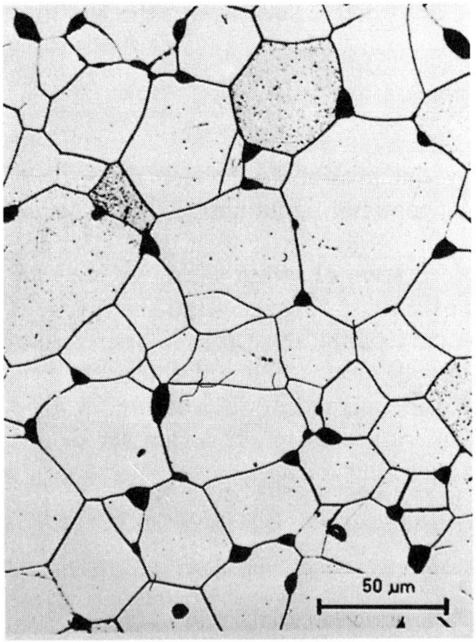

Fig. 4.13 Grain structure of an Mn–Zn ferrite (Perduijn and Peloschek, 1968).

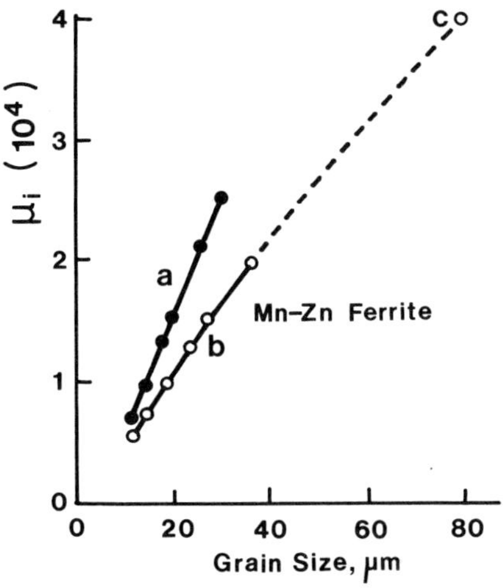

Fig. 4.14 Relationship between the initial permeability μ_i and grain size for various Mn–Zn ferrites (a, Perduijn and Peloschek, 1968; b, Roess, 1966; c, Beer and Schwartz, 1966).

the ferrites and hence reduce eddy current losses (Wagner, 1977, 1981). Unfortunately, these grain boundary phases are non-ferrimagnetic and reduce the permeability, so that the amounts of SiO_2 and CaO added are determined by a compromise between the required properties.

Ideally, we require a fully dense material with large pore-free grains like that shown in Fig. 4.13. These materials give the highest permeabilities because the large pore-free grains contain fewer obstacles to domain wall motion, and a low total grain boundary area also minimizes the resistance to domain wall movement over the boundaries. The relationship between permeability and grain size is shown in Fig. 4.14. Although the Mn–Zn ferrite contains pores at the grain boundaries, this is much less deleterious to the magnetic properties than would be the case if they were in the grains themselves.

4.9.2 Effect of sintering atmosphere

The final properties of sintered ferrites also depend on the sintering atmosphere. The oxygen stoichiometry in the ferrite influences the valences of the transition metal ions Fe, Co, Ni and Mn, which may be either divalent or trivalent or occasionally quadrivalent. The valence state of the ion will therefore have pronounced effects on both the magnetic and electrical properties. For example, the coexistence of ferrous and ferric ions gives rise to higher electrical conductivity due to low-activation-energy electron transfer from the Fe^{2+} ion to form Fe^{3+}:

$$Fe^{2+} \rightarrow Fe^{3+} + e^-.$$

If manganese ferrite is sintered in an oxidizing atmosphere such as air, the Mn ions are oxidized to Mn^{3+}, i.e.

$$2MnO + \tfrac{1}{2}O_2 \rightarrow Mn_2O_3.$$

This results in a large decrease in the permeability, due possibly to the fact that Mn_2O_3 is antiferromagnetic so that small regions of the material containing Mn_2O_3 act as obstacles to domain wall motion.

If the presence of excess oxygen results in the formation of cation vacancies, we may expect that specific arrangements of the cations over these vacancies and the A and B sites will lead to local regions with induced magnetocrystalline anisotropy. These regions may therefore act as traps or barriers to domain wall motion. Thus the permeability, which depends on the ease of wall movement and magnetization rotation, will be decreased. The sensitivity of the ferrites to the sintering atmosphere and other details of the preparation will of course depend on the particular ferrite.

4.10 PERMEABILITY OF FERRITES

The high permeabilities of ferrites in combination with their high resistivities are certainly their most important properties. In order to obtain high permeability, we require the maximum possible saturation magnetization and the lowest possible coercivity. Thus it is essential that the domain walls can be moved and ultimately removed from the material by very low applied fields, so that ideally the material should contain no obstacles to domain wall motion or magnetization rotation. These conditions are best satisfied by selecting materials with the lowest possible values of the magnetocrystalline anisotropy and magnetostriction, thus minimizing the domain wall energy and the sensitivity of the materials to stress. Since the magnetocrystalline anisotropy energy of the ferrites is given by the Akulov expression

$$E = K_0 + K_1(\alpha_1^2\alpha_2^2 + \alpha_2^2\alpha_3^2 + \alpha_3^2\alpha_1^2) + K_2\alpha_1^2\alpha_2^2\alpha_3^2,$$

with $K_1 < 0$ and K_2 small, the easy directions of magnetization are parallel to $\langle 111 \rangle$. Some typical values of K_1 and λ_s for various ferrites are shown in Table 4.3. The

Table 4.3 Typical values of the magnetocrystalline anisotropy coefficient and saturation magnetostriction for various ferrites.

Ferrite	K_1 (10^3 J m^{-3})	$\lambda_s(10^{-6})$
$MnFe_2O_4$	-4	-5
$NiFe_2O_4$	-7	-26
$FeFe_2O_4$	-12	40
$CoFe_2O_4$	200	-110
$MgFe_2O_4$	-4	-6
γ-Fe_2O_3	-5	—
$Mn_{0.62}Zn_{0.41}Fe_{1.97}O_4$	0.2	—

exceptionally high values of K_1 and λ_s for $CoFe_2O_4$ are, as we should expect, due to the large spin–orbit coupling in the Co^{2+} ion. Note also the exceptionally low value of K_1 for Mn–Zn ferrite. We shall see below that various Mn–Zn ferrites are excellent high-permeability materials. Since the modulus of K_2 is usually very much less than K_1, values of K_2 can usually be neglected. In polycrystalline materials the easy axes of magnetization within individual grains will be randomly orientated with respect to the applied field, so that for complete saturation of the material some rotation of the magnetization vector will be necessary. However, since the anisotropy field is approximately $2K_1/\mu_0 M_s$, the relatively small values of K_1 mean that any magnetization rotation processes will occur in relatively low fields. For example, even in the absence of domain walls, it should be possible to saturate the ferrite $NiFe_2O_4$ with a field of about $K_1/\mu_0 M_s$, i.e. 4×10 A m^{-1}. Experimental values are very much lower than this in spite of the impedance of domain wall movement and magnetization rotation by chemical heterogeneities, pores, grain boundaries and other crystal defects. Since the main mechanism of magnetization change is by domain wall displacement rather than rotation of the magnetization vector, the material can usually be very nearly saturated by fields of about 5000 A m^{-1}.

By varying the composition, it is possible to prepare ferrites with very low values of K_1 and coercivity, although much of this work depends on empirical techniques. However, as we shall see in Section 4.10.1, very successful high-permeability ferrites have been developed and are used on a very large commercial scale.

4.10.1 Permeability of Mn–Zn and Ni–Zn Ferrites

The Mn–Zn and Ni–Zn ferrites are by far the most important commercial soft ferrites, and have therefore been investigated in considerable detail. Since the magnetic and electrical properties are very strongly dependent on the composition, preparation conditions, grain size and porosity, the range of properties that can be obtained is very large. In view of this, it is quite impossible to present comprehensive data for all frequencies and temperatures, or even a summary of them. For such detailed information the reader is referred to the books by Heck (1974) and Snelling (1988), which also include detailed discussions of applications and design. Some representative properties are given in Table 4.4, from which it can be seen that the permeabilities of the Mn–Zn ferrites are invariably higher than those of the Ni–Zn ferrites. However, since the resistivities of the Mn–Zn ferrites are rather low compared with those of the Ni–Zn ferrites, the former are used in applications requiring low frequencies (<500 kHz). At

Table 4.4 Properties of Mn–Zn and Ni–Zn ferrites.

Property	Mn–Zn	Ni–Zn
Initial permeability μ_i(rel)	500–20 000	10–2000
Saturation magnetic polarization J_s (T)	0.3–0.5	0.1–0.36
Curie temperature (°C)	100–250	100–500
Coercivity H_c (A m^{-1})	4–100	16–1600
Resistivity ρ (Ω m)	0.02–20	$10–10^7$
Density (10^3 kg m^{-3})	4.6–4.8	4.8–4.9
Total power loss (W m^{-3})	50–200	Very low

higher frequencies, i.e. 500–50 MHz, where the energy losses due to eddy currents become larger than those due to hysteresis, it is preferable to use the Ni–Zn ferrites, for which the high resistivity reduces the eddy current losses. The magnetization curve for a typical Mn–Zn ferrite is shown in Fig. 4.15.

By very careful preparation, Perduijn and Peloschek (1968) have obtained Mn–Zn ferrites with permeabilities as high as 25000. This was achieved by choosing the composition 52.4 mol% Fe_2O_3. 25 mol% MnO and 22 mol% ZnO, which has a value of K_1 close to zero at room temperature. Since the value of K_1 depends on the ratio of the manganese ions to the ferrous ions, the preparation conditions must be very carefully controlled. For materials in which K_1 is very close to zero we should expect the domain walls to be very wide, and it may be, as Perduijn and Peloschek (1968) have suggested, that the magnetization process in this case is not just a simple displacement of thc walls in the ordinary sense, but rather involves the movement of practically free spins similar to the waving and reversal of stalks in a windswept cornfield. The permeability is not expected to be infinitely large, because many clusters of spins will be pinned at various crystallographic imperfections, and in particular the permeability will be limited by the presence of grain boundaries and pores. The increase in permeability with grain size has already been shown in Fig. 4.14. The magnetostriction in Mn–Zn ferrites is relatively small, but the presence of local internal stresses at pores, crystal defects and grain boundaries will also limit the ease of magnetization change and hence the permeability.

If we assume that the magnetization changes occur by spin rotation (this also occurs during domain wall propagation, though it is incoherent) then the rotational susceptibility is

$$\chi_{rot} = \frac{M_s}{H_A},$$

where H_A is the anisotropy field. Since

$$B = \mu_0(M + H),$$

$$\mu = \frac{B}{H} = \mu_0(\chi + 1),$$

the relative permeability due to rotational processes is

$$\mu_{rot} = \frac{M_s}{H_A} + 1 = \mu_0 \frac{M_s^2}{2K_1} + 1.$$

In order to obtain high permeabilities, we require high values of M_s and low values of K_1. If K_1 is low, the domain wall energy will also be low, since $\gamma = 4(AK_1)^{1/2}$, but since the wall width $\delta = \pi(A/K_1)^{1/2}$, we should expect that very wide low-energy walls are much less likely to interact with small crystallographic defects or grain boundaries. The individual grains will therefore tend to react independently to applied fields. The structure, energy and behaviour of domain walls in ferrimagnetic materials is further

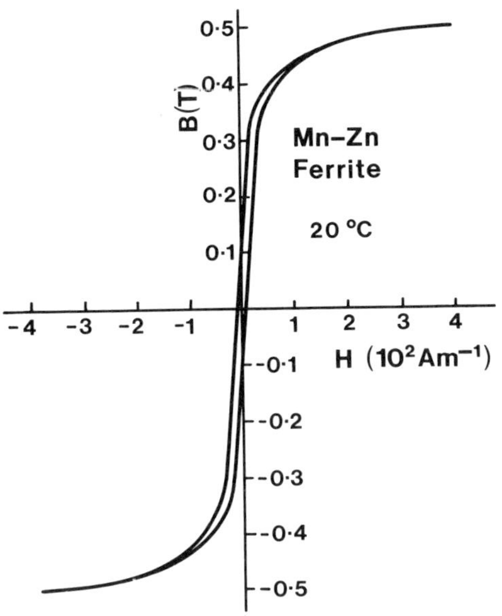

Fig. 4.15 Typical hysteresis loop for an Mn–Zn ferrite.

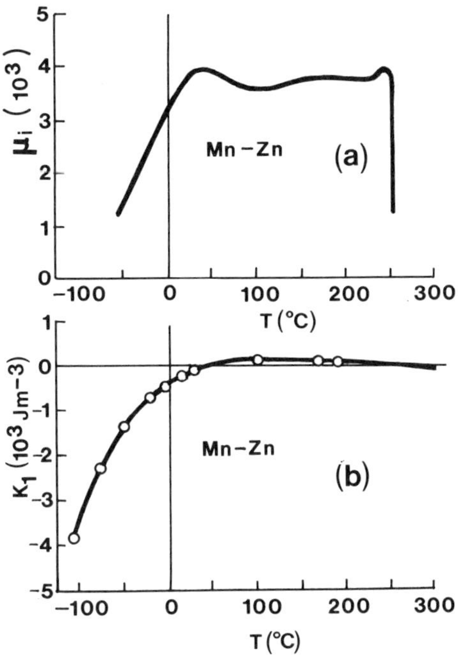

Fig. 4.16 Dependence of (a) initial permeability μ_i and (b) magnetocrystalline anisotropy constant K_1 on temperature for an Mn–Zn ferrite (after Ohta, 1963).

complicated by the fact that there are two magnetic sublattices. The dependence of the initial permeability of an Mn–Zn ferrite on K_1 and temperature has been determined by Ohta (1963), and is shown in Fig. 4.16.

The design of a ferrite material for a particular application is a very complicated problem, since a large number of electrical, magnetic and mechanical properties (e.g. wear characteristics of recording heat materials) must be simultaneously optimized.

4.11 MAGNETIC ANNEALING OF FERRITES AND INDUCED UNIAXIAL MAGNETOCRYSTALLINE ANISOTROPY

If a cubic ferrite is annealed above its Curie temperature in a magnetic field and subsequently cooled in the field, it may develop an additional uniaxial magnetocrystalline anisotropy. The preferred axis of magnetization is usually parallel to the field direction, so that the hysteresis loops have an increased rectangularity in this direction. The effect is particularly pronounced in ferrites containing Co^{2+} ions, in which there is a strong spin–orbit coupling and the uniaxial magnetocrystalline anisotropy is therefore thought to be due to the formation of interacting Co^{2+} ion pairs, though there is no precise theory. Kato and Takei (1933) have shown that the effect in the ferrite

$3(CoFe_2O_4)+Fe_3O_4$ is so large that they were able to produce a permanent magnet from the sintered powder. From Table 4.3, we note that the anisotropy constant for $CoFe_2O_4$ is itself very high, about 2×10^5 J m^{-3}, so that we should expect high values of remanence and coercivity, even in the absence of the magnetic annealing. These properties have led to the development of the permanent magnet material Vectolite, which has the composition $26Co_2O_3.30Fe_2O_3.44Fe_3O_4$. The maximum energy product, remanence and flux coercivity are 4.8 J m^{-3}, 0.16 T and 80 kA m^{-1} respectively, though the material is no longer used as a permanent magnet.

4.11.1 Isoperm Hysteresis Loops

Isoperm hysteresis loops can be observed in Mn–Zn ferrites containing about 51–55 mol% iron oxide, where both K_1 and λ_s are very small. Isoperm B–H curves (see Fig. 2.11) can also be produced by cobalt doping combined with a magnetic anneal with the field perpendicular to the required direction of magnetization.

4.11.2 Perminvar Hysteresis Loops

Perminvar hysteresis loops are characterized by their wasp-waisted or butterfly-wing shapes (see Fig. 2.9). In low fields the permeabilty is low but constant, with very low hysteresis loss. Perminvar properties can be obtained in both Ni–Zn and Mn–Zn ferrites, but only for iron contents of about 60 mol%, especially with cobalt additions. The addition of 1 mol% of cobalt can reduce hysteresis losses so that they are useful at higher frequencies.

4.11.3 Rectangular or 'Square' Hysteresis Loops

Many ferrites with various cobalt additions can be given thermomagnetic heat treatment to produce rectangular hysteresis loops. These ferrites are used in a variety of applications such as in switched mode power supplies. Squareness can also be induced by the application of stress, especially in ferrites that have low K_1 and large λ_s.

4.12 STRESS-INDUCED ANISOTROPY IN FERRITES

In the presence of a stress the magnetization vector attempts to orientate itself so as to minimize the stress energy. The effect is especially pronounced in ferrites with low K_1 and large λ_s. If, for example, a tensile stress is applied to a material with a negative magnetostriction such as $MnFe_2O_4$ or $CoFe_2O_4$, the stress energy will be lowered if the magnetization rotates away from the tensile stress axis, because magnetization in a direction perpendicular to this axis will be accompanied by a contraction in length, thus reducing the total stress energy. For materials with positive magnetostriction the magnetization vector will prefer to be parallel to the tensile stress axis. This stress-induced anisotropy must of course compete with the normal magnetocrystalline anisotropy, so

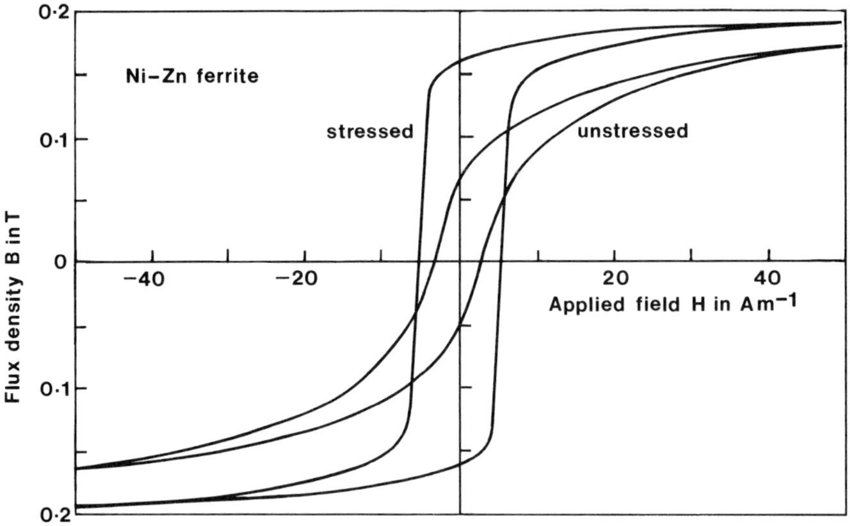

Fig. 4.17 Increase in permeability of a Ni–Zn ferrite due to compressive stresses (after Williams *et al.*, 1953).

that the equilibrium direction of magnetization will be that which minimizes the total energy. The reverse of these effects is observed if the stress is compressive. Since the overall domain structure will be affected by the presence of stresses, this may affect the shape of the hysteresis loop and its associated parameters. Rathenau (unpublished results) has shown that if the ferrite $Ni_{0.34}Zn_{0.66}Fe_2O_4$ in the form of a ring with $\lambda_s = -4 \times 10^{-6}$ is coated with a glass of higher thermal expansion and then cooled to room temperature, the resulting compressive stresses give a much more rectangular hysteresis loop.

The encapsulation of ferrite cores in various resins and plastics is now common practice, and if there are any resulting radial compressive stresses then may produce a marked increase in the permeability, provided, of course, that the material has a negative magnetostriction. The results obtained by Williams *et al.* (1953) for an Ni–Zn ferrite are shown in Fig. 4.17.

4.13 GRINDING OF FERRITES

It is difficult to control the shrinkage of ferrites that occurs during sintering. In applications such as transformers where flat surfaces are required or in inductors that require accurate air gaps, grinding and polishing is usually necessary. According to Knowles (1970), grinding puts the surface into compression but the bulk into tension, thus affecting the magnetic properties. Snelling (1974) has shown that a stress of about one-tenth of the tensile strength increases the losses and affects the permeability temperature characteristics.

4.14 COMPOSITION AND TEMPERATURE DEPENDENCE OF MAGNETIC PROPERTIES

The magnetic properties of the Mn–Zn and Ni–Zn ferrites are very strongly composition- and temperature-dependent. At room temperature the saturation magnetic polarization and coercivities range from 0.3 to 0.5 T and 4 to 100 A m^{-1} respectively. The corresponding ranges for Ni–Zn ferrites are 0.12–41 T and 16–1500 A m^{-1}.

The temperature variation of the saturation magnetic polarization for various Ni–Zn ferrites is shown in Fig. 4.18. Note that the Curie temperature decreases as the zinc content increases and that the saturation magnetic polarization reaches a maximum at a particular zinc content, namely $\delta = 0.5$ for Ni–Zn ferrites.

Since the saturation magnetization and magnetocrystalline anisotropy coefficient are both strongly dependent on the temperature, the temperature dependence of the initial permeability is expected to be complex. The anisotropy energy is usually proportional to the square or higher power of the saturation magnetization, so the initial permeability is expected to rise with increasing temperature and then to decrease rapidly to zero at the Curie temperature. The temperature variations of the initial permeabilities of various Ni–Zn ferrites are shown in Fig. 4.19. To some extent, the value of the magneto-crystalline anisotropy and the saturation magnetization can be controlled by careful control of the oxygen partial pressure during sintering. If the magnetocrystalline anisotropy energy is reduced or made zero, the μ_i versus temperature curve exhibits a secondary maximum that can in principle be produced at any temperature below the Curie temperature. Thus, by a careful choice of process control, a very wide range of μ_i versus T curves can be engineered to meet specific requirements.

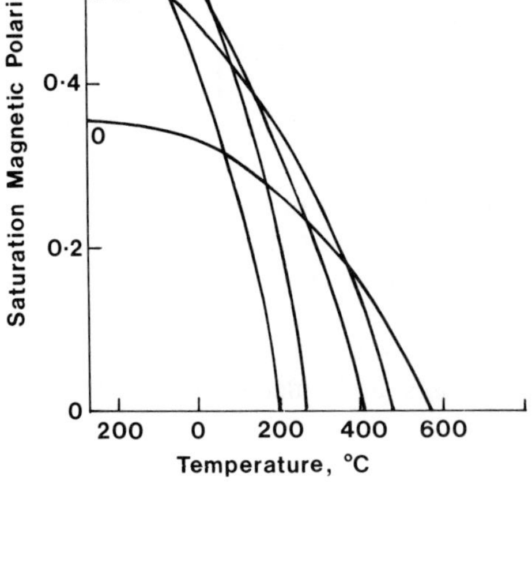

Fig. 4.18 Temperature variation of saturation magnetic polarization for various Ni–Zn ferrites (after Snelling, 1988).

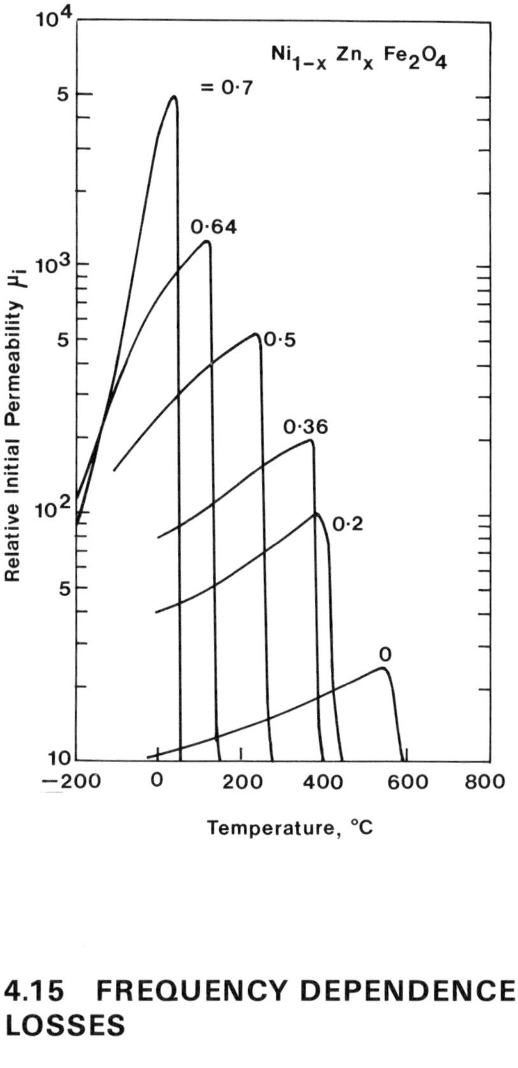

Fig. 4.19 Temperature variation of the initial permeabilities of various Ni–Zn ferrites (after Snelling, 1988).

4.15 FREQUENCY DEPENDENCE OF PERMEABILITY AND LOSSES

Although the ferrites are regarded as insulators, their resistivity, though high, is finite, so that when they are used as transformer or inductor cores, there are associated power losses. These losses can be expressed in terms of a frequency-dependent complex permeability μ containing a purely inductive component or real part μ' and a purely resistive component or imaginary part μ''; that is, $\mu = \mu' - j\mu''$, where $j = \sqrt{-1}$. The eddy current losses can be expressed by the loss tangent, where $\tan\delta = \mu''/\mu'$. For a time-dependent field $H = H_0 \exp(j\omega t)$, where ω is the angular frequency, the energy loss can be described by a phase difference δ between the field H and the resulting flux density B. The complex permeability $\mu(\omega)$

$$\mu(\omega) = \frac{B_0 \exp[j(\omega t + \delta)]}{H_0 \exp(j\omega t)} = \mu' - j\mu''.$$

Typical permeability spectra at room temperature for various Ni–Zn ferrites are shown in Fig. 4.20. The real part μ' is related to the energy stored by the component of B in phase with H, and the imaginary part μ'' describes the energy loss due to the component of B that is out of phase with H. The loss factor or loss tangent

$$\tan\delta = \mu''/\mu'.$$

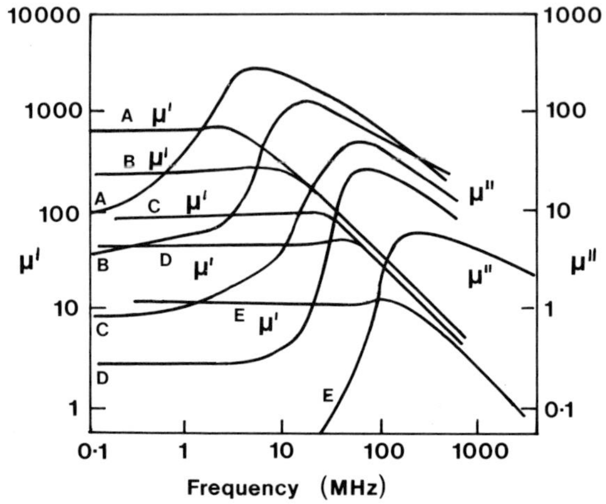

Fig. 4.20 Typical permeability spectra at room temperature for various Ni–Zn ferrites as a function of frequency: NiO:ZnO ratios are (A) 17.5:33.2; (B) 24.9:24.9; (C) 31.7:16.5; (D) 39.0:9.4; (E) 48.2:0.7, balance Fe_2O_3 (after Smit and Wijn, 1954).

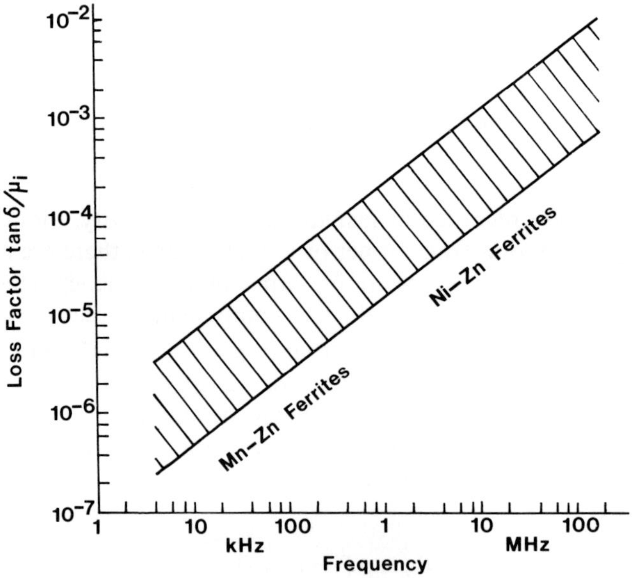

Fig. 4.21 Variation of loss factor with frequency for Mn–Zn and Ni–Zn ferrites.

The quality factor Q can be defined as the reciprocal of tan δ. A figure of merit for the material is defined as $(\tan\delta)/\mu$, sometimes referred to as the normalized loss factor:

$$\frac{\tan\delta}{\mu}=\frac{1}{\mu Q}=\frac{\mu''}{\mu\mu'}=\frac{\mu''}{\mu'^2}.$$

Mn–Zn ferrites have values of $(\tan\delta)/\mu<10^{-6}$ at 100 kHz, and may be used up to frequencies of about 2 MHz. For higher frequencies up to about 500 MHz the Ni–Zn ferrites are preferred because they have much lower normalized loss factors. The variation of the loss factor with frequency for Mn–Zn and Ni–Zn ferrites is shown in Fig. 4.21. For detailed discussions and data on ferrites the reader is referred to the excellent reference books by Snelling (1988) and Heck (1974), and to the review by Slick (1980).

4.16 APPLICATIONS OF SOFT FERRITES

The majority of soft or high-permeability ferrites are used in television deflection yokes, cores for television line output/eht transformers, and switched-mode power supplies. They are also used in telecommunications for inductor cores, particularly those used for *LC* filters, frequency division multiplexers, low-power wideband transformers and pulse transformers. Summaries of the principal applications of Mn–Zn ferrites and Ni–Zn ferrites, classified according to the required frequencies and permeabilities are given in Tables 4.5 and 4.6 respectively.

Ferrites are also widely used in transductors (variable inductors), which enable current control at power frequencies, and in controlling the inductance in high-frequency circuits both remotely and automatically. They are thus of great importance in communications engineering, especially for tuning, automatic frequency control, frequency modulation switching, coupling transformers and variable-frequency filters. The operation of these devices depends on the incremental or reversible permeability.

$$\mu_\Delta=\frac{1}{\mu_0}\frac{\mathrm{d}B}{\mathrm{d}H},$$

Table 4.5 Mn–Zn ferrites.

Application	Frequency (MHz)	Permeability μ_i
Inductor cores	0.2	800–4000
Inductor cores, antennas for medium and long-wave broadcasting	0.1–2	500–1000
Pulse transformers (low-power) wide-band transformers	1	3000–20 000
Transformers, chokes for switched-mode power supplies, scanning transformers for television receivers (power losses at 100 kHz and 25°C are about 200 mW cm^{-3})	0.01–1	2000–5000

Table 4.6 Ni–Zn ferrites.

Application	Frequency (MHz)	Permeability μ_i
Wide-band transformers	1–300	1000
Pulse transformers, antennas, cores for electromagnetic interference suppression	1–300	500–1000
Antennas for medium- and long-wave broadcasting, power transformer cores for electromagnetic interference suppression	0.5–5	150–500
Inductor cores, antennas for short-wave broadcasting, power transformer cores for electromagnetic interference suppression	2–30	70–150
Inductor cores, cores for electromagnetic interference suppression	10–60	12–70

i.e. the slope of the *B–H* curve, which can be varied from the initial permeability in a very small field to a value approaching unity when the applied field is large enough to saturate the material. The variation is achieved by means of a control field that is magnetically isolated from the high frequency windings.

The discovery of ferrites such as $Mg_{0.85}Mn_{0.45}Fe_{1.7}O_4$, $Mg_{0.4}Ni_{0.5}Mn_{0.1}Fe_2O_4$, $Mg_{0.7}Cu_{0.3}Fe_2O_4$, $Cu_{0.25}Mn_{0.75}Fe_2O_4$, $Li_{0.47}Ni_{0.06}Fe_{2.47}O_4$ and $Co_{0.02}Mn_{0.18}Fe_{0.8}Fe_2O_4$ with rectangular hysteresis loops led to their application in computers for the rapid storage and retrieval of digital information by switching of the direction of magnetization in very small (approximately 0.5 mm diameter) toroidal cores (Heck, 1974). In the early 1960s, many millions of these tiny toroidal cores, interwoven with read, write and select wires in planar arrays were used in mainframe computers, but this application has been superseded by semiconductor memories except in a few very specialized applications.

4.16.1 Ferrites for Use at Microwave Frequencies

Microwave ferrites can be defined as those used at frequencies in the range 100 MHz–500 GHz. The fundamental physical effect that is required for a microwave ferrite is that of gyromagnetic resonance, which occurs when the material is subjected to the electromagnetic fields in the presence of a static applied field. The effect is non-reciprocal, and is due to the fact that the permeability must be described by a tensor. The saturation magnetic polarization is of fundamental importance, since in the absence of any applied field it determines the natural resonance frequency through the magnitude of the internal field which it produces. The sharpness of the resonance, i.e. the linewidth, is also critically important, and is determined by spin-wave and magnetocrystalline anisotropy effects. Crystallographic defects and macroscopic faults such as porosity are also important in determining the sharpness of the resonance. Thus in the most demanding applications single crystals are usually required.

Ferrites are widely used in a variety of microwave devices because the transmission or absorption of electromagnetic waves depends on the frequency and the sense and magnitude of the internal magnetic field. When an electromagnetic wave is incident on a spontaneously magnetized material, the aligned magnetic moments due to electron spin may be made to precess at a frequency that depends on the net internal magnetic field. If a plane-polarized electromagnetic wave is transmitted through a magnetized material, the plane of polarization is rotated through an angle θ; the effect is of course the microwave equivalent of the well-known Faraday effect. If the same wave is internally reflected, it undergoes a further rotation through θ in the same sense, so that the total rotation is 2θ. This shows that the Faraday effect is non-reciprocal. A circularly polarized electromagnetic wave will stimulate resonant precession and hence absorption only if its polarization rotation is in the correct sense with respect to the direction of the static magnetic field inside the material. Thus the transmission or absorption of the wave can be controlled by the direction of the field inside the ferrite. This non-reciprocal property of ferrites has led to a variety of microwave devices such as waveguide and stripline isolators, resonance absorption isolators, Faraday isolators, gyrators, circulators, switches, modulators and limiters. For a discussion of microwave materials and devices the reader is referred to the books by Lax and Button (1962), Von Aulock (1965) and Heck (1974). For a review of microwave ferrites, see Nicolas (1980).

All of these devices are based on isolators or circulators. An isolator is a section of a waveguide in which the incident electromagnetic wave can propagate without any substantial attenuation, while the reverse waves are highly attenuated. A circulator is a device that has n ports, which can be inputs or outputs, such that a wave incident at port 1 emerges at port 2; a wave incident at port 2 emerges at port 3, and so on; a wave incident at port n emerges at port 1. All other propagation paths are inhibited. Thus the circulator provides easy isolation between different channels in a microwave system such as a radar or radio linkage.

The most important groups of materials are Mg–Mn ferrites (which have very high resistivities, of order 10^6–10^8 Ωm), Mg–Mn–Al, Ni–Zn–Al ferrites, the hexagonal ferrites based on $BaFe_{12}O_{19}$ and the ferrimagnetic garnets based on yttrium iron garnet $Y_3Fe_5O_{12}$ with substitutions of Gd, Ho and Al.

The lithium ferrite $LiFe_5O_8$ or $Li_{0.5}Fe_{2.5}O_4$ has a saturation magnetic polarization of 0.36 T and a relatively high Curie temperature of 645°C. Since the magnetization is due to trivalent iron ions Fe^{3+}, lithium ferrites have very narrow spin-wave resonance linewidths of about 250 A m^{-1}, and are suitable for low microwave power levels with very good temperature stability.

The $Mn_xMg_yFe_zO_4$ ferrites can be produced with a range of saturation magnetic polarizations (0.2–0.24 T), linewidths (of about 30 k A m^{-1}) and Curie temperatures (270–300°C), but the preparation has to be carefully controlled because Mn^{2+}, having the same electronic configuration as Fe^{3+}, can occupy both tetrahedral and octahedral sites in the spinel structure. Furthermore, the manganese can also exist in an Mn^{2+} ionic state. The properties therefore depend on the heat treatment and the heat-treatment atmosphere. Useful compositions, for low-microwave-power applications in the 7–15 GHz range have $0.10<x<0.15$, $0.4<y<0.5$ and $0.7<z<0.9$. The magnetization can be reduced by substituting aluminium (0–10%) for some of the iron.

As a result of their versatility, $MgFe_2O_4$ and related compounds have widespread applications in microwave devices. This is due largely to their low dielectric, magnetic and eddy current losses. The magnetization, which may be varied by as much as 250%, depends very critically on the firing temperature and the cooling or quenching rate. This is due of course to the degree of inversion that can be achieved in the magnetic structure, i.e. the occupation of A and B sites. In general, the ferrite can be represented by the formula

$$\overset{\text{A}}{Mg_x^{2+}Fe_{1-x}^{3+}} \quad \overset{\text{B}}{[Mg_{1-x}^{2+}Fe_{1+x}^{3+}]} \quad O_4$$

If E is the activation energy required to move a magnesium ion from a B to an A site then for a quenching temperature T

$$\frac{x(1+x)}{(1-x)^2} = \exp\left(-\frac{E}{kT}\right).$$

When $x=1$ a normal spinel structure is obtained, and when $x=0$ the inverse spinel structure is obtained. With mixed ferrites, a wide range of magnetic properties is obtainable that depend very strongly on the compositions and heat-treatment schedules.

Ferrites containing magnesium have very high electrical resistivity, up to $10^6-10^8\ \Omega m$ for $MgFe_2O_4$. Very good microwave properties can be obtained with mixed Mg–Mn ferrites over a range of compositions close to $Mg_{0.6}Mn_{0.1}Fe_{0.6}O_{1.6}$, The properties of Mg–Mn ferrites depend on their heat treatment and rate of cooling after sintering. When the ferrite $MgMn_{0.1}Fe_{1.6}O_{4.3}$ is slow-cooled from 1300°C, its saturation magnetic polarization $J_s=0.25$ T; whereas when it is quenched, $J_s=0.29$ T. Thus it should be borne in mind that changes in the magnetic properties may occur if the ferrites are allowed to overheat in the waveguide or microwave device.

The saturation magnetic polarization of nickel ferrite, $NiFe_2O_4$, is 0.32 T, and its high Curie temperature (about 570°C) makes it much more stable than the Mn–Mg ferrites. However, the fact that the Ni^{2+} ion has an orbital magnetic moment results in a broader linewidth, which can be reduced by the addition of a small quantity of cobalt Co^{2+}. Small additions of manganese are also necessary to reduce dielectric losses, yielding linewidths of about 20 kA m^{-1} at 9 GHz.

Small substitutions of aluminium have also proved useful to produce $NiFe_{2-x}Al_xO_4$ with $0<x<0.2$, and so too have substitutions of Zn to produce the widely used $Ni_{1-x}Zn_xFe_2O_4$ ferrites with $0<x<0.5$.

4.16.2 Hexagonal Ferrites for Use at Microwave Frequencies

The uniaxial ferrimagnetic materials $BaFe_{12}O_{19}$ and $SrFe_{12}O_{19}$ are very important permanent magnets and are produced in large quantities. Their structure and magnetic properties are described in Chapter 5. There are several materials structurally related to $BaFe_{12}O_{19}$ and $SrFe_{12}O_{19}$ that are useful in microwave devices operating at very high frequencies. The natural resonance frequency of $BaFe_{12}O_{19}$ is 48 GHz, but by substitution of Co, Al, Ti and Zn for some of the Ba, the resonant frequency can be varied from about 40 to 130 GHz. The hexagonal microwave ferrites are usually designated as follows.

$BaFe_{12}O_{19}$	M
$Me_2BaFe_{16}O_{27}$	Me_2W
$Me_2Ba_2Fe_{12}O_{22}$	Me_2Y
$Me_2Ba_3Fe_{24}O_{41}$	Me_2Z
$Me_2Ba_2Fe_{28}O_{46}$	Me_2X
$Me_2Ba_4Fe_{36}O_{60}$	Me_2U

where Me is Mn, Co, Zn, Ni, Ti, Mg, Al or Fe or a mixture of them. The structural relationships between the most interesting and useful materials, namely Me_2W, Me_2Y and Me_2Z, have been described by Morrish (1965). In the Me_2Y compound, unlike $BaFe_{12}O_{19}$, the hexagonal axis is the hard axis of magnetization, so that the magnetization vector lies in a plane perpendicular to the hexagonal axis, i.e. the basal plane. There is a small magnetocrystalline anisotropy in this plane, which has sixfold symmetry. Materials with an easy basal plane of magnetization are known commercially as Ferroxplana (Philips). Thus these ferrites may have a reasonable permeability, because, in the presence of a varying electromagnetic field, the magnetization vector can rotate in the easy plane with relative ease. The rotation is confined to this easy plane because of the high uniaxial magnetocrystalline energy perpendicular to it. These materials have resonant frequencies that are an order of magnitude greater than those for the cubic spinel ferrites such as Ni–Zn ferrite. The materials $Fe_{0.5}Co_{0.75}Zn_{0.75}W$, Co_2Z and $CoZn_{0.5}Fe_{0.5}Z$ also possess an easy plane of magnetization perpendicular to the hexagonal axis.

Many of the uniaxial M- or $BaFe_{12}O_{19}$-based compositions with various substitutions for the Ba are potentially useful and have been investigated. Although the Me_2Y materials have easy planes of magnetization perpendicular to the hexagonal axis, the same property can be produced in many of the other materials by appropriate substitution. Clearly, a very wide range of compositions with appropriate properties could be produced. The real and imaginary components of the permeabilities for Ni–Zn and Co_2Z ferroxplana are compared in Fig. 4.22, from which it can be seen that the peak in the losses indicated by the value of μ'' occurs at much higher frequencies in Co_2Z ferroxplana, so that the latter material can be used efficiently at much higher frequencies than Ni–Zn ferrites. The resistivities of the ferroxplana materials are typically in the range 10^2–10^8 Ωm.

The Me_2Z type is obtained by alternation of the M and Y units of the $BaFe_{12}O_{19}$ magnetoplumbite structure so that the easy direction of magnetization may be either parallel to the hexagonal axis or in a plane perpendicular to it. The Co–Z compositions have very high magnetocrystalline anisotropy, and are therefore potentially very useful at microwave frequencies. The Zn_2Z material has an easy direction of magnetization parallel to the hexagonal axis, whereas the Co_2Z material has an easy plane of magnetization perpendicular to the hexagonal axis. It has been found that increased permeabilities can be obtained from solid solutions of the two, for example, $Co_{0.6}Zn_{1.4}Z$ (i.e. $Ba_{3.0}Co_{0.6}Zn_{1.4}Fe_{24}O_{41}$) has a permeability maximum with a minimum in the net magnetocrystalline anisotropy. Even higher permeabilities can be obtained if, prior to sintering, the powder is orientated in a planar field, i.e. two DC fields at right-angles. This orientates the hexagonal platelets of the ferrite so that they are aligned in a plane

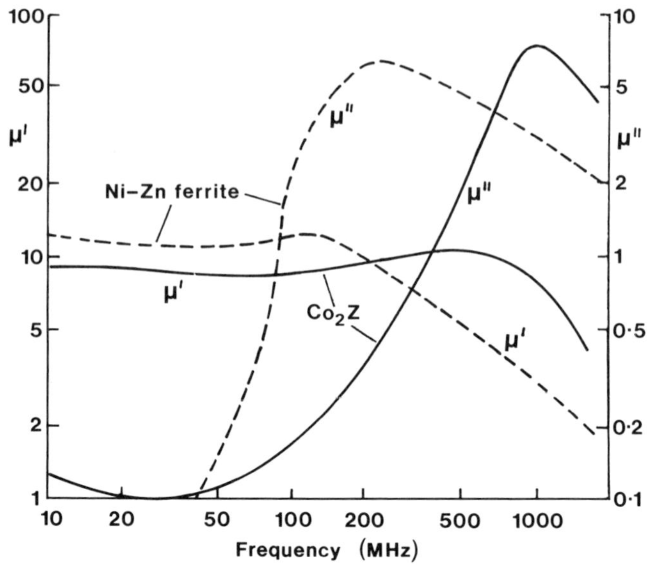

Fig. 4.22 Comparisons of real and imaginary components of permeabilities for Ni–Zn ferrite and Co_2Z ferroxplana (after Jonker *et al.*, 1956).

with increased permeability in this plane. For good performance at high frequencies the compositions and sintering conditions must be very carefully controlled so as to minimize the number of voids and to increase the electrical resistivity.

The M-type and W-type structures have easy directions of magnetization along the hexagonal axis, and since the uniaxial magnetocrystalline anisotropy is often very high, they have relatively low permeabilities – but they are still potentially useful materials.

Some magnetic properties of M, W and ferroxplana W, Y, Z materials are shown in Table 4.7.

The high-frequency magnetic properties of hexagonal ferrites have been reviewed by Sugimoto (1982) and the fundamental properties by Kojima (1982).

4.16.3 Garnets

Garnets, such as yttrium iron garnet, $Y_3Fe_5O_{12}$, are produced using conventional ceramic techniques, i.e. by sintering appropriate mixtures of the finely powdered oxides Fe_2O_3 and Y_2O_3. A narrow-linewidth garnet can be obtained only if the composition and sintering atmosphere (oxygen) are carefully controlled. Yttrium iron garnets with substitutions of Gd, Ho and Al are among the most important microwave materials, because they have very narrow resonance linewidths – typically of about 4 kA m^{-1} in the polycrystalline material and as low as 40 A m^{-1} in single crystals. Garnets are virtually indispensible at ultra-high frequencies. The very narrow linewidths in single crystals can be attributed to the absence of grain boundaries and crystallographic defects – although an optical quality surface finish is usually required to ensure the very narrowest linewidths. The magnetic properties of some microwave garnets are given in Table 4.8.

Table 4.7 Microwave ferrites and ferrimagnetics.

Material	J_s (T)	ΔH (kA m^{-1})	T_c (°C)
$NiFe_2O_4$	0.3	350	575
$MgFe_2O_4$		70	
Ni–Zn Fe_2O_4	0.5	120	375
$MnFe_2O_4$		50	
Ni–Co Fe_2O_4	0.31	140	590
Ni–Co–Al Fe_2O_4	0.15	330	450
$NiAl_{0.35}Fe_{1.65}O_4$	0.12	67	430
$NiAlFe_2O_4$	0.05	32	1860
$Mg_{0.9}Mn_{0.1}Fe_2O_4$	0.25	56	290
$Ni_{0.5}Zn_{0.5}Al_{0.8}Fe_{1.2}O_4$	0.144	17	
$Y_3Fe_5O_{12}$	0.178	55	280
$Y_{3-x}Al_xFe_5O_{12}$	0.12	80	250
$Y_{3-x}Gd_xFe_5O_{12}$	0.06	150	250
Single-crystal $Y_3Fe_{10}O_{12}$	0.178	0.5	292
Single-crystal $Y_{3-x}Al_xGaFeO_{12}$	0.06	0.5	160
Hexagonal ferrites:			
$BaFe_{12}O_{19}$	0.45	1.5	430
$Ba_3Co_2Fe_{24}O_{41}$	0.34	12	410
$Ba_2Zn_2Fe_{12}O_{22}$	0.28	25	130
$Ba_3Co_{1.35}Zn_{0.65}Fe_{24}O_{41}$		16	390
$Ba_2Ni_2Fe_{12}O_{22}$	0.16	8	500

Table 4.8 Magnetic properties of microwave garnets.

Garnet	J_s (T)	ΔH (kA m^{-1})	T_c (°C)
$Y_3Fe_5O_{12}$	0.178	4	280
$(Y, Al)_3Fe_5O_{12}$	0.120	6.4	250
$(Y, Gd)_3Fe_5O_{12}$	0.06	12	250

4.17 ELECTRICAL PROPERTIES OF FERRITES

Although the magnetic properties of ferrites are inferior to those of metals and alloys, their high electrical resistivity (Parker, 1975) is of great importance in applications where induced eddy currents must be minimized. Since the ferrites are oxide materials, they are either semiconductors or insulators. The resistivities of the ferrites cover a very wide range of values from 10^{-4} Ωm in magnetite to 10 Ωm in Ni–Zn ferrite, while a value as high as 10^9 Ωm can be obtained with a small addition of Mn.

Since the ferrites contain transition metal oxides, the presence of cation or anion vacancies may give rise to various valence states of the transition metal ions. The presence of Fe^{2+} ions among the Fe^{3+} ions results in n-type conductivity, since the extra electron may migrate from the Fe^{2+} to the Fe^{3+} ion. The process will of course be thermally activated, so we should expect an increase in the conductivity with increase in temperature. The presence of Ni^{3+} ions instead of Ni^{2+} ions in nickel ferrite $Ni^{2+}Fe_2^{3+}O_4$ is equivalent to an electron hole, so that the material shows p-type conductivity. For a material in which both n- and p-type carriers are present, the conductivity can be written as

$$\sigma = e(n_e\mu_e + n_h\mu_h),$$

where e is the electron charge, n_e and n_h are the concentrations of the electrons and holes, and μ_e and μ_h are their respective mobilities. The temperature dependence of the conductivity can be approximately described by the relation

$$\sigma = \sigma_0 \exp(-E/kT),$$

where E is the activation energy of the conduction process. Thus a graph of the function

$$\ln \sigma = \ln \sigma_0 - E/kT,$$

i.e. $\ln \sigma$ versus $1/T$, is often linear with slope $-E/k$, although there is usually some deviation for temperatures close to the Curie temperature. Unfortunately, detailed interpretation of the electrical properties of ferrites is very difficult because of the complexity of the materials and uncertainties in their exact physical and chemical states. However, as a general rule, in order to obtain high electrical resistivity, the presence of Fe^{2+} ions must be avoided. In the case of nickel ferrites the addition of a few mol% Mn increases the resistivity substantially. The function of the Mn can be explained as follows. In stoichiometric $Ni^{2+}Fe_2^{3+}O_4$ undesirable Ni^{3+} and Fe^{2+} ions may form according to the dissociation reaction

$$Ni^{2+} + Fe^{3+} \rightleftharpoons Ni^{3+} + Fe^{2+}$$

thus providing a mechanism for both electron and hole conduction, with a consequent lowering of the resistivity. In the presence of Mn it may be energetically favourable for the reactions

$$Ni^{3+} + Mn^{2+} \rightarrow Ni^{2+} + Mn^{3+}$$

$$Fe^{2+} + Mn^{3+} \rightarrow Fe^{3+} + Mn^{2+}$$

to occur, thus restoring the nickel and iron ions to the required valence states. The presence of Mn ions in two valence states will itself provide a mechanism for conduction, but this has a higher activation energy.

In considering the behaviour of ferrites in real devices, the magnetic and dielectric loss factors must also be taken into account.

4.18 STRUCTURE AND PROPERTIES OF THE FERRIMAGNETIC GARNETS

4.18.1 Crystal Structure

The ferrimagnetic garnets (Winkler, 1981) have the general formula $M_3Fe_2Fe_3O_{12}$, which can be written in the form $\frac{1}{2}[5Fe_2O_3 3M_2O_3]$, where M is yttrium or a metal of

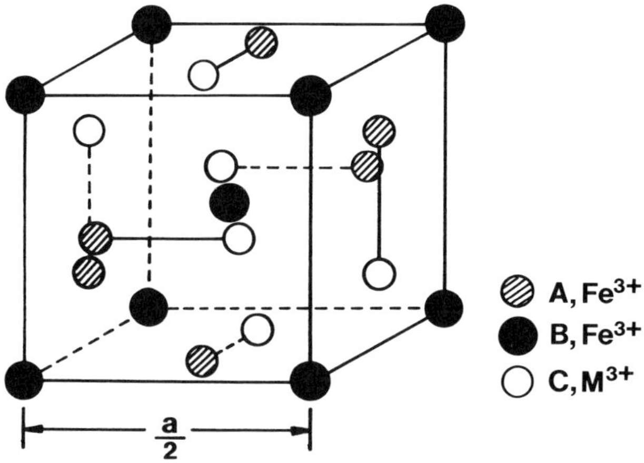

Fig. 4.23 One octant of the ferrimagnetic garnet structure. The oxygen ions are omitted for clarity (after Tebble and Craik, 1969).

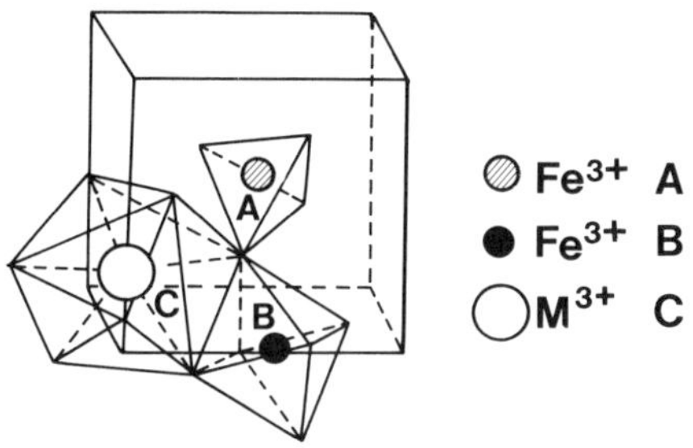

Fig. 4.24 Relative positions of the cations and oxygen anions in tetrahedral (A), octahedral (B), and polyhedral (C) sites of the ferrimagnetic garnet structure (after Gilleo, 1980).

the lanthanide group: Sm, Eu, Gd, Tb, Dy, Ho, Er, Tm, Yb or Lu. The crystal structure is cubic with 160 atoms per unit cell, which contains eight formula units of $M_3Fe_2Fe_3O_{12}$. In view of the large number of atoms per unit cell, it is more instructive to omit the 96 oxygen ions and to represent the structure by the cation lattice only. One octant of this lattice is shown in Fig. 4.23, where the ions are distinguished by the type of crystallographic site which they occupy. In the garnet structure there are three types of interstitial sites: tetrahedral (A), octahedral (B) and skewed cube polyhedral (C) sites. The relative positions of the cations and oxygen anions in these three sites are shown in Fig. 4.24. Each oxygen ion lies at a vertex that is common to four polyhedra of oxygen, one tetrahedron, one octahedron and two skewed cube polyhedra, although for simplicity only one of the latter is shown in Fig. 4.24. The orientations of these polyhedra, none of which is completely regular, do of course vary throughout the unit cell, but the crystal field symmetry around each of the A, B and C sites remains the same. The unit cell therefore consists of 24 A-site ions, 16 B-site and 24 C-site ions.

4.18.2 Magnetic Structure and Properties

As in the case of the ferrimagnetic spinels, the superexchange interaction via the oxygen ions is strong and basically antiferromagnetic. The magnetic moments of the A- and

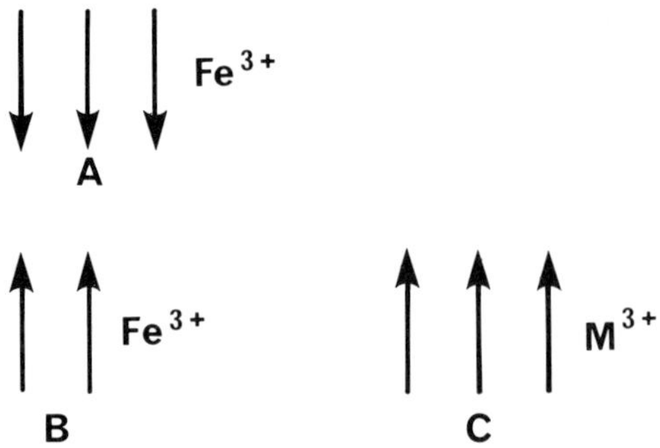

Fig. 4.25 Magnetic structure of the ferrimagnetic garnets, where M^{3+} is Y, Pr, Nd, Sm, Eu, Gd, Tb, Dy, Ho, Er, Tm, Yb or Lu.

B-site ions are antiferromagnetically aligned, while those of the C-site ions are antiferromagnetically aligned with respect to those on the A sites. The magnetic structure is shown in Fig. 4.25, and can be conveniently represented by writing the garnet formula as

$$\overset{\text{C}}{\longleftarrow}\quad\overset{\text{B}}{\longleftarrow}\quad\overset{\text{A}}{\longrightarrow}$$
$$[M_3^{3+}]\ [Fe_2^{3+}]\ [Fe_3^{3+}]\ O_{12}$$

The net magnetic moment per formula unit is

$$\mu = 2\times 5\mu_B + 3m_c - 3\times 5\mu_B$$

$$\mu = 3m_c - 5\mu_B,$$

where m_c is the magnetic moment of the M^{3+} ion. The magnetic properties of some simple ferrimagnetic garnets are summarized in Table 4.9. Although yttrium is not a lanthanide metal, it is included in the table because the corresponding ferrimagnetic garnet is one of the most important and has been investigated in considerable detail. Since the Y^{3+} ion has the closed electron shell structure of Kr, i.e. $4s^2\,4p^6$, it

Table 4.9 Properties of lanthanide garnets, $M_3Fe_2Fe_3O_{12}$.

M	Theoretical moments (μ_B)		Experimental moment (μ_B)	J_s (T) at 293 K	T_c (K)	T_{comp} (K)
	$3(2S)-5$	$3(L+2S)-5$				
Y	5	5	5.01	0.175	553	—
Sm	10	5	5.43	0.170	578	—
Eu	13	14	2.78	0.116	566	—
Gd	16	16	16.00	0.017	564	286
Tb	13	22	18.20	0.005	563	226
Dy	10	25	16.90	0.054	563	226
Ho	7	25	15.20	0.098	567	137
Er	4	22	10.20	0.129	556	83
Tm	1	16	1.20	0.138	549	—
Yb	2	7	0.00	0.163	548	≈6
Lu	5	5	5.07	0.175	549	—

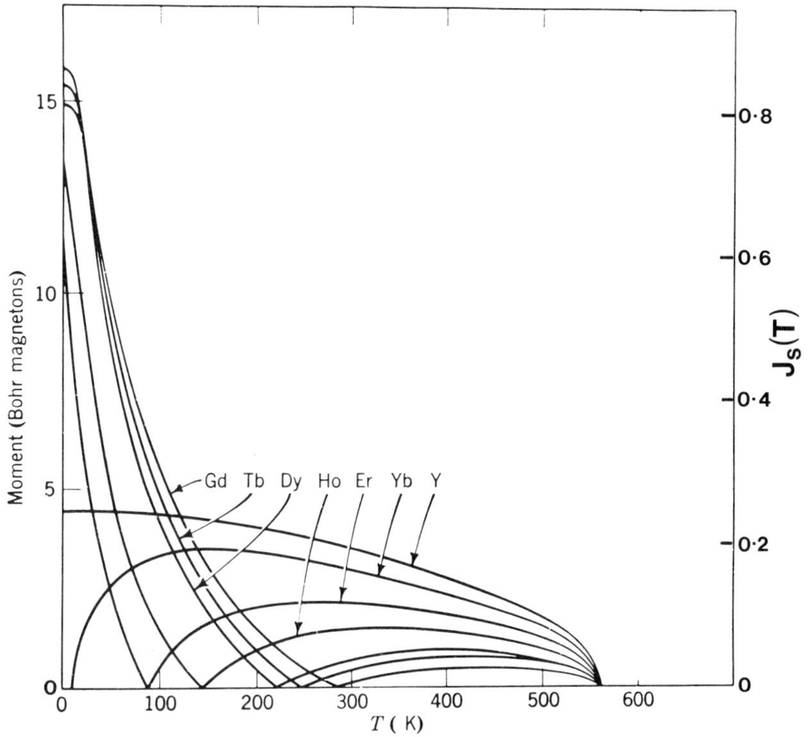

Fig. 4.26 The spontaneous magnetic moment and saturation magnetic polarization of the ferrimagnetic garnets as a function of temperature. Note the compensation temperatures, above which the direction of magnetization is reversed (after Pauthenet, 1956).

has no net magnetic moment and is therefore diamagnetic. The magnetic moment of $Y_3Fe_2Fe_3O_{12}$ (often referred to as YIG, i.e. yttrium iron garnet), is due to the uncompensated ions on the A sites, so that the magnetic moment per formula unit is $5\mu_B$ (obtained by setting $m_c(Y^{3+})$ in the above equation equal to zero). This is in excellent agreement with the experimental value of $5.01\mu_B$. However, if we compare the theoretical and experimental values in Table 4.9, we see that the latter generally lie between the spin-only and spin–orbit values. This suggests that the orbital angular momentum is partially quenched by the crystal field or that the magnetic moments include angles other than 180° or 0°.

The spontaneous magnetization of the ferrimagnetic garnets is plotted as a function of temperature in Fig. 4.26, from which we can see that there is a compensation temperature in every case except YIG, although that for thulium iron garnet is only just above absolute zero, though this is not shown in Fig. 4.26.

4.18.3 Magnetic Properties of Mixed Garnets

Since the garnets have a uniquely defined cation distribution, we can substitute for both the M^{3+} and Fe^{3+} ions directly without the problems of site inversion that arise in the mixed ferrites. The general formula for such a mixed garnet can be written as

$$\overset{\text{C}}{\longleftarrow} \quad \overset{\text{B}}{\longleftarrow} \quad \overset{\text{A}}{\longrightarrow}$$

$$[M^{3+}_{3-x}M'^{3+}_{x}] \; [Fe^{3+}_{2-y}M''^{3+}_{y}] \; [Fe^{3+}_{3-z}M''^{3+}_{z}] \; O_{12}$$

where M′ and M″ are the added ions, which may be paramagnetic (e.g. Gd^{3+}) or diamagnetic (e.g. Al^{3+}) respectively. Note that the M″ ions are divided between A and B sites, so that $z=1-y$, whereas the M′ ions (i.e. lanthanide ions) occupy the larger C sites.

The magnetic moment per formula unit at 0 K is

$$\mu=(3-x)m+xm'+(2-y)5\mu_B+ym''-(3-z)5\mu_B-zm'',$$

where m, m' and m'' are the magnetic moments of the M^{3+}, M'^{3+} and M''^{3+} ions respectively. However, the observed saturation magnetization and its temperature variation will depend on the particular pair of substitutions that is made, and for certain pairs the sublattice magnetizations may no longer be parallel but include angles other than 180° and 0°. For a detailed discussion of magnetic garnets the reader is referred to the book by Winkler (1981) and the review by Gilleo (1980).

4.19 STRUCTURE AND PROPERTIES OF THE FERRIMAGNETIC ORTHOFERRITES: LANTHANIDE ORTHOFERRITES

The orthoferrites have the general formula $MFeO_3$, where M is any of the lanthanide elements or yttrium. The structure of the orthoferrites is basically a distorted form of the perovskite structure shown in Fig. 4.27, and is easily described by the orthorhombic unit cell shown in Fig. 4.28, in which the distorted perovskite subcells are still discernible (Geller, 1956). The structure and properties of these materials have been studied in considerable detail by Bobeck (1967).

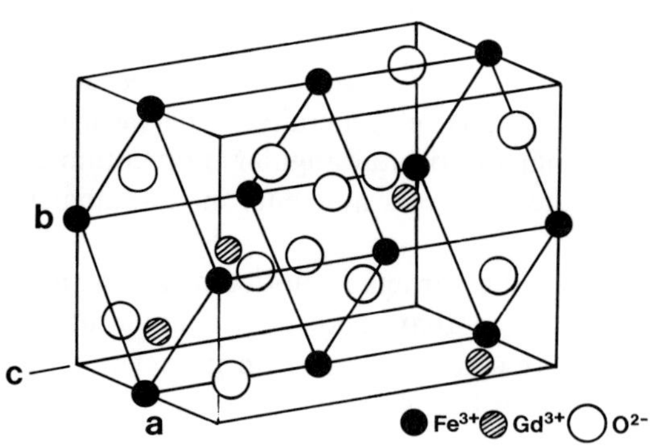

Fig. 4.27 Crystallographic structure of the lanthanide orthoferrites $MFeO_3$, where M is Y, Pr, Nd, Sm, Eu, Gd, Tb, Dy, Ho, Er, Tm, Yb or Lu (after Geller and Wood, 1956).

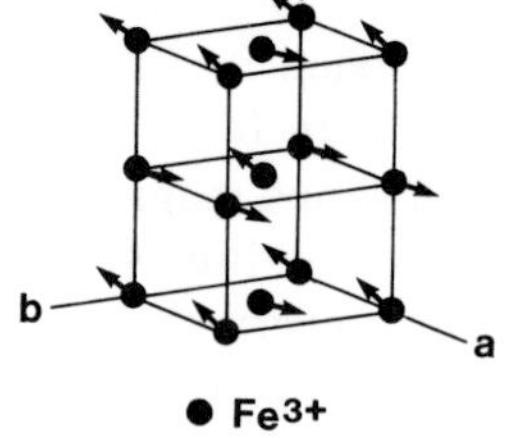

Fig. 4.28 Orthorhombic unit cell of the orthoferrites $MFeO_3$, showing the canted structure of the Fe^{3+} ions

The orthoferrites are basically antiferromagnetic, but there is a weak ferromagnetism due to slight canting (about 0.5°) of the antiferromagnetically aligned Fe^{3+} ions. The magnetic structure, which has a unit cell identical with that of the molecular crystallographic unit cell, is shown in Fig. 4.28. The Fe^{3+} ionic magnetic moments are aligned antiparallel along the axis, and, from Fig. 4.28, we can readily see that the canting of the spins produces a net ferromagnetic moment perpendicular to this, i.e. along the *c* axis. The only exception to this behaviour is the orthoferrite $SmFeO_3$, which has a net magnetic moment at room temperature parallel to the *a* axis. The lanthanide ions show only paramagnetic behaviour, since they make no contribution to the spontaneous magnetization. Some of the basic properties of the lanthanide orthoferrites are summarized in Table 4.10. Note that although the saturation magnetizations are very low, the Curie temperatures are all about 350–460°C. The dependence of the spontaneous magnetization on temperature for two orthoferrites is shown in Fig. 4.29.

The orthoferrites have some remarkable magnetic properties. For example, when they are magnetized to saturation, reverse fields of several hundred k A m^{-1} are often required in order to reverse the magnetization. The reversal field is that required to nucleate a reverse domain, and is about ten times the intrinsic flux density in the material. However, once the reverse domain has been nucleated, it can be moved by a field as small as 100 A m^{-1}, i.e. the wall coercivity or pinning field $H_c \leqslant 100$ A m^{-1}.

The lanthanide orthoferrites are all optically transparent near the red end of the spectrum, so that the domains can be observed by means of the Faraday effect provided the crystals are less than about 100 μm thick. Since the spontaneous magnetization is so low and parallel to the *c* axis, the magnetostatic energy associated with the magnetization in this direction is small so that domains that are magnetized perpendicular to the plane of a thin platelet can readily be observed. In practice, the specimens are either grown as single-crystal platelets or cut from larger crystals such that the *c*-axis is perpendicular to the plane of the platelet. The sizes and shapes of the domains are determined by minimization of the magnetostatic and wall energies.

Table 4.10 Magnetic properties of lanthanide orthoferrites.

M	J_s (mT)	Canting angle α (mrad)	T_c (K)
Y	10.5	8.9	643
La	8.29	9.1	738
Pr	7.09	8.5	707
Nd	6.19	8.5	693
Sm	8.39	8.2	673
Eu	8.36	8.0	663
Gd	9.40	9.8	661
Tb	13.7	7.8	652
Dy	12.7	8.0	645
Ho	9.1	8.2	643
Er	8.09	8.1	641
Tm	14.0	8.0	631
Yb	14.2	8.9	632
Lu	11.9	10.7	622

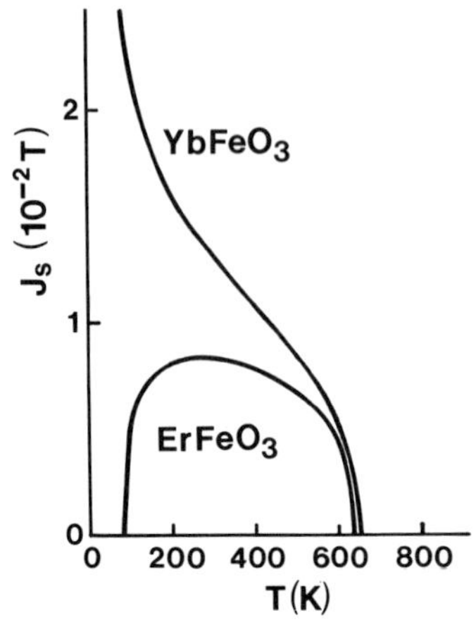

Fig. 4.29 Temperature dependence of the spontaneous magnetic polarization of the orthoferrites $YbFeO_3$ and $ErFeO_3$ (after Gorodetsky and Treves, 1965).

4.20 APPLICATION OF GARNET AND ORTHOFERRITE DOMAIN STRUCTURES AS INFORMATION STORAGE DEVICES

Under suitable conditions, it is possible to nucleate isolated domains in garnets and orthoferrites that are cylindrical and have the same radii, as shown in Fig. 4.30. These reverse domains are mutually repulsive, and can be moved through the material by fields as small as 100 A m^{-1}, provided that the crystal platelet is defect-free so that the wall energy is independent of position and hence there is no tendency for the walls to be pinned at particular points in the crystal. If the crystal is divided into a matrix of positions that have Cartesian coordinates (x_n, y_n) then the presence of a reverse domain at say (x_1, y_3) could be used to signify the stored binary digit 1. The absence of a reverse domain at any point can then be used to represent zero. Thus a single-crystal platelet of any garnet or orthoferrite could be used as the fundamental element of a computer store. Since the reverse domains can be made very small (about 5 μm diameter), one square millimetre of garnet, for example, could store about 100 000 bits of information. Bobeck (1967) has shown that stable cylindrical domains with radii comparable to the plate thickness can be generated if a stabilizing bias field is applied perpendicular to the plane of the plate. If this field is removed, the reverse domains become unstable, resulting in stripe-out. The domain radius also depends on the bias field, and if this is too large then the cylindrical domains may collapse.

The most suitable materials for a bubble memory so far developed consist of a layer of a ferrimagnetic garnet deposited by liquid-phase epitaxy with an induced uniaxial anisotropy onto a gadolinium gallium garnet slice, $Gd_3Ga_5O_{12}$.

In order for this material to function as a memory, some means of generating, storing, moving and reading the information must be devised. This has been achieved by first

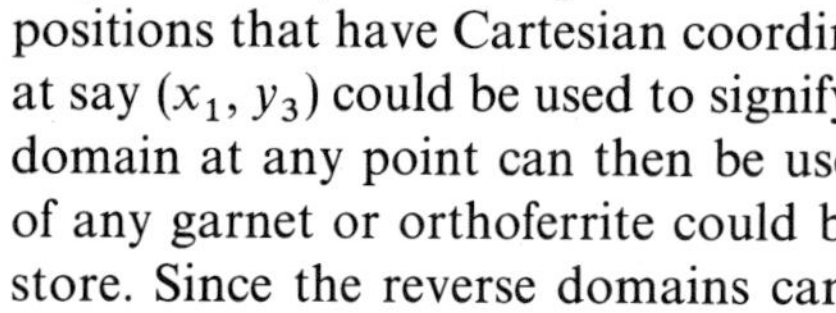

Fig. 4.30 Basis of an information store containing cylindrical domains of reversed magnetization ('bubble domains') in a thin platelet of an orthoferrite or garnet crystal.

depositing a layer of SiO_2 about 1 μm thick onto the bubble material, followed by an overlay pattern of ferromagnetic Permalloy (with a composition close to 80% Ni–Fe). The Permalloy overlay pattern may be in the form of 'T' and 'I' bars or chevrons, or other more complex shapes, with width about half the bubble diameter and with length about five times their width. The bubbles can be generated by special Permalloy discs at the beginning of the bubble track (Perneski, 1969). The generation and control of the bubbles is achieved by a rotating field in the plane of the film. The bubbles are guided into their appropriate storage positions by the changing polarity of the 'T' and 'I' bars or chevrons.

The detection of bubbles (i.e. reading) can be carried out by means of carefully positioned magnetoresistive sensors. The magnetoresistance effect is the change in the resistance of a material in the presence of a magnetic field. In this case the field is provided by the stray field from the bubble domains. Although the change in resistance due to the stray field is very small, it is sufficient to generate a reliable electrical signal.

4.20.1 Garnets with Growth-Induced Uniaxial Magnetocrystalline Anisotropy

Although the orthoferrites have been used to demonstrate the possible use of domains as information stores, their relatively low switching speeds and bit storage makes them unsuitable for commercial applications.

Since the ferrimagnetic garnets have a cubic structure and hence cubic magnetocrystalline anisotropy, these were not at first considered suitable for magnetic bubble domain stores. Bobeck *et al.* (1970) have shown that it is possible to produce suitable bubble domain materials from bulk flux-grown garnets containing two or more lanthanide metals, e.g. Er, Eu, Gd. The growth process itself induces sufficient uniaxial magnetocrystalline anisotropy, a property that is necessary for the formation of stable cylindrical bubble domains in thin platelets. Excellent bubble domain crystals can be grown by liquid-phase epitaxy on non-magnetic substrates, but in order to grow a thin magnetic layer of the correct orientation (i.e. $\{100\}$, $\{111\}$ or $\{110\}$ surface planes), a non-magnetic substrate with a matching lattice spacing is required. One of the most suitable substrate materials is the non-magnetic garnet $Gd_3Ga_5O_{12}$. In general, the ferrimagnetic garnets that are suitable for magnetic bubble storage elements have rather complicated compositions. The reasons for this can be explained as follows.

If, for example, some of the Fe^{3+} ions in $Y_3Fe_2Fe_3O_{12}$ are replaced by Ga^{3+} ions, the latter tend to prefer the smaller tetrahedral A sites, thus replacing the Fe^{3+} ions on these sites, with a consequent reduction in the magnetization per unit cell. The garnet $Y_3Ga_{1.1}Fe_{3.9}O_{12}$ has a suitable magnetization (i.e. the magnetostatic energy for magnetization parallel to the c-axis is not too high compared with the magnetocrystalline anisotropy energy), and can be grown with growth-induced magnetocrystalline anisotropy parallel to the c-axis. Unfortunately, it is not suitable for epitaxial growth on $Gd_3Ga_5O_{12}$ because the lattice parameter mismatch is too large. However, if some of the yttrium is replaced by europium on the C sites, suitable matching of the crystal and substrate can be obtained without altering the magnetic properties substantially.

These requirements mean that bubble domain garnets have complex compositions. A very large number of potentially suitable garnets are available by appropriate choice of composition, e.g. $(Eu, Y)_3(Ga, Fe)_5O_{12}$, $(Er, Eu)_3(Ga, Fe)_5O_{12}$, $(Er, Gd)_3(Ga, Fe)_5O_{12}$,

$(Eu, Gd)_3(Ga, Fe)_5O_{12}$, $(La, Sm, Lu)_3(Ga, Ge)_5O_{12}$, $(Gd, Y)_3(Fe, Mn, Ga)_5O_{12}$ and $(Bi, Y, Ca)_3(Fe, Si)_5O_{12}$.

Metallic films, either crystalline or amorphous (e.g. Gd–Co), can also support magnetic bubbles. Since these bubbles are very much smaller than those in garnets, their potential information storage density is very much greater.

4.21 BASIC PROPERTIES OF MAGNETIC BUBBLES

4.21.1 Bubble Formation and Bubble Diameter

The principal requirement for the formation of bubble domains in a thin film of a ferrimagnetic garnet is that there should be a strongly preferred axis of magnetization perpendicular to the plane of the film. This is a necessary but insufficient condition, since, in general, the domain structure that will be formed will be the labyrinth or stripe domain pattern as shown in Fig. 4.31(a). The stripe domains are magnetized alternately into and out of the plane of the film, and the domains themselves are separated by vertical domain boundaries. If a magnetic field is applied perpendicular to the plane of

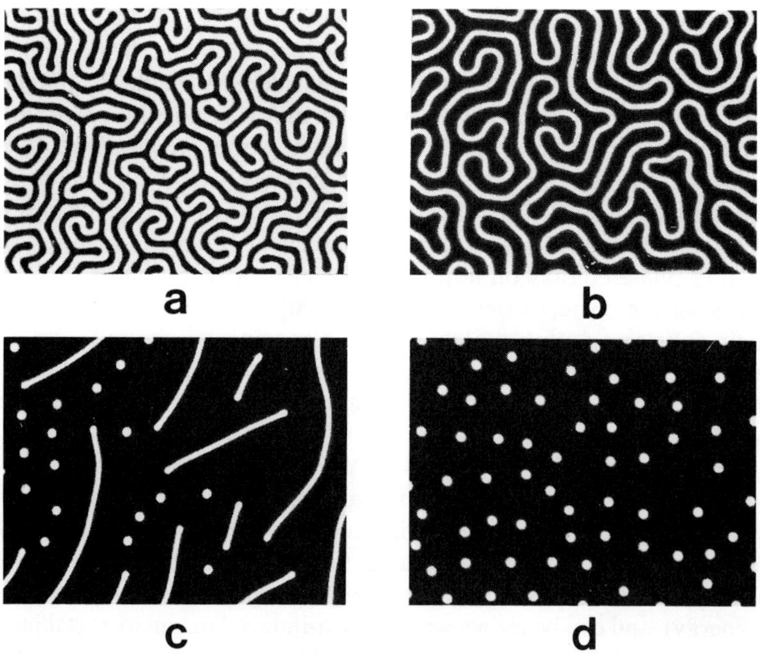

Fig. 4.31 Domains in an epitaxial garnet crystal of $Bi_{0.6}Tm_{0.3}Y_{2.1}Fe_{4.1}Ga_{0.9}O_{12}$ (6 μm thick): (a) in zero applied field; (b) in a small field applied perpendicular to the plane of the crystal (note the growth in the domains for which the magnetization is parallel to the field direction); (c) formation of 'bubbles' in a larger field; (d) formation of 'bubbles'. The bubbles can be removed by the application of a sufficiently large field so that the crystal is then fully saturated perpendicular to the plane of the paper and in the field direction. The domains were obtained by using the Faraday effect. Magnification: 300 × (courtesy, Plessey Company).

the paper (this is known as the bias field), the domains whose magnetization is parallel to the field expand at the expense of those magnetized in the opposite direction (Figs 4.31b, c). If the bias field is increased still further, the remaining stripe domains may shrink so as to form isolated cylindrical domains, known rather inaccurately as 'magnetic bubbles' (Fig. 4.31d). Thus magnetic bubbles are simply cylinders of magnetization bounded by a vertical domain wall surrounded by a region of reverse magnetization, and sustained by a bias field.

If we assume that the thickness of the garnet film is very small compared with the planar dimensions of the film, the demagnetizing factor perpendicular to the plane of the film $D_z \approx 1$, i.e. $D_x \approx D_y \approx 0$. The demagnetizing field associated with magnetization in this direction is

$$H_D = D_z M_s = M_s.$$

The magnetocrystalline anisotropy field H_A is given by

$$H_A = \frac{2K_u}{\mu_0 M_s},$$

where K_u is the induced magnetocrystalline anisotropy energy. Thus for stable domains magnetized perpendicular to the plane of the film

$$H_A > M_s,$$

$$\frac{2K_u}{\mu_0 M_s} > M_s;$$

that is,

$$K_u > \tfrac{1}{2}\mu_0 M_s^2.$$

Thus M_s should be relatively low so that the demagnetizing field associated with magnetization perpendicular to the plane of the film is also low. If $H_A < M_s$, the bubbles cannot exist because the preferred direction of magnetization will then be in the plane of the film. Furthermore, it is an advantage, particularly for garnets, to use a material with a relatively low anisotropy field H_A ($=2K_u/\mu_0 M_s$), since this increases the bubble mobility and also decreases the coercivity.

Thus if the condition $H_A > M_s$ is satisfied, the stable domain structure can be subdivided into stable bubbles by applying a bias field perpendicular to the plane of the film. It is convenient to express the suitability of a bubble material for the storage of information in terms of a quality factor Q, which is defined as the ratio of the induced uniaxial magnetocrystalline anisotropy energy to the magnetostatic energy associated with magnetization perpendicular to the plane of the film, i.e.

$$Q = \frac{K_u}{\tfrac{1}{2}\mu_0 M_s^2} = \frac{2K_u}{\mu_0 M_s^2}.$$

The value of Q for a suitable bubble domain material is typically about 2. It must always be greater than 1; but if Q is too large, it becomes difficult to nucleate the reverse bubble domains during the writing process. For stable bubbles the bubble length h (equal to the thickness of the film) should not be significantly greater than the bubble diameter D. At present, materials with $D \approx h \approx 6\ \mu\text{m}$ and a domain wall thickness of about 0.1 μm are proving to be satisfactory.

It is convenient to define another parameter λ, known as the characteristic length, i.e.

$$\lambda = \frac{\gamma}{\mu_0 M_s^2} = \frac{4(AK)^{1/2}}{\mu_0 M_s^2},$$

where γ is the domain wall energy per unit area. Note that λ is proportional to $Q^{1/2}/M_s$, a fact that further emphasizes the importance of the quality factor Q. The value of λ is important in that the choice of bubble diameter D, film thickness h and bubble domain wall width δ must be considered in relation to λ; that is, D/λ, h/λ and δ/λ are parameters of practical and theoretical significance. For most practical bubble domain materials, i.e. substituted garnets, λ is usually in the range 0.1–1.0 μm. For a garnet film of thickness $h = 4\lambda$ the bubble diameter will be $D = 8\lambda$, although $D = 9\lambda$ is often used in practical devices.

The stability of a single bubble in a bias field H_B perpendicular to the plane of the film is determined by three contributions to its total energy:

(1) the domain wall energy $E_w = \pi D_\gamma h$;
(2) the magnetic energy of the bubble in the bias field H_B, i.e. $E_H = \frac{1}{8}\mu_0 M_s H_B \pi D^2 h$;
(3) the magnetostatic energy of the film and the bubble.

In general, bubbles can be stable in a range of bias fields H_B between two extremes: the field H_{CO} that causes the bubble to collapse, and the stripe-out field H_{SO} that is low enough to produce elliptical instability so that the bubble expands into a long serpentine stripe. Examples of bubbles and stripes can be seen in Fig. 4.31(c). The bubble diameters at collapse and stripe-out are usually specified in terms of λ, i.e. as D_{CO}/λ and D_{SO}/λ. The bubble diameter usually varies by a factor of about three over the stable range, corresponding to a range in H_B of about 10%, but the exact values depend on h/λ.

4.21.2 Dynamic Properties of Bubbles

Another very important parameter in the application of magnetic bubbles as information stores is their mobility; this is defined as $\mu = v/\Delta B$, where $\Delta B = D\,\mathrm{d}B/\mathrm{d}x$ and $\mathrm{d}B/\mathrm{d}x$ is the flux gradient across the bubble. The quantity ΔB is known as the drive field, and if the bubble is to move, the drive field must be greater than the coercivity of the bubble wall. Thus an efficient bubble domain material must have a very low wall coercivity, and it is for this reason that the bubble domain materials must have a high degree of crystallographic perfection, so that domain wall pinning is kept to a minimum. Typically, an efficient bubble material will have a crystallographic defect density of less than $5 \times 10^4\ \text{m}^{-2}$.

Bubble domain motion is also impeded by viscous drag or damping of the spin movement within the domain walls. A simple theoretical treatment shows that the bubble mobility μ is independent of the drive field, i.e.

$$\mu = \frac{\gamma' \delta}{\alpha \pi},$$

where γ' is the gyromagnetic ratio, α is the domain wall damping constant and δ is the domain wall width. This relationship implies that the bubble velocity v is proportional to the drive field ΔB. However, most real bubble materials show a nonlinear relationship between v and ΔB.

For high mobility we require materials with a high gyromagnetic ratio γ', and it is for this reason that nearly all practical bubble materials contain one or more of the lanthanide metals (e.g. Eu, Er, Yb, Sm, Tm, Lu), for which there is often a substantial contribution to the magnetic moment from the orbital angular momentum, thus increasing the gyromagnetic ratio. Ideally, the damping constant α should be as low as possible.

Another factor that affects the velocity of the wall is the 'in-plane' magnetocrystalline anisotropy, i.e. the intrinsic magnetocrystalline anisotropy of the garnet K_1, since this determines the ease of rotation of the spin in the bubble domain walls and hence affects the resistance to domain wall motion. Thus K_1 should be as low as possible, while the induced anisotropy K_u perpendicular to the plane of the film should be greater than $\frac{1}{2}\mu_0 M_s^2$, as discussed earlier. A second quality factor Q' can be defined as the ratio $2K_1/\mu_0 M_s^2$, and it has been suggested that the ratio Q/Q' provides a good indication of the maximum wall velocity.

The mobility of a bubble is strongly affected by the spin configuration in the surrounding wall. In a 'normal' bubble the spin configuration is the same as that in a 180° Bloch wall (see Chapter 1), with all the spins rotating in the same sense. Certain bubbles persist in much higher bias fields before collapsing than normal bubbles, and their collapse radius is also very small. These bubbles are known as hard bubbles, and it has been suggested that their domain walls contain several segments separated by Bloch lines, i.e. regions over which the sense of the spin rotation changes. The persistence of these bubbles is due to the fact that as the bias field is increased, the Bloch lines resist compression and hence inhibit bubble collapse. The existence of Bloch lines in very thin cobalt films has been confirmed by Grundy *et al.* (1972), who used the techniques of Lorentz electron microscopy. The velocity of a hard bubble down a field gradient is very much lower than that for 'normal' bubbles. Another important difference in the behaviour of 'normal' and hard bubbles is that the latter have a velocity component perpendicular to the field gradient, whereas normal bubbles always move parallel to the latter. Thus for the accurate storage of information the presence of hard bubbles must be avoided. The most successful method for avoiding the formation of hard bubbles in garnets is the implantation of high-energy ions such as H, He or Ne. The implanted layer expands and produces stresses that, via magnetostriction, cause the magnetization within the affected layer to rotate into the plane of the film. The bubbles are therefore capped so that the formation of Bloch lines is suppressed. Changes in the bubble structure sometimes occur during propagation in a drive field, and may also result in a nonlinear relationship between v and ΔB.

Another parameter that is of importance in determining the suitability of a bubble material is the inertial time constant τ^*. This is a measure of the time interval that elapses before the bubbles respond to changes in the propagating fields. Changes in the bubble diameters occur when they are moving, so their dynamic behaviour is very complex (Malozemoff and Slonczewski, 1979). Ideally, we require τ^* to be as small as possible. For further information on magnetic bubble materials and properties, the reader is referred to the books by Bobeck and Della Torre (1975), O'Dell (1974), Smith (1974), Malozemoff and Slonczewski (1979), Eschenfelder (1981) and Jouve (1986), and to the reviews by Bobeck (1975) and Eschenfelder (1980). Some of the basic requirements for a field access bubble domain memory are listed in Table 4.11. The values of these parameters for some potentially useful magnetic bubble materials are shown in Table 4.12.

The ferrimagnetic garnets are thc preferred materials for magnetic bubble devices, because their properties can more easily be tailored to specific requirements by chemical substitution and appropriate epitaxial crystal growth of the thin films. The magnetic bubbles in orthoferrites tend to be larger and less stable and to have relatively low mobilities.

The hexaferrites $BaFe_{12}O_{19}$ and $SrFe_{12}O_{19}$ have been considered for possible application in magnetic bubble information storage devices, but their high uniaxial magnetocrystalline anisotropy results in the formation of very small bubbles, which are difficult to nucleate and to move sufficiently rapidly.

The potential use of amorphous Gd–Co films for magnetic bubble domain devices has been discussed very briefly in Chapter 3.

It has been suggested that Bloch lines within domain walls (i.e. where there is a change in the sense of rotation of the individual ionic magnetic spin moments) could themselves be used to store digital information (Konishi, 1983; Hidaka *et al.* 1983), but the problem of writing and reading the digits reliably at high speed has not yet been solved. The information storage density would be very large indeed (Suzuki, 1986).

All aspects of magnetic bubble technology have been very thoroughly discussed in the book by Eschenfelder (1981) and the review by Eschenfelder (1980), while more recent developments have been discussed in the book edited by Jouve (1986).

Table 4.11 Basic requirements for a field-access bubble domain memory: approximate values of parameters.

Saturation magnetic polarization J_s	<250 mT
Anisotropy field $H_A = 2K_u/\mu_0 M_s$	$>1.5M_s$
Quality factor $Q = H_A/M_s$	ideally about 1.5
Bubble diameter D	6 μm
Thickness h	6 μm
Mobility μ	$>2\times10^4$ m s^{-1} T^{-1}
Characteristic length $\lambda = \gamma/\mu_0 M_s^2$	1 μm
Coercivity H_c	<25 A m^{-1}
Inertial time constant τ^*	<100 ns
Crystal defect density x	$<5\times10^4$ m^{-2}

Bubble size and shape should be insensitive to rotating in-plane fields and temperature fluctuations.

Table 4.12 Magnetic properties of bubble domain ferrimagnetic garnets (after Eschenfelder, 1981).

Ferrimagnetic garnet	H (μm)	J_s (mT)	Q	K (10^3 J m^{-3})
$Sm_{1.2}Lu_{1.8}Fe_5O_{12}$	0.5	175	2.5	30
$Sm_{0.5}Lu_{0.7}Y_{1.0}(CaGe)_{0.8}Fe_{4.2}O_{12}$	1.7	51	3.5	3.6
$Sm_{0.4}Lu_{0.5}Y_{1.2}(CaGe)_{0.9}Fe_{4.1}O_{12}$	3.0	33	5.3	2.3
$Sm_{0.1}Y_{1.9}(CaGe)_{1.0}Fe_{4.0}O_{12}$	6.0	16	4.1	0.4
$Eu_{1.0}Tm_{2.0}Fe_5O_{12}$	0.5	138	1.7	12
$Eu_{1.0}Tm_{2.0}Ga_{0.6}Fe_{4.4}O_{12}$	2.0	75	3.0	7
$Eu_{0.65}Y_{2.35}Ga_{1.2}Fe_{3.8}O_{12}$	7	16	8	0.8
$Sm_{0.85}Tm_{2.15}Fe_5O_{12}$	0.5	138	2.5	19
$Sm_{0.3}Tm_{0.75}Y_{1.2}(CaGe)_{1.75}Fe_{4.25}O_{12}$	0.9	50	3.1	3
$Sm_{0.2}Tm_{0.2}Y_{1.7}(CaGe)_{0.9}Fe_{1.4}O_{12}$	3.3	29	3.9	1.3
$La_{0.2}Sm_{0.6}Lu_{1.6}(CaGe)_{0.6}Fe_{4.4}O_{12}$	1.1	78	2 9	7.1
$La_{0.6}Sm_{0.3}Lu_{2.1}Ga_{0.9}Fe_{4.1}O_{12}$	1.3	51	2.6	2.7

4.22 OXIDES FOR MAGNETIC RECORDING

Magnetic recording of audio, visual and digital information is now one of the most important applications of magnetic materials – indeed, the expenditure on magnetic tapes exceeds that on transformer core materials, although of course the quantity by weight is considerably smaller. The recorded information is stored in the form of the remanent magnetization of the material such that the magnitude of the remanence is (ideally) proportional to the amplitude of the original electromagnetic input signal. Thus a material with a linear remanent magnetization is required. Since we wish to retrieve the recorded information, the remanence must be sufficiently large to give a strong noise-free playback signal. A material with a high saturation magnetization is required, and if this is to be combined with high remanence values then the material must also have a relatively high coercivity – this is usually in the range 8–40 kA m^{-1}, so that adequate magnetization of the tape can be obtained with a low-field recording head.

Another important requirement (which is not immediately obvious) is that the slope of the initial magnetization curve must be small enough at low fields to avoid 'print-through' magnetization of the tape from adjacent layers on the reel on which recordings have already been made. A remanent magnetic polarization curve suitable for magnetic recording is shown in Fig. 4.32(a). However, in practice, magnetic recording is carried out in the presence of an AC bias field with a frequency much higher than the highest frequency in the signal. The amplitude of the AC bias field is chosen so that it is at least equal to or greater than the remanence coercivity of tape material. The resulting remanance curve (Fig. 4.32a) is known as the anhysteretic remanence curve. It is approximately linear for small signal fields, and has a relatively low signal-to-noise ratio. The mechanism of production of the anhysteretic remanence M_{ar} in a signal field H_s is shown in Fig. 4.32(b). The recording process is very complex and involves many subtleties to achieve high-quality recording and reproduction. Thorough accounts are given in the handbook edited by Mee and Daniel (1990) and by Mallinson (1987).

The recording process is further complicated by the magnetic properties of the

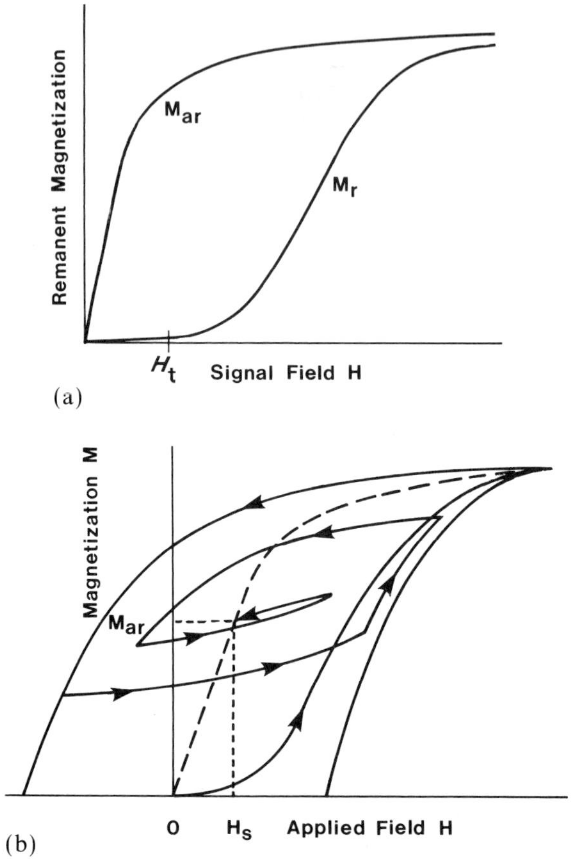

Fig. 4.32 (a) Remanent (M_r) and anhysteretic remanent (M_{ar}) magnetization curves for a material suitable for magnetic recording. Note that there is a threshold field H_t below which the remanent magnetization is very small. (b) Magnetization route to obtain the associated anhysteretic remanent magnetization curve (M_{ar}) shown as a dashed line and as a full line in (a).

recording head material, its dimensions and geometry. All of these factors introduce nonlinear effects, thus adding to the complexity of the recording and reproduction process (see e.g. Jones and Mee, 1990).

The recording head consists of a magnetically soft core of laminated Ni–Fe or ferrite blocks, which are magnetized by surrounding coils that carry the signal. The information is stored as small strips of remanent magnetization in the plane of the tape parallel to its longitudinal direction. Magnetization perpendicuar to the plane of the tape is strongly opposed by the demagnetizing field. Unfortunately, the fields produced by the recording head, which contains a small gap of about 25 μm, to allow flux leakage, are nonlinear.

The oxides γ-Fe_2O_3, Fe_3O_4 and CrO_2 have the all required properties mentioned above, but so far only γ-Fe_2O_3 and Co-doped modifications are the most widely used materials. However, it has been shown that CrO_2 tapes have many advantages that have not yet been fully exploited, although they are available commercially on a limited scale.

4.22.1 Maghemite γ-Fe_2O_3

(*i*) *Magnetic properties of γ-Fe_2O_3*

The very important ferrimagnetic oxide γ-Fe_2O_3 has a defect inverse spinel structure with lattice vacancies on octahedral sites. The magnetic structure has already been discussed in section 4.5.2 and is shown in Fig. 4.9. Since the formula for γ-Fe_2O_3 is equivalent to $FeFe_{5/3}O_4$ the net magnetization is due to the uncompensated iron ions on the B sites, i.e.

$$\mu = (\tfrac{5}{3} - 1)5\mu_B = \tfrac{10}{3}\mu_B = 3.33\mu_B$$

per formula unit, i.e. $J_s = 0.535$ T. This compares reasonably well with the experimental values of 0.5–0.52 T. Since maghemite transforms to the very weakly ferrimagnetic oxide α-Fe_2O_3 (rhombohedral structure) in the range 400–600°C, its Curie temperature cannot be measured directly, although it has been estimated to be about 590°C.

(*ii*) *Preparation of elongated single-domain γ-Fe_2O_3 particles*

Fortunately, elongated signal domain particles of γ-Fe_2O_3 can be prepared relatively easily. The starting material is ferric oxyhydroxide α-FeOOH in the form of small acicular (i.e. needle-shaped) crystals. The process can be summarized by the following equations:

$$6\alpha\text{-FeOOH (400°C)} = 3\alpha\text{-}Fe_2O_3 + 3H_2O$$

$$3\alpha\text{-}Fe_2O_3 + H_2 = 2Fe_3O_4 + H_2O$$

Careful oxidation of the resulting acicular magnetite particles at about 250°C produces acicular particles of γ-Fe_2O_3 i.e.

$$2Fe_3O_4 + \tfrac{1}{2}O_2 = 3\gamma\text{-}Fe_2O_3.$$

Several methods of preparing acicular γ-Fe_2O_3 particles have been described by Bate (1975, 1980, 1986). Most detail is given in the 1980 reference.

Unfortunately, the dehydration of α-FeOOH results in the formation of pores within the particles that have very significant effects on the particle coercivities and hence on the switching field distribution. The tendency of particles to sinter into irregular shapes reduces the shape anisotropy and the efficiency of alignment and packing of the particles, all of which reduce the recording quality. The reduction and variability of the particle coercivities arises because the magnetization reversal mechanism for pore-containing or irregularly shaped particles may occur by incoherent processes such as fanning (Luborsky, 1961).

Improvements have been made by using $FeCl_2$ to precipitate acicular γ-FeOOH (lepidocrocite), which can then be converted to acicular γ-Fe_2O_3 particles, which have few pores and smoother surfaces – although it is difficult to make the particles small, i.e. with $l \approx 0.5$ μm.

Further improvements can be obtained by using $Fe(OH)_3$ as the starting material and treating it hydrothermally in the presence of a growth-controlling additive to give fine acicular γ-Fe_2O_3 particles, which are free from pores and surface blemishes, such as dendrites. They show little tendency to sinter and have a narrow range of coercivities. This process is known as the NP process (non-polar) because there are no internal magnetic poles, which would be present if the particles contained pores. The magnetic poles produce demagnetizing fields, which assist and complicate the demagnetization process, with a consequent reduction in the coercivity. Non-polar γ-Fe_2O_3 particles have been used in high-quality audio tapes, but the difficulty in producing small particles has limited their widespread application.

(iii) Magnetic recording properties of elongated single-domain γ-Fe_2O_3 particles

The magnetic properties of acicular γ-Fe_2O_3 particles, suitable for magnetic recording, are summarized in Table 4.13. Note that the coercivity and reduced remanence depend on the degree of alignment in the sample used for measurement. In the recording tapes themselves it is highly desirable to have as high a degree of alignment as possible. This is achieved by aligning the particles in a field of about 0.1 T during manufacture, though it is very difficult to obtain values of J_r/J_s greater than about 0.8, because of particle agglomeration and interaction effects.

Since the particles have a high degree of acicularity, we should expect the coercivity due to shape anisotropy to be (Chapter 1)

$$H_c = 0.48(D_z - D_x)M_s,$$

assuming that the particles are randomly orientated, non-interacting and that magnetization reversal occurs by coherent rotation. Since the axial ratio of length to diameter $l/d \approx 7$, the coercivity will be very close to the maximum possible value (Chapter 2), i.e.

$$H_c = 0.48 \times 0.5 \times 4 \times 10^5 \approx 10^5 \text{ A m}^{-1}.$$

However, since this is very much greater than the experimental values, it is clear that our assumptions concerning the magnetization reversal mechanism are incorrect. Furthermore, we have not yet considered the effects of magnetocrystalline anisotropy on the coercivity. According to Takei and Chiba (1966), the easy axis of magnetization in γ-Fe_2O_3 is [110], which coincides with the direction of elongation of the particles. Assuming, as before, that the particles are randomly orientated and that reversal occurs by coherent rotation, the coercivity for particles with cubic magnetocrystalline anisotropy

Table 4.13 Properties of γ-Fe_2O_3 particles.

J_s	0.5 T
T_c	590°C
K_1	-4.6×10^3 J m^{-3}
ρ	4.6×10^3 kg m^{-3}
H_c	2×10^4–3×10^4 A m^{-1}
J_r/J_s	0.8
Dimensions	$l = 0.2$–0.7 μm
	$l/d = 7$

is, according to Néel (1947a, b),

$$H_c = 0.64 K_1 / J_s,$$
$$= 0.64 \times 4.6 \times 10^3 / 0.5 \approx 6 \times 10^3 \text{ A m}^{-1}.$$

Thus if we include the magnetocrystalline anisotropy, the coercivity of a random array of γ-Fe_2O_3 particles should be even higher. However, since the coercivity of γ-Fc_2O_3 particles does not vary significantly with temperature, Johnson and Brown (1958) concluded that the coercivity is due principally to shape anisotropy.

We now have to account for the fact that the observed coercivities are very much smaller than those predicted by the Stoner–Wohlfarth (1948) theory, which assumes coherent magnetization reversal. From electron micrographs it can be seen that the axial ratios of the particles are at least 5, whereas from magnetic measurement Johnson and Brown (1958) estimated the ratio at about 1.6, suggesting that, magnetically, the particles behaved as though they were less acicular than they appeared in the electron micrographs. The particles are irregular in shape and contain many cavities. It has therefore been suggested that the low coercivities of γ-Fe_2O_3 can be attributed to incoherent mechanisms of magnetization reversal such as curling and fanning. Jacobs and Bean (1955) suggested that γ-Fe_2O_3 particles as observed in the electron microscope could be compared approximately with a chain of spheres, and calculated the coercivities for two mechanism of magnetization reversal known as symmetric fanning and coherent rotation. The coercivity for symmetric fanning was shown to be lower, and therefore was expected to be the preferred mechanism, as was essentially confirmed by Luborsky (1961), who compared the theoretical and experimental results. Magnetization reversal by the curling mechanism (Luborsky, 1961) is not expected to occur in irregularly shaped acicular single-domain particles such as γ-Fe_2O_3. As we shall see in Section 4.22.2, it is believed to occur in the more perfectly formed acicular CrO_2 single-domain particles. The operative mechanism and the associated reduction depends on the particle size, length-to-diameter ratio, shape, degree of perfection and basic magnetic properties.

The coercivity is also reduced due to particle interactions, the magnitude of which depends on the packing fraction of the particles – a typical recording tape coating contains about 40% by volume of γ-Fe_2O_3 particles, so that the expected reduction in the coercivity would be from H_c to $H_c(1-p)$, with $p=0.4$, i.e. about 60%. Experimental results obtained by Morrish and Yu (1955) are shown in Fig. 4.33. The combined effects of particle interactions and incoherent modes of magnetization reversal are very complex and are still not clearly understood.

(iv) *Cobalt-modified* γ-Fe_2O_3

The magnetic recording properties of γ-Fe_2O_3 particles can be significantly improved by the addition of cobalt, either by particular substitution or by surface doping. The principal effect of the cobalt is to increase the coercivity from about 30 kA m^{-1} to 50 kA m^{-1} by increasing the magnetocrystalline anisotropy. Since shorter wavelengths and bit lengths are required in modern high-density recording, the associated increase in the demagnetizing fields must be counterbalanced by an increase in the coercivity. Thus cobalt-modified γ-Fe_2O_3 tapes are now rivalling γ-Fe_2O_3 as the most widely used recording medium; indeed, they are the predominant material for video tapes.

One method of preparing cobalt-modified γ-Fe_2O_3 particles is to disperse the γ-Fe_2O_3

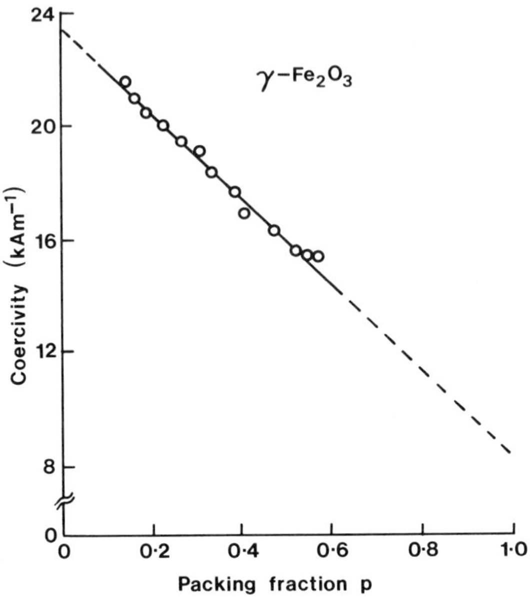

Fig. 4.33 Coercivity of γ-Fe_2O_3 single-domain particles as a function of the packing fraction p (after Morrish and Yu, 1955).

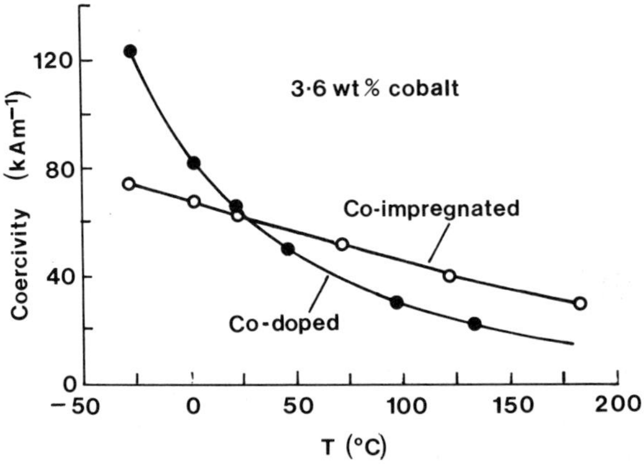

Fig. 4.34 Effects of doping and impregnation of γ-Fe_2O_3 particles with cobalt on the temperature dependence of the coercivity (after Kishimoto *et al.*, 1984).

particles in an aqueous solution of $CoSO_4$ and to precipitate $Co(OH)_2$ onto the surface of the particles by adding NaOH. After warming the particles for a short time at 45°C, the $Co(OH)_2$ decomposes so that cobalt ions penetrate the surfaces of the γ-Fe_2O_3 particles, possibly forming a layer of cobalt ferrite $CoFe_2O_4$ or a mixture of ferrites. Particles of γ-Fe_2O_3, Fe_3O_4 and α-FeOOH can also be used as starting materials. Kishimoto *et al.* (1984) have shown that it is important that the cobalt ions be confined to the surfaces of the γ-Fe_2O_3 particles rather than to be uniformly distributed throughout the particles. They showed that the coercivity for the surface doped particles was less sensitive to temperature variations than that for the bulk-doped particles (Fig. 4.34). The magnetic behaviour and properties of cobalt-modified γ-Fe_2O_3 particles are still not fully understood, but it is clear that the nature of the particle surfaces plays a critical role in determining the coercivity. Knowles (1984) has suggested that a domain wall theory of coercivity may be more appropriate than one based on coherent or incoherent modes of magnetization reversal.

4.22.2 Chromium Dioxide CrO_2

(*i*) *Preparation of (CrO_2) particles*

Chromium dioxide can be prepared by the thermal decomposition of chromyl chloride CrO_2Cl_2, chromium trioxide CrO_3 or aqueous CrO_3, although the latter two methods require oxygen under considerable pressure (about 500–3000 atm). The first method yields acicular particles with lengths of about 5–10 μm and diameters of about 0.5–1.0 μm. Acicular particles with these dimensions are unlikely to be single domains. The most frequently used method is the oxidation of Cr_2O_3 with excess CrO_3 under pressure in the presence of water:

$$Cr_2O_3 + CrO_3 \rightarrow 3CrO_2$$

Swoboda *et al.* (1961) have shown that the particle size can be decreased by the addition of catalysts to the CrO_3 starting material. The coercivity of large particles, $l = 3$–10 μm, $d = 1$–3 μm, prepared without a catalyst was only 5×10^3 A m^{-1}, whereas the addition of 2% Sb_2O_3 gave particles with an average coercivity of 3×10^4 A m^{-1} and dimensions $l = 0.1$–2.0 μm and $d = 0.04$–0.08 μm. According to Darnell (1961), the largest minor dimension for single-domain particle behaviour is about 0.46 μm, and particles with an average axial ratio of about 5:1 and diameter of about 0.2 μm would be suitable for magnetic recording.

Commercial acicular CrO_2 particles are usually produced by adding Sb_2O_3 and Fe_2O_3, which enable lower temperatures and pressures to be used and also have the beneficial effect of raising the Curie temperature. Unfortunately, the additives reduce the saturation magnetization by about 25%. The particles are very regular in shape, smooth and highly acicular with axial ratios of about 20:1. Typical CrO_2 particles for magnetic recording are shown in Fig. 4.35. Particle lengths of order 0.1 μm can be obtained with coercivities as high as 224 kA m^{-1}, but commercially available particles have coercivities of about 44 kA m^{-1} with saturation magnetic polarizations 0.46 T. Iron, antimony, rhodium, iridium and osmium can be added to the particles, and act as growth inhibitors to reduce the particle diameters with a consequent increase in coercivity, as would be expected in regularly shaped pore-free particles, in which magnetization reversal takes place by the curling mode. Although doping the particles with iridium results in a very significant improvement in the magnetic properties (Kullmann *et al.*, 1984), iridium is not used commercially because it is too expensive. Iron is still the only commercially employed dopand and enables the coercivity to be varied over a wide range, namely 30–75 kA m^{-1}, i.e. similar to that obtainable in Co-modified γ-Fe_2O_3.

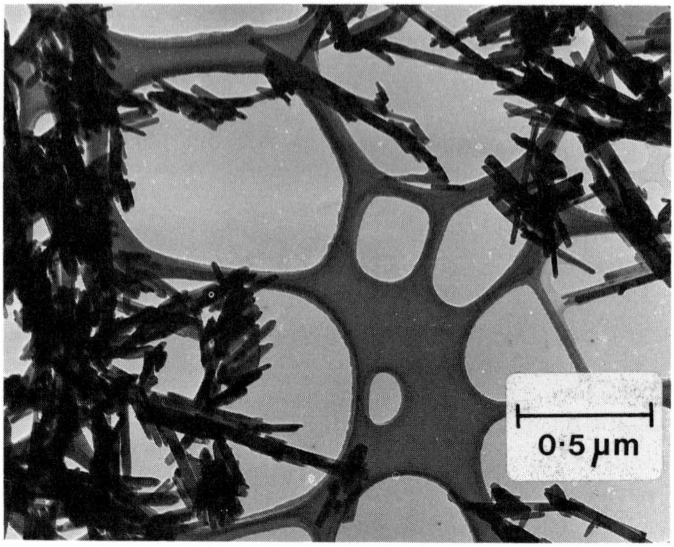

Fig. 4.35 Electron micrograph of acicular particles of CrO_2 suitable for magnetic recording. The length-to-diameter ratio $l/d \approx 10$.

(*ii*) *Magnetic properties of chromium dioxide* (CrO_2)

Chromium dioxide, CrO_2, is an unusual oxide material in that it is ferromagnetic rather than ferrimagnetic. It has the rutile structure shown in Fig. 4.36, with lattice parameters $a = 0.4423$ nm and $c = 0.2917$ nm. The formula CrO_2 suggests that the chromium ions are in the Cr^{4+} state with the electronic configuration [Ar]$3d^2$, so we should expect a magnetic moment of $2\mu_B$ per ion, assuming that the orbital angular momentum is quenched by the crystal field. Since the spins are ferromagnetically aligned and there are two Cr^{4+} ions per unit cell, the magnetic moment per unit cell should be

$$\mu = 2 \times 2\mu_B;$$

that is

$$J_s = 4\mu_B\mu_0/a^2c = 0.65 \text{ T},$$

which is in reasonable agreement with the experimental results. The saturation magnetic polarization at room temperature (20°C) is 0.615 T; there is some variation in this value, depending on the method of preparation. However, some of this variation may be due to the fact that CrO_2 is difficult to saturate. Darnell and Cloud (1965) have shown that the magnetization is only 94–98% of the saturation magnetization, even in a field of 8×10^5 A m^{-1}, though this is much larger than the anisotropy field

$$\frac{2K_1}{\mu_0 M_s} = \frac{2K_1}{J_s} = 8 \times 10^4 \text{ A m}^{-1}.$$

Darnell and Cloud (1965) have attributed the difficulty in saturation to increased anisotropy from surface chromium ions. According to Rodbell *et al.* (1967), CrO_2

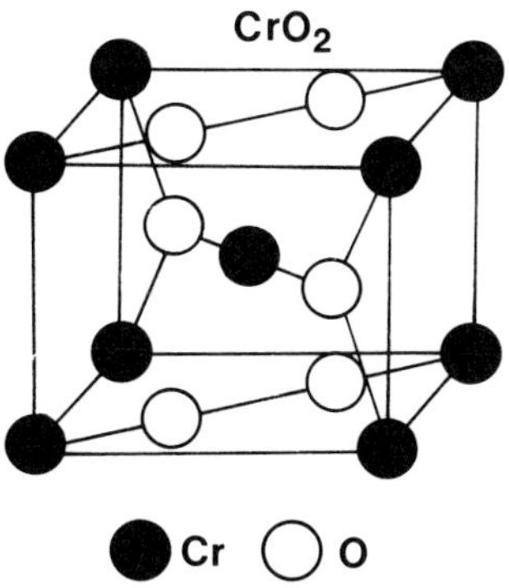

Fig. 4.36 Rutile structure of ferromagnetic CrO_2.

has uniaxial magnetocrystalline anisotropy, with the c-axis as the easy axis and $K_1 = 2.5 \times 10^4$ J m^{-3}. The Curie temperature of CrO_2 is 113.5°C.

From measurements of the temperature dependence of the coercivity, Darnell (1961) concluded that the shape anisotropy was the principal factor, i.e. the coercivity variation could be correlated with the variation of $M(T)$ rather than $K_1(T)/M_s(T)$. For coherent rotation of the magnetization vector the expected coercivities due to shape and magnetocrystalline anisotropy for randomly orientated particles are respectively

$$\begin{aligned} H_c &= 0.479(D_z - D_x)M_s \\ &= 0.479 \times 0.5 \times 490 \times 10^3 = 1.17 \times 10^5 \text{ A m}^{-1} \end{aligned}$$

and

$$\begin{aligned} H_c &= 0.479 \frac{2K_1}{\mu_0 M_s} \\ &= \frac{0.958 \times 2.5 \times 10^4}{4\pi \times 10^{-7} \times 490 \times 10^3} = 4 \times 10^4 \text{ A m}^{-1}. \end{aligned}$$

Since the observed coercivitics are in the range 2.4×10^4–4.0×10^4 A m^{-1}, it has been suggested that the difference between these and the theoretical values can be attributed to the incoherent mechanism of magnetization reversal known as curling. Darnell (1961) has also shown that the value of the rotational hysteresis integral R

$$R = \int_0^\infty \frac{W_r}{J_s} \, d\left(\frac{1}{H}\right)$$

(where W_r is the rotational hysteresis energy) lies between 1.10 and 1.36, which also suggests that magnetization reversal occurs by the curling mechanism.

(*iii*) *Recording properties of CrO_2*

It has been shown that the regular shapes of CrO_2 particles (Fig. 4.35) make them easier to orientate than γ-Fe_2O_3, so that remanence ratios as high as 0.90 can be obtained, thus increasing the signal outputs. Also, the high coercivity should improve the storage density by making the particles less susceptible to demagnetization by neighbouring particles. These expectations have largely been confirmed by Speliotis (1968). The relatively low Curie temperature $T_c = 113°C$, is a disadvantage, because any significant rise in the ambient temperature will decrease the magnetization and hence the remanence. However, since the demagnetizing fields are themselves proportional to the magnetization, their effect on the particles with reduced coercivity at high temperatures will be similar to that at room temperature, provided of course that the temperature does not rise too much above about 50°C. Thus CrO_2 is potentially a very useful magnetic recording material, and has already found limited commercial production, especially in high-quality audio, video, data tapes and disks where high recording densities are required. Coercivities as high as 200 kA m^{-1} have been obtained, but for commercially available particles $H_c \approx 45$ kA m^{-1} with $J_s \approx 0.48$ T. The high coercivities are necessary to overcome the higher demagnetizing fields resulting from the shorter wavelengths or bit lengths used in the high-density recording.

CrO_2 particles are more reactive and abrasive than γ-Fe_2O_3, but one advantage they have over the latter is that their relatively low Curie temperature of 115°C enables thermographic tape duplication from a master tape with a much higher Curie temperature. This is done by heating the two tapes while they are in contact. The process is both fast and cheap.

4.22.3 Barium Ferrite Particles $BaFe_{12}O_{19}$

Barium ferrite particles crystallize as thin hexagonal platelets in which the easy axis, i.e. the hexagonal axis, is perpendicular to the plane of the platelet. Since the uniaxial magnetocrystalline anisotropy is high, the coercivities of the single-domain particles are typically 160–240 kA m^{-1}. These coercivities are much too high for magnetic recording, but they can be reduced by the addition of CoO and TiO_2; thus $BaFe_{10.5}Co_{0.75}Ti_{0.75}O_{19}$ particles have a coercivity of about 72 kA m^{-1}, but there is no significant reduction in the saturation magnetic polarization.

The particles are prepared by the 'glass crystallization' method, in which a mixture of BaO, Fe_2O_3 and B_2O_3 is melted and then very rapidly cooled between steel rollers. The boron oxide provides a glassy matrix in which the $BaFe_{12}O_{19}$ platelets form, but is later removed with hot acetic acid. The resulting particles are very small, with diameters of about 0.08 μm and thicknesses of about 0.03 μm. The demagnetizing field due to the shape anisotropy is parallel to the hexagonal axis, and so assists the applied field during magnetization reversal. For a random assembly of uniaxial single domain particles the coercivity is

$$H_c = 0.48\left(\frac{2K_1}{\mu_0 M_s} - D_z M_s\right),$$

where D_z is the demagnetization factor parallel to the hexagonal axes of the particles. For $BaFe_{10.4}Co_{0.8}Ti_{0.8}O_{19}$ the experimental and theoretical values are in reasonable agreement.

There are three modes in which information can be recorded on magnetic tape: analogue in-plane, digital in-plane and digital perpendicular. For the latter mode materials with a sufficiently high magnetocrystalline anisotropy are required with $K_1 > \frac{1}{2}\mu_0 M_s^2$ so as to overcome the demagnetizing field perpendicular to the plane of the tape. Barium ferrite particles with modified composition and properties are quite suitable for this, and although they are not yet widely used for in-plane or perpendicular recording, they are used in magnetic security coding strips on tickets, bankers cards, credit cards and bank cheques.

A comparison of the magnetic recording properties of various particulate materials is shown in Table 4.14. Although iron nitride (Fe_4N) particles appear to have very promising magnetic recording properties and higher corrosion resistance than iron particles, the latter property was not observed in practical tapes.

4.22.4 Oxidized Iron Particles

The higher saturation magnetic polarizations and coercivities of acicular metal or alloy particles make them particularly attractive as magnetic recording media. If the anisotropy is predominantly due to shape, the coercivity should be proportional to the saturation magnetization, so that iron particles should have a coercivity about 2.5 times that of γ-Fe_2O_3. Commercially available acicular iron particles have coercivities in the range 88–136 kA m^{-1}, which are significantly higher than the expected value.

Acicular iron particles are usually produced by reducing acicular α-FeOOH at 375–450°C. Since the pure iron particles are highly reactive, they are stabilized and protected by coating with chromates or sodium oleate. The surface oxidation of the particles results of course in a reduction of the net saturation magnetization.

Further improvements in the coercivity and hysteresis loop squareness M_r/M_s can be achieved by heating the particles at about 700°C. Acicular lepidocrocite γ-FeOOH particles can also be used as the starting material.

Iron particles with coercivities of about 88 kA m^{-1} are now being used in high-quality audio tapes and 8 mm video tapes because of their high signal-to-noise ratio. Particles with lengths as small as 0.1 μm can be prepared with length-to-diameter ratios up to about 10, and coercivities as high as 160 kA m^{-1}. Since the coercivity is determined by shape anisotropy, the coercivity changes with temperature at the same rate as the saturation magnetization M_s, which is relatively small at room temperature.

As Bate (1986) has pointed out, the advantages of using magnetic tape are that the properties of the tape and the particles can be controlled and measured separately in order to optimize the final behaviour of the tape. Furthermore, there is at present a range of suitable particles (see Table 4.14) and this may well be extended with further research.

Particles also have the advantage that they can be highly oriented by a magnetic field or not, as required. Tapes and hard disks have highly oriented particles – hard disks are individually coated while they are spinning. Floppy disks, however, require randomly oriented particles because they are punched from a wide tape, thus giving a magnetically isotropic disk. The properties of various particulate media and their magnetic recording applications are given in Table 4.15.

The disadvantages of using coated magnetic tapes are that the magnetic particles represent only about 40% of the volume on the coating and only 20% on rigid disks. Thus the magnetic properties are diluted, with an accompanying reduction in recording performance. It is difficult to avoid particle agglomeration or to obtain a uniform switching field.

Reviews of magnetic recording materials have been published by Bate (1975, 1980, 1986), Sharrock (1989) and Köster and Arnoldussen (1990). The reader is also referred to the book edited by Mee and Daniel (1990) and that by Mallinson (1987).

A summary of the applications of ferrites is given in Table 4.16.

Table 4.14 Properties of magnetic recording materials (after Bate, 1986).

Property	γ-Fe_2O_3	Co–Fe_2O_3	CrO_2	$BaFe_{12}O_{19}$+Co, Ti	Fe	Fe_4N
J_s (T)	0.42	0.45	0.46	0.3–0.46	1.1–1.38	0.8–0.95
J_r/J_s	0.5	0.5–0.8	0.5	0.6–0.7	0.23–0.52	0.47–0.53
T_c (°C)	590	590	115–126	320	768	490
K_1 (J m^{-3})	-4.6×10^3	-5–100×10^3	2.5×10^4	3.3×10^5	4.4×10^4	—
H_c (kA m^{-1})	20–28	44–60	36–54	26–158	30–132	51–88
Switching-field distribution $\Delta H/H_c$	0.26–0.61	0.3–0.6	0.35–0.6	0.16–0.6	0.5–0.75	0.5–0.7
ρ (10^6 kg m^{-3})	4.6	4.8	4.88–4.95	5.28	5.8	5.8
Particle size (μm)[a]	l=0.3 d=0.06	l=0.3 d=0.06	l=0.5 d=0.05	Diameter=0.1 Thickness=0.02	l=0.3 d=0.06	l=0.5 d=0.07
Crystal structure	Cubic	Cubic	Tetragonal	Hexagonal	bcc	fcc
Magnetic structure	Ferrimagnetic	Ferrimagnetic	Ferromagnetic	Ferrimagnetic	Ferromagnetic	Ferromagnetic

[a]Particles are needle-shaped, apart from $BaFe_{12}O_{19}$, which are hexagonal platelets.

Table 4.15 Magnetic material, saturation remanence $M_r(\infty)$, coercivity H_c, switching-field distribution Δh_r and number of particles per unit volume, N, of various particulate magnetic recording media (after Table 3.9 of Köster and Arnoldussen, 1990).

Application	Material	$M_r(\infty)$ (kA m^{-1})	H_c (kA m^{-1})	Δh_r	N (10^3 μm^{-3})
Reel-to-reel audio tape	γ-Fe_2O_3	100–120	23–28	0.30–0.35	0.3
Audio tape:					
IEC I	γ-Fe_2O_3	120–140	27–32	0.25–0.35	0.6
IEC II	CrO_2	120–140	38–42	0.25–0.35	1.4
	γ-Fe_2O_3 + Co	120–140	45–52	0.25–0.35	0.6
IEC IV	Fe	230–260	80–95	0.30–0.37	3
Professional video tape	γ-Fe_2O_3	75	24	0.4	0.1
	CrO_2	110	42	0.3	1.5
	γ-Fe_2O_3 + Co	90	52	0.35	1
Home video tape	CrO_2	110	44–58	0.35	2
	γ-Fe_2O_3 + Co	105	52–74	0.35	2
	Fe	220	110–125	0.38	4
Instrumentation tape	γ-Fe_2O_3	90	27	0.35	0.6
	γ-Fe_2O_3 + Co	105	56	0.50	0.8
Computer tape	γ-Fe_2O_3	87	23	0.30	0.16
	γ-Fe_2 + Co	90	44	0.33	0.8
	CrO_2O_3	120	40	0.29	1.4
Flexible disk	γ-Fe_2O_3	56	27	0.34	0.3
	γ-Fe_2O_3 + Co	60	50	0.34	0.5
	Fe	200	120	0.34	2
Computer disk	γ-Fe_2O_3	56	26–30	0.30	0.3
	γ-Fe_2O_3 + Co	60	44–55	0.30	0.5

Table 4.16 Applications of ferrites.

Application	Ferrite
Inductor and transformer cores	Mn–Zn and Ni–Zn ferrites up to 300 MHz
Magnetic recording heads	Mn–Zn, Ni–Zn ferrites
Permanent magnets	$BaFe_{12}O_{19}$ and $SrFe_{12}O_{19}$
Microwave devices	Li, Ni, Ni–Al, Ni–Zn and Mn–Mg spinel ferrites; garnets such as YIG, GdIG, YAG; hexaferrites with an easy plane of magnetization (ferroxplana)
Information storage, magnetic recording including magneto-optical and bubble memories	Spinels such as γ-Fe_2O_3, Co-doped γ-Fe_2O_3; single-crystal orthoferrites and garnets; Mg–Mn, Cu–Mn, Li–Ni, Co–Mn ferrite cores
Electromechanical transducers	Spinels such as $CoFe_2O_4$
Magnetoelastic wave devices	Single-crystal garnets
Magneto-optical devices	Single-crystal garnets

4.23 MAGNETIC THIN FILMS FOR IN-PLANE MAGNETIC RECORDING

Most in-plane thin-film magnetic recording materials are based on Co–Ni alloys with additions of phosphorus, which is used to control the coercivity. The films are prepared by electroplating or by an electroless plating process. Apart from the presence of phosphorus, which increases the coercivity, the magnetic properties are also strongly affected by the pH of the electroplating solution and the film thickness.

Co–P and Co–Ni–P films can be produced with coercivities in the range 15–130 kA m^{-1}. The lowest coercivities are obtained with Co–P alloys with a negligibly small P content, while alloys with the highest coercivities have a high P content and a Co:Ni ratio of about 4:1. The high coercivities are believed to result from (a) a grain structure, in which the grains are randomly oriented, with a high in-plane magnetocrystalline anisotropy, and are partially decoupled from each other by non-ferromagnetic phosphorus-rich grain boundaries; and (b) the high stacking fault concentration, which results from a mixture of fcc Ni-rich grains and hcp Co-rich grains. This microstructure impedes domain wall nucleation, and provides pinning sites for the domain walls that are nucleated.

Many other additions apart from phosphorus are effective in segregating the grain boundaries in cobalt alloys. Plated Co–Ni–W and Co–Ni–Cr have similar magnetic properties to Co–Ni–P. Indeed, plating Co–Ni–P films onto Cr, W or Mo films improves the control of the magnetic properties.

There are a large number of potentially suitable alloys for thin-film magnetic recording, including Co–Pt and Co–Re (see Table 4.17), but the choice of an alloy for commercial application depends on several other factors such as corrosion and wear resistance.

Reviews of thin-film magnetic recording materials have been published by Arnoldussen (1986) and Howard (1986).

Table 4.17 Typical properties of magnetic film recording media (after Table 2 of Arnoldussen, 1986). Copyright 1986 IEEE.

Material	Deposition process[a]	Orientation[b]	Dominant crystal structure[c]	M_s (kA m^{-1})	H_c (kA m^{-1})	Measures of loop squareness S, S^*
Co	OIE	IPA	hcp	1100–1400	61–120	
	NIE, SP	IPI	hcp	1100–1400	30–60[d]	
Fe	OIE	IPA	bcc	1600	60–90	
	SP	IPI	bcc	1600	10	
Ni	OIE	IPA	fcc	400	20–28	
	SP	IPI	fcc	400		
Co–Ni	OIE	IPA	hcp/fcc	800–1200	30/70	
Co–Fe	OIE	IPA	bcc	1400–1600	60–120	0.9, 0.9
Co–Sm	NIE[e]	IPA	Amorphous	500–1000	33–55	1.1
Co–P	EL, EP	IPI	hcp	800–1100	36–96	0.9, 0.9
Co–Re	SP	IPI	hcp/fcc	500–750	18–58	0.9, 0.9
Co–Pt	SP	IPI	hcp/fcc	800–1400	→140	
Co–NiP	EI, EP	IPI	hcp/fcc	600–1000	40–120	0.8, 0.8
Co–(30 at%) Ni:N_2	SP	IPI	hcp/fcc	650	80	0.95
Co–Ni:O_2	OIE	IPA	hcp/fcc	300–400	80	0.7–0.8
Co–Ni–Pt	SP	IP*I	hcp/fcc	800–900	60–70	0.9, 0.97
Co–Ni–W	SP	IPI	hcp/fcc	450	30–50	0.8, 0.8
Fe_3O_4	NIE, SP	IPI	IS	400	17–32	
γ-Fe_2O_3:Co	SP	IPI	IS	200–250	40–100	0.8, 0.8
γ-Fe_2O_3:Os	SP	IPI	IS	240	160	0.8, 0.8
Co–(18 at%)Cr	SP	⊥	hcp	300–550	80–100(⊥)	
Co–(20 at%)Cr	SP	⊥	hcp	400	65–95(⊥)	
Co–(22 at%)Cr	SP	⊥	hcp	300–340	80–105(⊥)	

[a] OIE, oblique-incidence evaporation; NIE, normal-incidence evaporation; SP, sputtered; EL, electroless-plated; EP, electroplated.
[b] IPA, in-plane anisotropic; IPI, in-plane isotropic; ⊥, perpendicular.
[c] hcp, hexagonal close-packed; fcc, face-centred cubic; bcc, body-centred cubic; IS, inverse spinel.
[d] Coercivities for Co films deposited on Cr or W underlayers.
[e] Co–Sm values for films deposited in the presence of an orienting field.

4.24 Co–Cr FILMS FOR PERPENDICULAR MAGNETIC RECORDING

Co–Cr films with uniaxial anisotropy and the easy axis of magnetization perpendicular to the plane of the film can be prepared by sputtering onto a polyamide or other substrate. Although cobalt itself has a high uniaxial magnetocrystalline anisotropy, it is necessary to add the chromium in order to reduce the saturation magnetization so that the anisotropy field $H_A > M_s$, i.e.

$$\frac{2K_1}{\mu_0 M_s} > M_s,$$

or

$$K_1 > \tfrac{1}{2}\mu_0 M_s^2,$$

assuming a maximum demagnetizing factor $D_z = 1$. If this condition is satisfied, the Co–Cr film can sustain a stable magnetization perpendicular to the plane of the film. The relatively high saturation magnetizations and coercivities that can be achieved make the alloy films very suitable for perpendicular recording, because a high recording resolution and output voltage can be obtained. Furthermore, the films have very good mechanical and chemical stabilities, with reproducible magnetic properties. They also have good adhesion to the substrate.

The magnetic properties are strongly dependent on the sputtering time and conditions, i.e. the temperature and argon pressure. The saturation magnetization decreases approximately linearly with increasing chromium content, and is zero at about 25 at%. The films have saturation magnetic polarizations of about 0.25–0.5 T.

From the hysteresis and torque curves, it can be shown that the columnar microstructures of the films have a uniaxial magnetocrystalline anisotropy, with the easy axis perpendicular to the plane of the film. The effective induced anisotropy coefficient K_u is therefore given by

$$K_u = K_i - \tfrac{1}{2}\mu_0 M_s^2,$$

where K_i is the intrinsic anisotropy coefficient and $\tfrac{1}{2}\mu_0 M_s^2$ is the shape anisotropy energy for magnetization perpendicular to the plane of the very thin film. For a film of thickness 1 μm and $M_s \approx 3 \times 10^5$ A m^{-1}, K_u is typically about 5×10^4 J m^{-3}, so that K_1 is certainly greater than $\tfrac{1}{2}\mu_0 M_s^2$, a condition that must be satisfied if the film is to be magnetized perpendicular to its plane in order to store high-density digital information of about 4000 bits mm^{-1}.

The alignment of the c axis in the columnar structure shows some dispersion, but can be reduced by depositing the Co–Cr onto a thin film of titanium. For further details see the reviews by Arnoldussen (1986) and Howard (1986).

REFERENCES

Arnoldussen, T. C. (1986). *Proc. IEEE* **74**, 1526.

Bate, G. (1975). *Magnetic Oxides*, Part 2, ed. D. J. Craik (Wiley, London), p .689.

Bate, G. (1980). *Ferromagnetic Materials: A Handbook on the Properties of Magnetically Ordered Substances*, Vol. 2, ed. E. P. Wohlfarth (North-Holland, Amsterdam), p. 2381.

Bate, G. (1986). *Proc. IEEE* **74**, 1513.

Beer, A. and Schwartz, T. (1966). *IEEE Trans. Magn.* **MAG-2**, 470.

Bobeck, A. H. (1967). *Bell Syst. Tech. J.* **46**, 1901.

Bobeck, A. H., Spencer, E. G., Van Uitert, L. G., Abrahams, S. C., Barns, R. L., Grodkiewicz, W. H., Sherwood, R. C., Schmidt, P. H., Smith, D. H. and Walters, E. M. (1970). *Appl. Phys. Lett.* **17**, 131.

Bobeck, A. H. (1975). *Magnetic Oxides*, Part 2, ed. F. J. Craik (Wiley, London), p. 743.

Bobeck, A. H. and Della-Torre, E. (1975). *Magnetic Bubbles* (North-Holland, Amsterdam).

Chol, G. (1971). *J. Am. Ceram. Soc.* **54**, 34.

Darnell, F. J. (1961). *J. Appl. Phys.* **32**, 1269.

Darnell, F. J. and Cloud, W. H. (1965). *Bull. Soc. Chim. Fr.* **4**, 1164.

Eschenfelder, A. H. (1980). *Ferromagnetic Materials: A Handbook on the Properties of Magnetically Ordered Substances*, Vol. 2, ed. E. P. Wohlfarth (North-Holland, Amsterdam), p. 297.

Eschenfelder, A. H. (1981). *Magnetic Bubble Technology*, 2nd edn (Springer-Verlag, Berlin).

Geller, S. (1956). *J. Chem. Phys.* **24**, 1236.

Geller, S. and Wood, E. A. (1956). *Acta Cryst.* **9**, 563.

Gilleo, M. A. (1980). *Ferromagnetic Materials: A Handbook on the Properties of Magnetically Ordered Substances*, Vol. 2, ed. E. P. Wohlfarth (North-Holland, Amsterdam), p. 1.

Gorodetsky, G. and Treves, D. (1965). *Proc. Int. Conf. Magnetism, Nottingham, 1964* (Int. Phys. and Phys. Soc. Lond.), p. 606.

Gorter, E. W. (1954). *Philips Res. Rep.* **9**, 295, 321, 403.

Gorter, E. W. and Schulkes, J. A. (1953). *Phys. Rev.* **90**, 487.

Grundy, P. J., Hothersall, D. C., Jones, G. A., Middleton, B. and Tebble, R. S. (1972). *Phys. Stat. Sol.* (a), **9**, 79.

Guillaud, C. (1951). *J. Phys. (Paris)* **12**, 239.

Heck, C. (1974). *Magnetic Materials and Their Applications* (Butterworths, London).

Hidaka, Y., Matsuyama, K. and Konishi, S. (1983). *IEEE Trans* **MAG-19**, 1841.

Howard, J. K. (1986). *J. Vac. Sci. Tech.* **A4**, 1.

Jacobs, I. S. and Bean, C. P. (1955). *Phys. Rev.* **100**, 1060.

Johnson, C. E. and Brown, W. F. (1958). *J. Appl. Phys.* **29**, 1699.

Jonker, G. H., Wijn, H. P. J. and Braun, P. B. (1956). *Philips Tech. Rev.* **18**, 145.

Jouve, H. (ed.) (1986). *Magnetic Bubbles* (Academic Press, London).

Kato, Y. and Takei, T. (1933). *J. Inst. Elec. Engrs Japan* **53**, 408.

Kishimoto, M., Amemiya, M. and Hayama, F. (1984). *J. Appl. Phys.* **55**, 2272.

Kittel, C. (1956). *Introduction to Solid State Physics*, 2nd edn (Wiley, New York), p. 447.

Kittel, C. (1986). *Introduction to Solid State Physics*, 6th edn (Wiley, New York), p. 475.

Knowles, J. E. (1970). *J. Phys. D: Appl. Phys.* **3**, 1346.

Knowles, J. E. (1984). *IEEE Trans. Magn.* **MAG-20**, 84.

Kojima, H. (1980). *Ferromagnetic Materials: A Handbook on the Properties of Magnetically Ordered Substances*, Vol. 3, ed. E. P. Wohlfarth (North-Holland, Amsterdam), p. 305.

Konishi, S. (1983). *IEEE Trans. Magn.* **MAG-19**, 1838.

Köster, E. and Arnoldussen, T. C. (1990). *Magnetic Recording Handbook*, ed. C. D. Mee and E. D. Daniel (McGraw-Hill, New York).

Krupička, S. and Novák, P. (1982). *Ferromagnetic Materials: A Handbook on the Properties of Magnetically Ordered Substances*, Vol. 3, ed. E. P. Wohlfarth (North-Holland, Amsterdam), p. 189.

Kullmann, U., Köster, E. and Meyer, B. (1984). *IEEE Trans. Magn.* **MAG-20**, 742.

Lax, B. and Button, K. J. (1962). *Microwave Ferrites and Ferrimagnetic Materials* (McGraw-Hill, New York).

Luborsky, F. E. (1961). *J. Appl. Phys.* **32**, 171S.
McGuire, R. T. (1955). *Phys. Rev.* **97**, 831.
Mallinson, J. C. (1987). *The Foundations of Magnetic Recording* (Academic Press, London).
Malozemoff, A. P. and Slonczewski, J. C. (1979). *Magnetic Domain Walls in Bubble Materials: Applied Solid State Science*, Suppl. I (Academic Press, New York).
Maxwell, L. R. and Pickart, S. J. (1953). *Phys. Rev.* **92**, 1120.
Mee, C. D. and Daniel, E. D. (eds) (1990). *Magnetic Recording Handbook* (McGraw-Hill, New York).
Morrish, A. H. (1965). *The Physical Principles of Magnetism* (Wiley, New York).
Morrish, A. H. and Yu, S. P. (1955). *J. Appl. Phys.* **26**, 1049.
Néel, L. (1947a). *C. R. Acad. Sci. Paris* **224**, 1488.
Néel, L. (1947b). *C. R. Acad. Sci. Paris* **224**, 1550.
Néel, L. (1948). *Ann. de Phys.* **3**, 137.
Nicolas, J. (1980). *Ferromagnetic Materials: A Handbook on the Properties of Magnetically Ordered Substances*, Vol. 2, ed. E. P. Wohlfarth (North-Holland, Amsterdam), p. 243.
O'Dell, T. H. (1974). *Magnetic Bubbles* (Macmillan, London).
Ohta, K. (1963). *J. Phys. Soc. Japan* **18**, 685.
Parker, R. (1975). *Magnetic Oxides*, Part 1, ed. D. J. Craik (Wiley, London), p. 421.
Pauthenet, R. (1956). *C. R. Acad. Sci. Paris* **243**, 263.
Perduijin, D. J. and Pelozchek, H. P. (1968). *Proc. Br. Ceram. Soc.* **10**, 263.
Perneski, A. J. (1969). *IEEE Trans. Magn.* **MAG-5**, 554.
Rodbell, D. S., DeVries, R. C., Barber, W. D. and DeBlois, R. W. (1967). *J. Appl. Phys.* **38**, 4542.
Roess, E. (1966). *Z. Angew. Phys.* **21**, 391.
Sharrock, M. P. (1989). *IEEE Trans. Magn.* **MAG-25**, 4374.
Slick, P. I. (1980). *Ferromagnetic Materials: A Handbook on the Properties of Magnetically Ordered Substances*, Vol. 2, ed. E. P. Wohlfarth (North-Holland, Amsterdam), p. 189.
Smit, J. and Wijn, H. P. J. (1954). *Adv. Elec. Elec. Phys.* **6**, 69.
Smit, J. and Wijn, H. P. J. (1959). *Ferrites* (Philips Technical Library, Eindhoven).
Smith, A. B. (1974). *Bubble Domain Memory Devices* (Artech, Dedham, MA).
Snelling, E. C. (1974). *IEEE Trans. Magn.* **MAG-10**, 616.
Snelling, E. C. (1988). *Soft Ferrites: Properties and Applications* (Butterworths, London).
Speliotis, D. E. (1968). *IEEE Trans. Magn.* **MAG-4**, 53.
Standley, K. J. (1972). *Oxide Magnetic Materials*, 2nd edn (Clarendon Press, Oxford).
Stoner, E. C. and Wohlfarth, E. P. (1948). *Phil. Trans. R. Soc. Lond.* **A240**, 599.
Sugimoto, M. (1980). *Ferromagnetic Materials: A Handbook on the Properties of Magnetically Ordered Substances*, Vol. 3, ed. E. P. Wohlfarth (North-Holland, Amsterdam), p. 393.
Suzuki, R. (1986). *Proc. IEEE* **74**, 1582.
Swoboda, T. J., Arthur, P., Cox, N. L., Ingraham, J. N., Oppergard, A. L. and Sadler, M. S. (1961). *J. Appl. Phys.* **32S**, 374.
Takei, H. and Chiba, S. (1966). *J. Phys. Soc. Japan* **21**, 1255.
Tebble, R. S. and Craik, D. J. (1969). *Magnetic Materials* (Wiley-Interscience, London).
Von Aulock, W. H. (1965). *Handbook of Microwave Ferrite Materials* (Academic Press, New York).
Wagner, U. (1977). *J. Magn. Magn. Mater.* **4**, 166.
Wagner, U. (1981). *J. Magn. Magn. Mater.* **23**, 73.
Williams, H. J., Sherwood, R. C., Goertz, M. and Schnettler, F. J. (1953). *AIEE Trans. Comm. Elec.* **72**, 531.
Winkler, G. (1981). *Magnetic Garnets* (Vieweg, Braunschweig/Wiesbaden).
Yafet, Y. and Kittel, C. (1952). *Phys. Rev.* **87**, 290.

Chapter 5

Permanent Magnet Materials

5.1 INTRODUCTION

Permanent magnet materials, or hard magnetic materials, are characterized by high coercivities and high remanent magnetizations, so that their hysteresis loops enclose the largest possible areas. Since permanent magnets form a large and very important group of materials, their structure and properties will be considered in detail, although the discussion will be concentrated on materials that are of present or potential technological value. The special technological importance of permanent magnets is their ability to produce attractive or repulsive forces without contact, to act through a vacuum or any non-ferromagnetic intervening material, and to provide a magnetic flux – all without the expenditure of energy and hence at no operating cost. They have a considerable advantage over electromagnets in that they can be very small and hence are valuable in applications where miniaturization is important. The fact that permanent magnets can provide a magnetic flux without a power supply means that they are very reliable, and also avoid the necessity for heat dissipation. The efficiency of permanent magnets as determined by the constancy and reliability of the external magnetic flux they provide is of course totally dependent on their resistance to demagnetization in service when subjected to opposing fields, to their own demagnetizing fields and to temperature changes. The resistance to demagnetization is referred to as magnetic hardness, and is determined by the magnitude of the flux coercivity ${}_{B}H_{C}$, i.e. the reverse field required to reduce the flux density in the magnet to zero.

The principal functions of a permanent magnet are to supply magnetic flux or magnetomotive force, i.e. a magnetic potential difference in a circuit. An example of the first function is the use of a permanent magnet in a loudspeaker to provide a flux round the speaker coil, which experiences a mechanical force when a current from the amplifier flows through the coil. When a permanent magnet is used in any kind of holding or attracting device, the soft magnetic 'keeper' or yoke is magnetized by the flux provided by the permanent magnet, so that an attractive force is set up between the magnet and the yoke. Sometimes permanent magnets for producing magnetic flux are also used in combination with a soft magnetic yoke, which acts as a flux enhancer or concentrator, and for this reason the yoke is usually made of a material with a high saturation magnetic polarization – ideally in combination with high permeability. In this case the yoke is also magnetized by the flux from the permanent magnet, or, strictly speaking, by the magnetomotive force. These two important properties of permanent magnets enable them to be used in a very wide range of electrical, magnetic, electromagnetic and mechanical devices.

Thus the most important property of a permanent magnet is that its magnetization should be as large and as stable as possible and remain so even when it is subjected to demagnetizing fields or temperature changes. A typical B–H loop for a permanent magnet is shown in Fig. 5.1(a). Since the flux density B inside the material is the sum of the flux due to the magnetization – and the internal field H,

$$B=\mu_0(M+H).$$

We recall from Chapter 1 that the flux coercivity ${}_BH_c$ is defined as the reverse field required to reduce the flux density B to zero, so that

$$B=0=\mu_0(M+H);$$

that is,

$$B=0 \quad \text{when} \quad {}_BH_c=-M.$$

Thus the magnetization of the material is finite at the coercive field point. Furthermore, it is important to realize that the flux density B continues to increase even after the material has reached its saturation magnetic polarization. This arises because there is

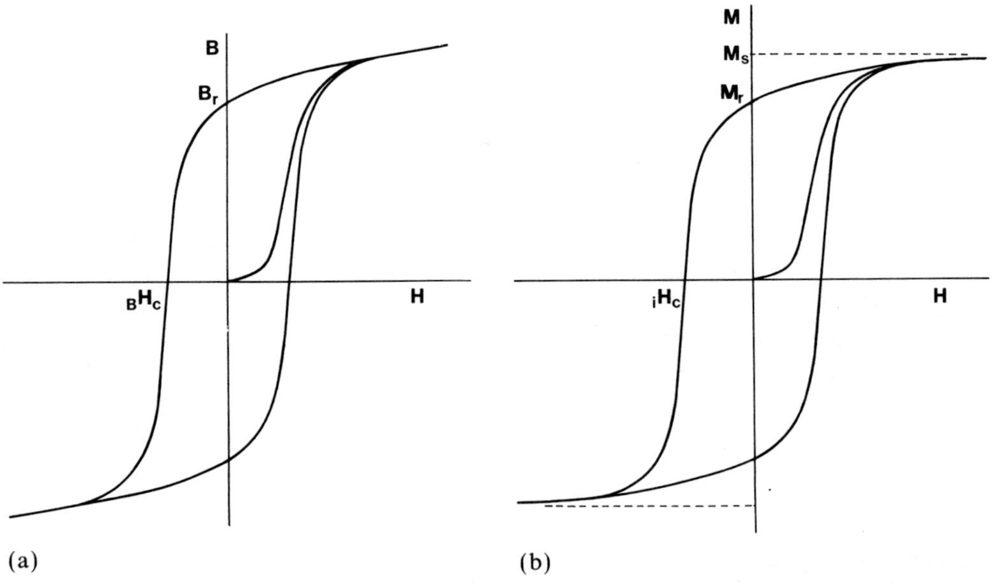

Fig. 5.1 (a) A typical B–H loop for a permanent magnet material. Note that the flux density B does not saturate, but continues to increase as H increases. (b) A typical intrinsic or M–H loop for a permanent magnet material, showing saturation magnetization M_s in high fields.

still a contribution to B due to the increase in the applied field H. For comparison, a typical intrinsic or $M-H$ loop is shown in Fig. 5.1(b). Note that in this case the intrinsic coercivity ${}_iH_c$ is defined as the reverse field required to reduce the magnetization to zero in the presence of the reverse field. M_r is the remanent magnetization which is usually significantly less than the saturation magnetization M_s.

5.2 DISTRIBUTION OF THE FIELDS *B, H* AND *M* IN A TOROIDAL PERMANENT MAGNET WITH AN AIR GAP

Consider a permanent magnet material in the form of a toroid, a section of which has been removed to leave an air gap of length l_g and a length l_m of the permanent magnet material as shown in Fig. 5.2. The presence of the air gap produces an internal demagnetizing field H_D, which is of course in the opposite direction to the magnetization M and the net internal flux B. The magnitude of the demagnetizing field depends on the ratio l_g/l_m and the geometry of the specimen, although in this case the demagnetizing field H_D is given approximately by

$$H_D = \frac{l_g}{l_m} M.$$

5.3 ENERGY PRODUCT OF A PERMANENT MAGNET MATERIAL

Consider a permanent magnet material in the form of a toroidal ring as shown in Fig. 5.2. Let the total length of the permanently magnetized material be l_m and that of the air gap l_g. Since there are no sources of electrical current around the ring, the magnetomotive force acting round the ring is zero (Ampère's circuital law), i.e.

$$\oint \boldsymbol{H} \cdot \mathrm{d}\boldsymbol{l} = 0;$$

hence

$$H_m l_m + H_g l_g = 0, \tag{5.1}$$

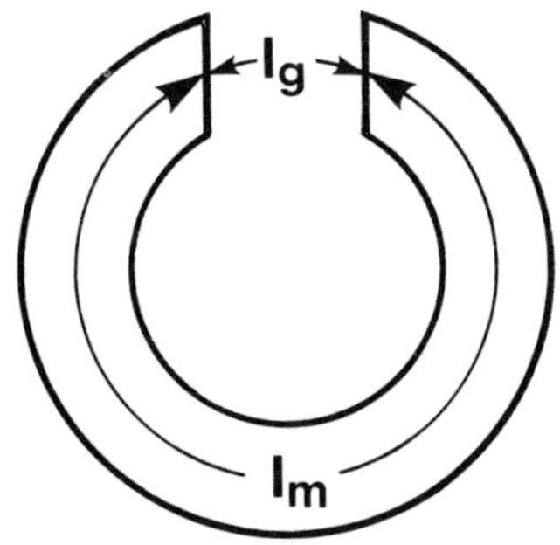

Fig. 5.2 Permanent magnet material in the form of a toroid from which a section has been removed to leave an air gap of length l_g.

where H_m and H_g are the magnetic fields inside and outside the material respectively. Let the cross-sectional areas of the material and the air gap be A_m and A_g respectively. Since the magnetic flux is continuous across the boundaries between materials,

$$\Phi_m = \Phi_g,$$

$$B_m A_m = B_g A_g,$$

$$B_m A_m = \mu_0 H_g A_g, \tag{5.2}$$

assuming that the permeability of air is equal to that of free space ($\mu_0 = 4\pi \times 10^{-7}$ H m^{-1}). If (5.2) is multiplied by H_g, we obtain

$$B_m A_m H_g = \mu_0 H_g^2 A_g$$

or

$$H_g^2 = B_m H_g A_m / \mu_0 A_g. \tag{5.3}$$

From (5.1),

$$H_g = -H_m l_m / l_g,$$

and on substitution for H_g in (5.3), we obtain (ignoring sign)

$$H_g^2 = \frac{B_m H_m V_m}{\mu_0 V_g}, \tag{5.4}$$

where V_m is the volume of the material and V_g is the volume of the air gap. Thus in order to produce a given field H_g in the air gap, the volume of the material required will be a minimum when the product $B_m H_m$ is a maximum. This can readily be appreciated by rearranging (5.4) as

$$(V_m)_{min} = \frac{\mu_0 H_g^2 V_g}{(B_m H_m)_{max}}.$$

Since H_m in the material is in the opposite direction to B_m, the product $B_m H_m$ represents the maximum BH product in the demagnetizing quadrant of the B–H loop for the material. The demagnetizing quadrant and the variation of the BH product for a typical permanent magnet material are shown in Fig. 5.3. It is important to appreciate the fact that the field H is the effective field in the specimen, and is produced by an externally applied field. The maximum value of the product BH, namely $(BH)_{max}$, is one of the most important parameters in assessing the quality of a permanent magnet material, and is equal to the area of the largest rectangle or square that can be inscribed in the demagnetizing quadrant of the B–H loop. The value $(BH)_{max}$ is closely related to the magnitude of the remanent flux density B_r and the flux coercivity $_BH_c$, but it also depends on the exact shape of the demagnetization curve. The product $(BH)_{max}$ is sometimes

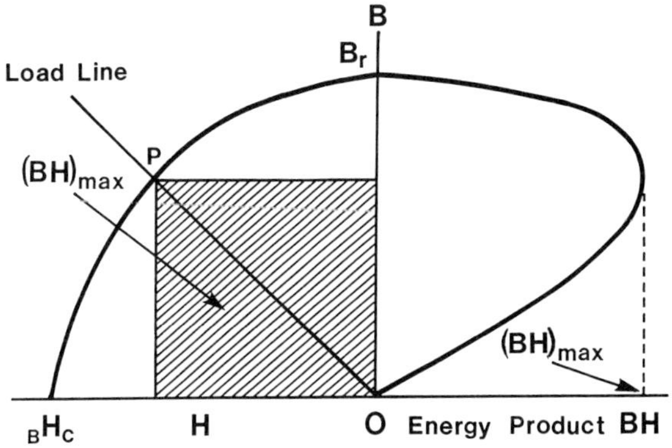

Fig. 5.3 The demagnetizing quadrant of a typical permanent magnet material and the variation of the energy product BH as a function of the demagnetizing field.

referred to as the maximum energy product, since its dimensions are J m^{-3}, but, strictly speaking, the maximum energy stored in a permament magnet material in the absence of any externally applied field is half $(BH)_{max}$, i.e.

$$E_{max} = \tfrac{1}{2}(BH)_{max}.$$

The factor $\frac{1}{2}$ appears because this is a self-energy, so that the magnetization of the material is both a source of magnetic flux and magnetic field. This can readily be understood as follows. When the demagnetization curve is measured, the specimen is usually in a form for which the demagnetizing field due to its own magnetization is negligibly small, so that changes in B are produced by the external field. In this case the energy of the permanent magnet is BH, and is a maximum when the area of the inscribed rectangle is a maximum, as shown in Fig. 5.3. However, when the permament magnet is being used as such with an air gap, the demagnetizing field inside the material is produced by, and is in the opposite direction to, its own magnetization M. The flux density in the material B is

$$B = \mu_0(M - H_D),$$

and the demagnetizing field

$$H_D = DM,$$

where D is the demagnetizing factor, which depends on the geometry and l_g/l_m, so that the magnetization is both a source of flux and field. Thus we can regard the flux as being produced by half the magnetization, and the demagnetizing field by the other half. The factor of $\frac{1}{2}$ appears so that the effects of the magnetization should not be counted twice. It should be emphasized that when measurements of B are being made in the presence of an externally applied field, the energy of the material is of course BH (J m^{-3}). Although permanent magnets are nearly always used in the presence of their own demagnetizing fields, there are applications (e.g. fail-safe lifting devices) where external fields are applied.

5.3.1 Working Point and Load Line

From (5.2), $B_m = \mu_0 H_g A_g / A_m$, and from (5.1), $H_g = -H_m l_m / l_g$, so substituting for H_g, we obtain

$$B_m = \mu_0 H_m l_m A_g / l_g A_m$$

(note that H_m is a negative field in the second quadrant). If we assume that $A_g = A_m$ then

$$B_m = \mu_0 H_m l_m / l_g. \tag{5.5}$$

Thus a graph of B_m versus H_m is a straight line with negative slope $\mu_0 l_m / l_g$; this line is the load line (Fig. 5.3), and its intersection at the point P with the demagnetizing B–H curve gives the working point of the magnet. Thus we see that the working point depends on the ratio of the length l_m of the magnetic material to the length l_g of the air gap in which the field H_g is required. Ideally, a permanent magnet (usually together with its soft iron yoke) is designed so that it operates at a point on the demagnetization curve that corresponds to maximum energy product and thus minimizes the volume of permanent magnet material required to produce a given magnetic field H_g in the air gap. In practice, allowance has to be made for the magnetic reluctance of the magnet and soft iron (or other high-permeability yoke) by means of a reluctance factor α, so that

$$H_g l_g = -\alpha H_m l_m, \qquad \alpha \leqslant 1.$$

Allowance also has to be made for flux leakage from the magnet system, particularly at the air gap, where the magnetic flux will spread significantly beyond the geometrical volume V_g of the latter; thus for a flux leakage factor β,

$$B_g A_g = -\beta B_m A_m.$$

For good designs $\alpha \approx 0.7$–0.95 and $\beta \approx 0.1$–0.8.

5.3.2 Theoretical Maximum Energy Product $(BH)_{max}$

Consider an ideal permanent magnet material with the rectangular hysteresis loop shown in Fig. 5.4 for which the magnetization is constant and equal to the saturation magnetization M_s unless a critical reverse field equal to the intrinsic coercivity ${}_iH_c$ is applied, after which the material is magnetized to saturation in the opposite direction. Since $\boldsymbol{B} = \mu_0(\boldsymbol{M} + \boldsymbol{H})$, the B–H loop is therefore a parallelogram with $B_r = \mu_0 M_s = J_s$ as shown in Fig. 5.5. If we now consider the demagnetizing quadrant, the flux density B inside the material is the magnitude of the vector sum of $\boldsymbol{M}_s$ and $\boldsymbol{H}$, i.e.

$$B = \mu_0(M_s - H),$$

and hence

$$BH = \mu_0(M_s - H)H. \tag{5.6}$$

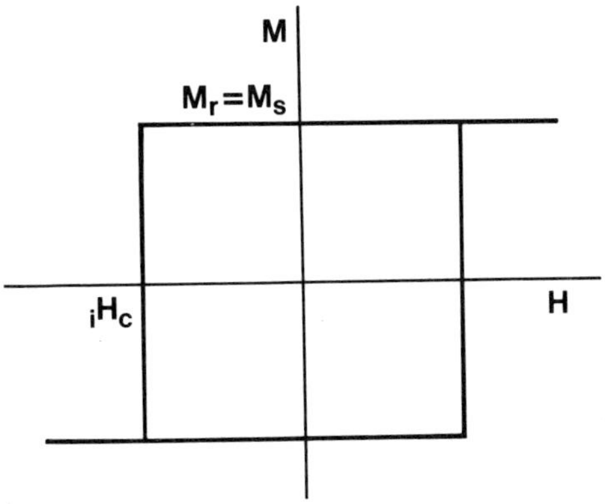

Fig. 5.4 An ideal permanent magnet material with a rectangular M–H hysteresis loop, with $M_r = M_s = J_s/\mu_0$.

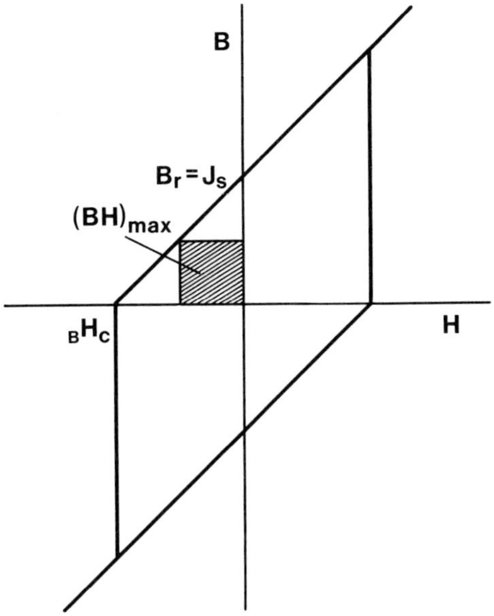

Fig. 5.5 The ideal B–H hysteresis loop for the permanent magnet material shown in Fig. 5.4. Note that in this example the flux coercivity ${}_BH_c$ is equal to the intrinsic coercivity iH_c.

For a maximum in BH

$$\frac{d(BH)}{dH} = 0 \quad \text{and} \quad \frac{d^2(BH)}{dH^2} < 0.$$

Therefore

$$\frac{d(BH)}{dH} = \mu_0 M_s - 2\mu_0 H = 0,$$

giving

$$H = \tfrac{1}{2} M_s,$$

and

$$\frac{d^2(BH)}{dH^2} = -2\mu_0 < 0.$$

Thus $H = \frac{1}{2} M_s$ at the maximum value of BH, so that after substituting in (5.6), we find

$$(BH)_{max} = \tfrac{1}{2}\mu_0 M_s^2 - \tfrac{1}{4}\mu_0 M_s^2;$$

that is,

$$(BH)_{max} = \tfrac{1}{4}\mu_0 M_s^2 = \frac{B_r^2}{4\mu_0},$$

since $B_r = \mu_0 M_s$.

This absolute maximum can be realized only if the flux coercivity ${}_BH_c \geqslant \tfrac{1}{2}M_s$, in which case $(BH)_{max}$ occurs at a flux density of $\tfrac{1}{2}B_r$ in a reverse field $\tfrac{1}{2}M_s$ as shown by the shaded area in Fig. 5.5. This absolute maximum can still be obtained even if the flux reversal occurs when B is still positive as shown in Fig. 5.6(a), provided that ${}_BH_c \geqslant \tfrac{1}{2}M_s$. Thus the ultimate maximum energy product of a permanent magnet material depends on the saturation magnetization M_s and the flux coercivity ${}_BH_c$, and therefore, ideally, both of these should be as large as possible. For an ideal permament magnet with an intrinsic coercivity ${}_iH_c > M_s$ flux reversal will occur only when B is negative as shown in Fig. 5.6(b).

For most permanent magnet materials the magnetization decreases with increase in the reverse field, so that the demagnetizing B–H curve is not linear with H, and the $(BH)_{max}$ value is therefore less than the theoretical maximum. In general, B_r is also less than $\mu_0 M_s$, i.e. $B_r = \mu_0 M_r$, where M_r is the remanent magnetization, so that the theoretical $(BH)_{max}$ product is reduced to $\tfrac{1}{4}\mu_0 M_r^2$. Since the flux density B decreases very rapidly with

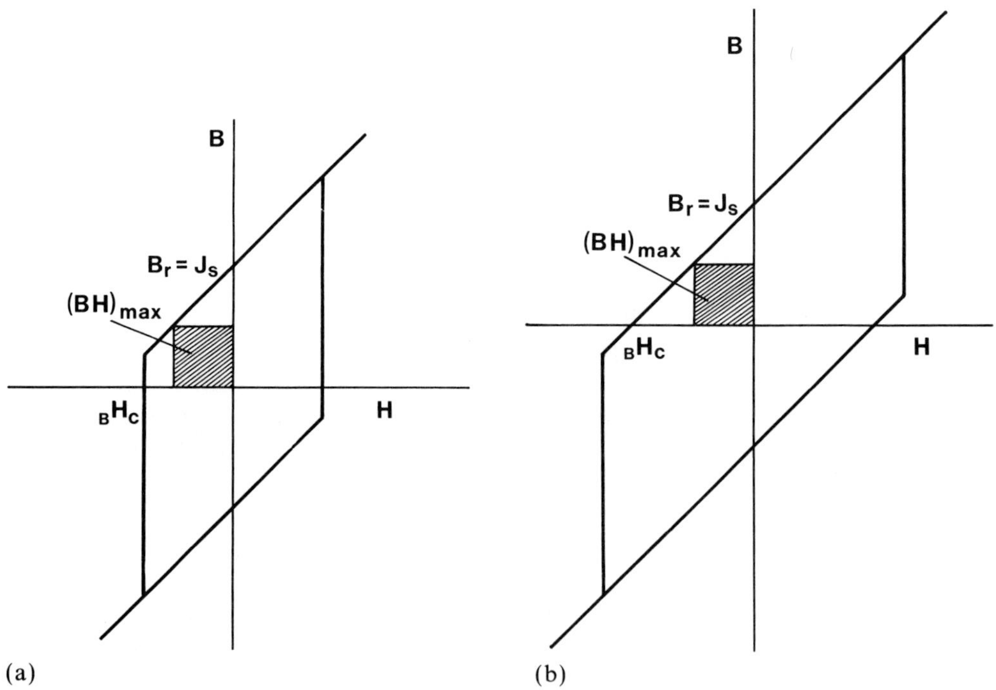

Fig. 5.6 (a) An ideal B–H hysteresis loop for which flux reversal occurs when B is still positive. (b) An ideal B–H hysteresis loop for which flux reversal occurs when B is negative; this behaviour is observed in materials for which the intrinsic coercivity $H_c > M_r$.

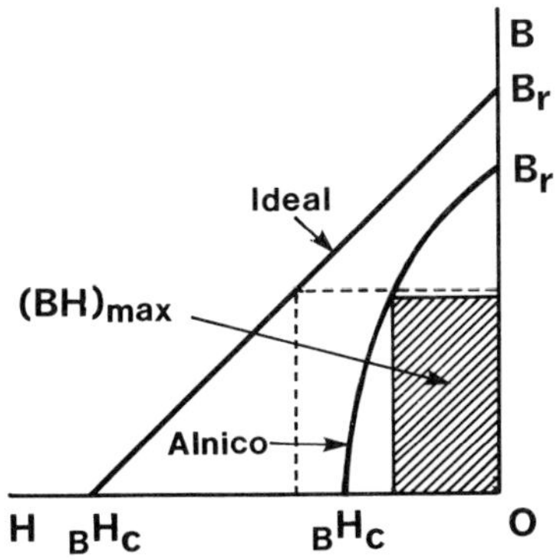

Fig. 5.7 Comparison of the demagnetizing quadrants of an ideal permanent magnet material and a commercial Alnico permanent magnet with the same saturation magnetic polarization.

reverse field, the $(BH)_{max}$ value is considerably less than $\frac{1}{4}\mu_0 M_r^2$, particularly if the flux coercitivy $_BH_c$ is less than $\frac{1}{2}M_s$, as is generally the case for many commercially important permanent magnet materials. A schematic comparison of the demagnetizing quadrants of an ideal permanent magnet material and an Alnico (Fe–Co–Ni–Al alloy) magnet with the same saturation magnetization M_s is given in Fig. 5.7, which clearly shows that the failure of the Alnico to realize the theoretical $(BH)_{max}$ is due to three causes:

(1) $B_r \neq \mu_0 M_s$;
(2) the B–H curve is nonlinear;
(3) $_BH_c < \frac{1}{2}M_r$ or $< \frac{1}{2}M_s$.

As we shall see later, there are several permanent magnet materials with maximum energy products that do approach the theoretical maximum – one of these is the sintered intermetallic compound $SmCo_5$, for which the $(BH)_{max}$ value is about 80% of the theoretical maximum.

5.4 BEHAVIOUR OF PERMANENT MAGNETS UNDER DYNAMIC OR RECOIL CONDITIONS

We have seen in Section 5.3 that when a permanent magnet is required in applications involving the conversion of electrical into mechanical energy, e.g. in small electric motors, dynamos, loudspeakers and eddy current brakes, it is the location of the working point on the demagnetizing curve that is of the greatest importance in determining the efficiency of the machine. Since in all of the above applications the function of the permanent magnet is to provide magnetic flux in an air gap, the efficiency is maximized if the working point is at or near the $(BH)_{max}$ point. However, there are many applications where the permanent magnet is required to produce a mechanical force in a soft ferromagnetic material, e.g. in magnetic chucks, clamps, couplings and clutches. In these applications we are concerned with the behaviour of the permanent magnet when the working point varies owing to changes in the demagnetizing field. When permanent magnets are used with varying working points, they are said to be operating under dynamic or recoil conditions.

If the application involves cyclical movement of the working point, it is essential that the flux density in the material is at least approximately reversible, so this imposes another requirement on the properties of the magnet. When the working point of a permanent magnet changes, there is an accompanying change in its magnetic free energy, which may be converted into mechanical work. Since, in the above-mentioned applications, the permanent magnets are generally used in conjunction with magnetizable materials that are capable of relative movement, a decrease in the magnetic free energy produces mechanical work, and conversely the expenditure of mechanical work produces an increase in the magnetic free energy. If the two energies are exactly equal then the permanent magnet is operating reversibly. Consider the demagnetizing quadrant of a permanent magnet shown in Fig. 5.8(a), and assume that initially the demagnetizing field of the air gap corresponds to the working point at P, and that the magnetizable material is at a considerable distance from the gap. If there is a slight change in this configuration by moving the magnetizable material closer to the magnet, the flux density and the field in the permanent magnet will be changed by the amounts dB and dH respectively. This change is accompanied by movement of the working point from P to Q as shown in Fig. 5.8(a), so that there is a change in the magnetic free energy of the magnet given by

$$\mathrm{d}F=\tfrac{1}{2}\int(\boldsymbol{H}\cdot\mathrm{d}\boldsymbol{B}-\boldsymbol{B}\cdot\mathrm{d}\boldsymbol{H})\,\mathrm{d}V. \tag{5.7}$$

By elementary geometry, it can easily be shown that this energy is given by the area of the sector OPQ, assuming that the working point moves reversibly along the demagnetizing curve. Consider the demagnetizing curve shown in Fig. 5.8(a). Suppose that the working point moves along the curve from P to Q, so that there is an infinitesimal increase in the flux density δB accompanied by a decrease in the demagnetizing field δH. The change in energy of the magnet is given by (5.7), which is equivalent to half the shaded area shown in Fig. 5.8(a). Since OPC is on the same base (i.e. PC) as APC

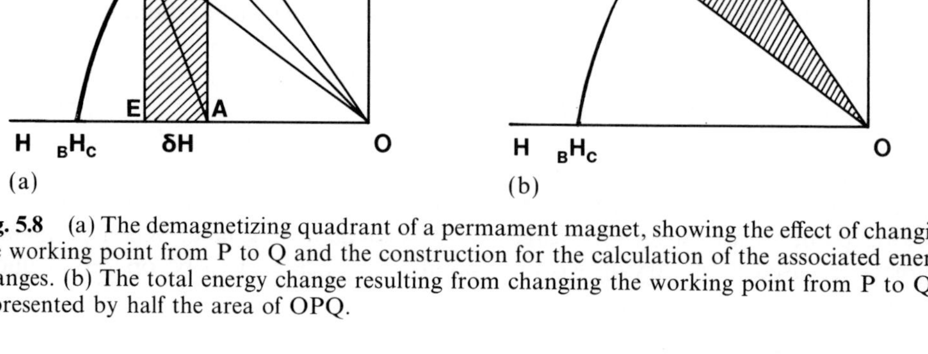

Fig. 5.8 (a) The demagnetizing quadrant of a permament magnet, showing the effect of changing the working point from P to Q and the construction for the calculation of the associated energy changes. (b) The total energy change resulting from changing the working point from P to Q is represented by half the area of OPQ.

and is of equal vertical height AC, it has the same area, which is also equal to half the area of the rectangle PCAE. Similarly, OQC has the same area as DCQ. Since the area of the sector OPQ is equal to the sum of the areas of OPC and OQC plus the area of PCQ, we can see immediately that the recoil energy of the magnet given by (5.7) is just equal to half the area of the sector OPQ, as shown in Fig. 5.8(b). Note that in evaluating (5.7) the signs of the changes in B and H and the signs of the quantitites themselves must be taken into account.

From Fig. 5.8(b), it is clear that the maximum recoil or dynamic energy is equal to half the total area enclosed when P is at the coercivity point ${}_BH_c$ and Q is at the B_r point. Note that when P is at the ${}_BH_c$ point, the flux density in the material is, by definition, zero, so that the mechanical force is also zero. When Q is at the remanence point B_r, the magnetic circuit is completely closed, and the force exerted is therefore a maximum. In this state the magnetic free energy is a minimum.

In most practical materials this degree of reversibility is seldom realized, and in general it is important to ensure that the point P is not too far down the demagnetization curve. Thus for dynamic applications a suitable permament magnet material is one for which the working point can move *reversibly* from the remanence point B_r over the greatest possible extent down the demagnetization curve. In an ideal permanent magnet for dynamic applications the demagnetization curve should be linear and reversible from the B_r point to the ${}_BH_c$ point, and for such a material the maximum theoretical mechanical work that can be obtained can readily be derived as follows. Consider the ideal demagnetization curve shown in Fig. 5.9. If after a change in the arrangement of the magnet system, the working point moves from the ${}_BH_c$ point to the B_r point, the total change in the magnetic free energy is given by half the total area under the B–H curve, i.e.

$$F=\tfrac{1}{2}\times\tfrac{1}{2}{}_BH_cB_r=\frac{1}{4\mu_0}J_s^2,$$

assuming that in an ideal material $B_r=J_s=\mu_0M_s$ and that ${}_BH_c=J_s/\mu_0=M_s$. The latter values correspond to those in the ideal rectangular hysteresis loop shown in Fig. 5.4. Thus for dynamic applications the maximum efficiency is obtained only when the flux coercivity is $\geqslant M_s$, whereas for static applications the maximum theoretical $(BH)_{max}$ is obtained when the coercivity ${}_BH_c\geqslant\frac{1}{2}M_s$. We therefore require very high coercivity

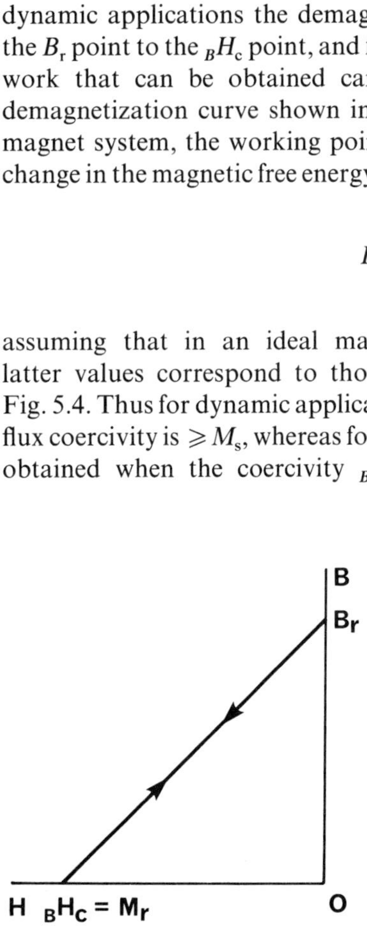

Fig. 5.9 The ideal demagnetization 'curve' of a permanent magnet material suitable for dynamic applications. Note that the flux coercivity ${}_BH_c=M_r$ and that the material will be even more efficient if $M_r=M_s$. Note also that the flux is completely reversible over the ${}_BH_c$–B_r range.

materials with high remanence B_r. The maximum mechanical energy is obtained when the maximum magnetic energy $J_s^2/4\mu_0$ is transformed.

From the above discussion, it can be concluded that the energy criteria for static and dynamic applications are somewhat different – though not unrelated. For static applications we require the area that can be inscribed in the demagnetizing quadrant of the B–H loop to be as large as possible, while for dynamic applications we require the total area enclosed by the demagnetizing quadrant to be as large as possible.

5.4.1 Recoil Permeability and Recoil or Dynamic Energy of a Permanent Magnet

Consider the toroidal permanent magnet shown in Fig. 5.2. Assume that the gap is closed by means of a piece of high-permeability material such as soft iron, i.e. the magnet has a keeper (Fig. 5.10a). If for simplicity we also assume that the contact of the keeper with the magnet is perfect, i.e. there is no high reluctance air gap, then the magnet has no demagnetizing field and is therefore at the remanence point B_r as shown in Fig. 5.11. When the keeper is removed, the magnet experiences a demagnetizing field that reduces the internal flux density to B_p, corresponding to the point P on the demagnetization curve. The magnet now produces magnetic flux in the air gap (Fig. 5.10b). If we allow for flux leakage, this is approximately equal to the flux density in the material, and so the magnet is said to be in a working condition and the point $(-H_D, B_P)$ or P is known as the working point of the magnet. When the keeper is replaced, the flux density does not in general return to the point $(0, B_r)$ via the demagnetization curve, but moves along the curve PQR to a lower remanent flux density B_R, as shown in Fig. 5.11.

If the keeper is removed again, the magnet returns almost exactly to the point $(-H_D, B_P)$ via the curve RSP so as to trace what is called a recoil hysteresis loop. In general, these recoil loops are very narrow, and it is often convenient to specify the flux on the loop by means of the straight line PR, which is known as the recoil line. The slope of this line $(B_R - B_P)/H_D$, is known as the recoil permeability μ_{rec}. The recoil permeability is an important parameter in the design of permanent magnets, particularly when they are to be used in dynamic applications where the working point is constantly

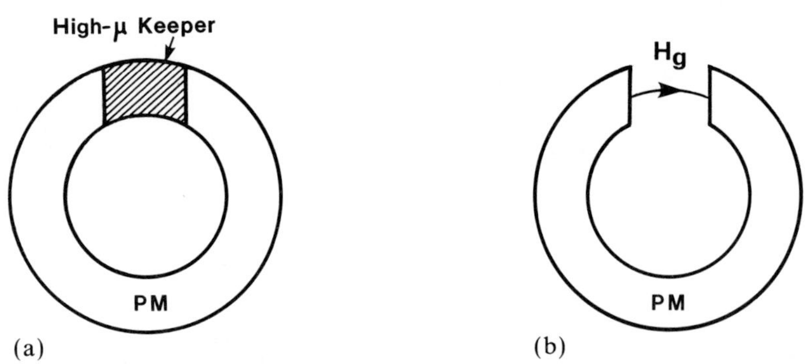

Fig. 5.10 (a) Toroidal permanent magnet with a high-permeability keeper; the mechanical contact between the keeper and the magnet is assumed to be perfect. (b) Toroidal permanent magnet with an air gap in which there is a field H_g.

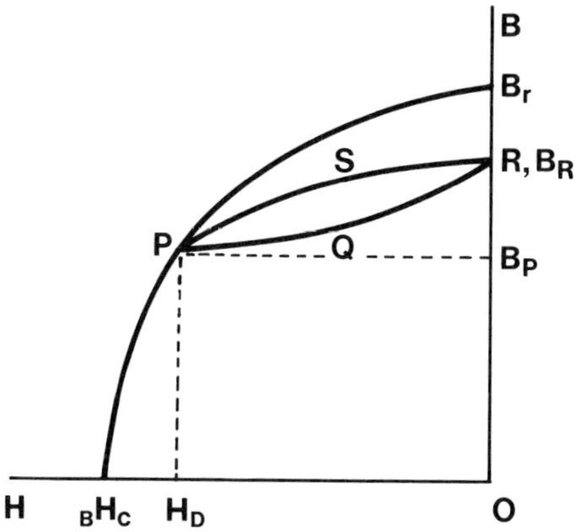

Fig. 5.11 The demagnetizing quadrant for the toroidal permanent magnet shown in Fig. 5.10(b), showing the effects of changes in the working point resulting from movement of a keeper and the tracing of a recoil hysteresis loop PQRSP.

changing. For some materials such as barium and strontium ferrites, $SmCo_5$ and Nd–Fe–B, the recoil lines are very close to the demagnetization curves, which are very close to the ideal for dynamic applications.

5.4.2 Effect of Flux Leakage under Dynamic Conditions

When a permanent magnet is used in a lifting or attracting device, this is usually done in combination with a high-permeability material in the working gap to form a magnetic circuit with a leakage gap, as shown in Fig. 5.12. The function of the permanent magnet is to provide a magnetomotive force, which magnetizes the high-permeability yoke and thus produces an attractive force and a field in the leakage gap. The associated demagnetizing field reduces the flux density in the magnet to the value corresponding to the working point P (Fig. 5.13), and the demagnetizing field itself is of course the

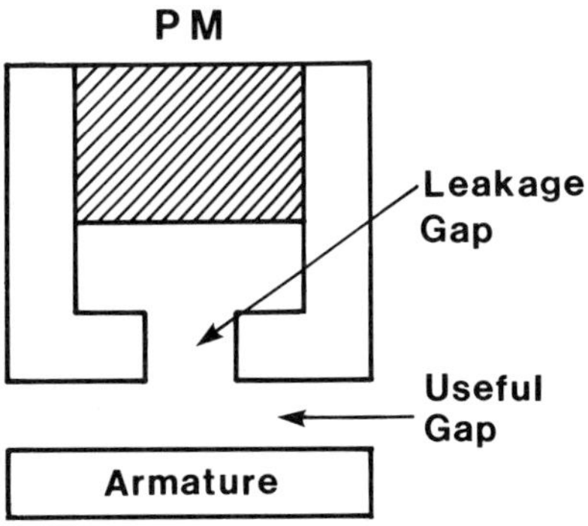

Fig. 5.12 Permanent magnet used in a lifting or attracting device in combination with a high-permeability material (the armature) to form a magnetic circuit with a leakage gap.

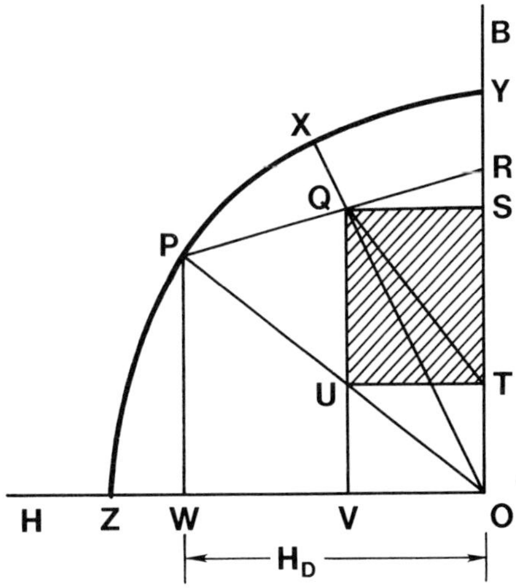

Fig. 5.13 Determination of recoil energy for a magnet with flux leakage working under dynamic conditions. The useful recoil energy is given by half the area of the rectangle QSTU.

field H_D, i.e. OW. If the armature approaches the leakage gap of the magnet, the flux in the gap divides itself between the leakage and useful gaps, depending on their relative magnetic reluctances (or permeances). Since this decreases the effective demagnetizing field acting on the permanent magnet, its flux density increases, corresponding to movement of the working point along the recoil line PQR. At the point Q, for example, the useful flux density in the gap between the armature and the magnet is QU, while the leakage flux density is reduced from PW to UV. The useful magnetizing field is given by UT=VO, i.e. the magnetomotive force is equal to VO$\times l_m$, where l_m is the length of the permanent magnet. The useful magnetic energy available for doing work on the armature is equal to half the area enclosed by QSTU, while the leakage energy is equal to half the area of UTOV. If the armature were of infinite permeability and made perfect magnetic contact with the pole pieces, there would be no leakage flux and hence no magnetomotive force. The useful flux density would then be a maximum equal to OR, and the force on the armature would also be a maximum, although in this case the leakage and useful energies would be zero. Since this ideal situation is never realized, because the permeability of the armature is less than infinite and the contact never perfect, an appreciable magnetomotive force is required to magnetize the armature. Depending on the magnetic properties of the material, this imposes constraints on the value of the magnetizing field OV. Since the force on the armature depends on both its level of flux density and hence on the magnetizing field, we expect this to be a maximum when the area of QSTU is a maximum for the particular open circuit working point P. From elementary geometry, the maximum area QSTU that can be inscribed in the triangle OPR is obtained when PQ=QR, from which it follows that QS=$\frac{1}{2}$OW and ST=$\frac{1}{2}$OR. Thus for the open circuit working point P the maximum useful recoil energy is given by half the area of QSTU, i.e.

$$E_{rec}=\tfrac{1}{2}\times\tfrac{1}{2}\mathrm{OR}\times\tfrac{1}{2}\mathrm{OW}=\tfrac{1}{8}B_rH_D.$$

Since by changing the design of the magnet the working points P can be changed, it is clear that the recoil energy depends on the reluctance of the air gap when the magnet is on open circuit. The maximum possible area of OPR is the total area of the demagnetizing quadrant, i.e. the area of the sector OZY. If, ideally, the demagnetizing quadrant is linear and completely reversible, as shown in Fig. 5.9, then the maximum possible recoil energy is given by

$$E_{rec,\ max}=\tfrac{1}{4}B_r\ {}_BH_c,$$

assuming that the working point P can recoil from the coercivity point ${}_BH_c$ along the demagnetizing curve to the remanence point B_r. Note that in many discussions of the behaviour and properties of permanent magnets the energy and recoil energy products are often used, and it should be emphasized that these are of course twice the actual magnetic energies. As a general rule, the maximum recoil energy for a permanent magnet is obtained when the working point P is slightly below the $(BH)_{max}$ point. This is so because as P moves down the demagnetizing curve, the recoil energy product increases more rapidly than the $(BH)_{max}$ product decreases. The precise location of the optimum point depends on the shape of the demagnetizing curve and the recoil permeability. From the above discussions we conclude that when a permanent magnet is required to work under recoil conditions, it is desirable that the magnetization should be as reversible as possible. Thus in a dynamic application such as permanent magnet motors where the magnet may be subjected to large and varying demagnetizing fields, the coercivity ${}_BH_c$ should of course be as large as possible.

The design of permanent magnet circuits is, in general, a complicated process, and in the foregoing discussion we have done little more than to illustrate the basic principles, although with modern materials such as $SmCo_5$ and Nd–Fe–B the design process has been simplified. For further details the reader is referred to Edwards (1962), Parker and Studders (1962), Parker (1990), Schüler and Brinkmann (1970) and Moskowitz (1976).

5.5 ALNICO PERMANENT MAGNET ALLOYS

The Alnicos are essentially heat-treated Fe–Co–Ni–Al–Cu alloys with additions of Cu and Ti. They can be divided into two main groups: (a) the isotropic Alnicos 1–4, which have a cobalt content in the range 0–20 wt%, and (b) the anisotropic Alnicos with a cobalt content in the range 22–40 wt% and a titanium content of 5–8.5 wt%.

5.5.1 Isotropic Alnicos 1–4

The magnets are prepared by cooling at about 30 K s^{-1} from the single-phase bcc state at 1250°C followed by tempering at 650°C for about 4 h. The first of the series to be discovered (Mishima, 1932) contained only iron, nickel and aluminium but is now known as Alnico 3. It was soon discovered that small additions of Co, Cu and Ti improved the properties significantly. The compositions and magnetic properties of the isotropic Alnicos are given in Table 5.1. Typical B–H demagnetization curves are shown in Fig. 5.14.

Origin of coercivity in isotropic Alnicos 1–4

During cooling, the high-temperature bcc α-phase undergoes a spinodal decomposition into two bcc phases α_1 and α_2. The α_1 is Fe- or Fe–Co-rich and is strongly ferromagnetic, whereas the α_2 phase is Ni–Al-rich and is weakly ferromagnetic. The two phases form a complex three-dimensional system of 'regularly' spaced slightly elongated particles in which the axes of elongation are approximately parallel to the $\langle 100 \rangle$ directions. Such a microstructure is characteristic of phase separation by spinodal decomposition. After tempering, the Fe- or Fe–Co-rich α_1 phase becomes more strongly ferromagnetic, while the Ni–Al-rich α_2 phase is believed to become very weakly ferromagnetic or paramagnetic. Thus the tempering is essential if the alloys are to develop their full permanent magnet properties, which then requires the difference in the saturation magnetic polarizations of the α_1 and α_2 phases to be as large as possible. In this state the coercivity can be attributed to the shape anisotropy of the interconnected 'single-domain particles', although the complexity of the microstructure makes any quantitative comparison with the Stoner–Wohlfarth (1948) theory impossible.

Table 5.1 Compositions and magnetic properties of isotropic Alnicos.

Characteristics	Alnico	Composition (wt%, balance is Fe)					Magnetic properties		
		Ni	Al	Co	Cu	Others	B_r (T)	$_BH_c$ (kA m^{-1})	$(BH)_{max}$ (kJ m^{-3})
Isotropic, Co-free	3	24–30	12–14	0	0–3	0–1 Ti	0.5–0.6	40–54	10
Isotropic, low-Co	1,4	21–28	11–13	3–5	2–4	0–1 Ti	0.55–0.75	36–56	11–12
Isotropic, medium-Co	2	16–20	9–11	12–14	3–6	0–1 Ti	0.65–0.8	40–50	13–14
Isotropic, extra-high-H_c	2	18–21	8–10	17–20	2–4	4–8 Ti 0–1 Nb	0.6–0.7	60–72	14–16

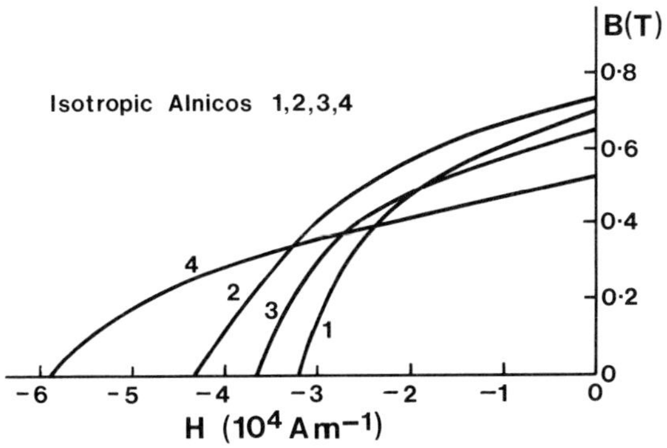

Fig. 5.14 Typical demagnetization curves of Alnicos 1–4.

5.5.2 Anisotropic Alnico 5

The energy products, coercivities and remanences can be increased significantly by the controlled cooling of the cobalt-containing alloys (20–25% Co) from 1200°C in a saturating magnetic field. This thermomagnetic treatment produces improved properties parallel to the direction of the applied field during cooling and was first introduced by Oliver and Shedden (1938) and Jonas and Meerkamp van Embden (1941).

The magnetic properties in the preferred direction are further improved by subsequent tempering for several hours at about 600°C. The application of a field during this treatment, however, has no effect on the final magnetic properties. As with the isotropic Alnicos, the effect of tempering is to increase the difference in the saturation magnetizations of the α_1 and α_2 phases. The coercivity of Alnico 5 before and after tempering as a function of the quenching temperature is shown in Fig. 5.15.

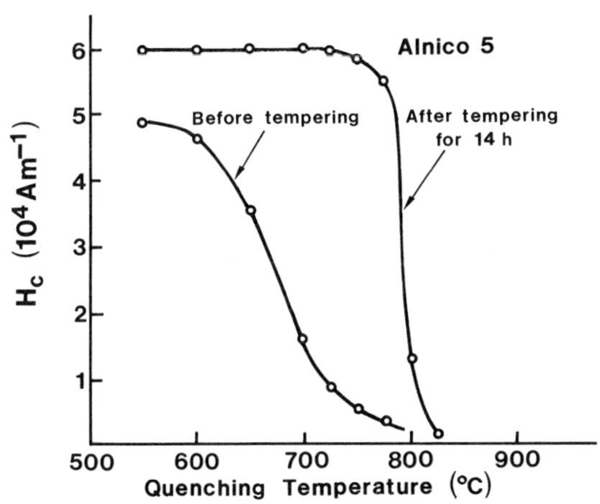

Fig. 5.15 Coercivity of Alnico 5 before and after tempering as a function of quenching temperature.

5.5.3 Grain-Oriented Alnico 5

Improvements in the magnetic properties of Alnico 5 can be made by grain-orienting the alloy before the thermomagnetic and tempering treatments. This is done by casting it in heated or exothermic moulds onto steel or copper slabs (preferably water cooled). On solidification, the grains tend to grow with their axes parallel to the [100] direction, which is perpendicular to the cold metal surface. Grain-oriented Alnico 5 is often referred to as Alnico 5 DG (DG = directed grain). Alnico 5 DG is certainly not fully columnar, but the preferred grain orientation is such that there is a significant improvement in the magnetic properties (Table 5.2). Optimum properties are of course obtained when the magnets are single crystals; the maximum energy product achieved so far is 80 kJ m^{-3}. Unfortunately, single-crystal Alnico 5 is difficult to produce on a commercial scale. Typical demagnetization curves for various forms of Alnico 5 are shown in Fig. 5.16.

High-coercivity forms of polycrystalline Alnico 5, known as Alnicos 6 and 7, can be produced by adding small quantities of titanium and niobium, but the higher coercivities are obtained at the expense of the remanence and the energy product.

5.5.4 Anisotropic Alnico 8

The superior properties of Alnico 5 DG compared with Alnico 5 are largely achieved by an increase in the remanence and B–H loop squareness; the coercivities of the two magnets are almost the same. The higher coercivities in polycrystalline Alnico 8 are obtained by the addition of 4–5 wt% titanium and an increased cobalt content. Typical coercivities are in the range 110–170 kA m^{-1}, with $(BH)_{max} = 40$–48 kJ m^{-3}. Unfortunately, the high coercivities are obtained at the expense of lower remanences, in the range 0.74–0.9 T, which are significantly lower than those for Alnicos 5 and 5 DG. Typical B–H demagnetization curves for Alnicos 8 and 9 are shown in Fig. 5.17. Even higher coercivities can be obtained by increasing the cobalt and titanium contents to about 38% and 8% respectively (Table 5.2).

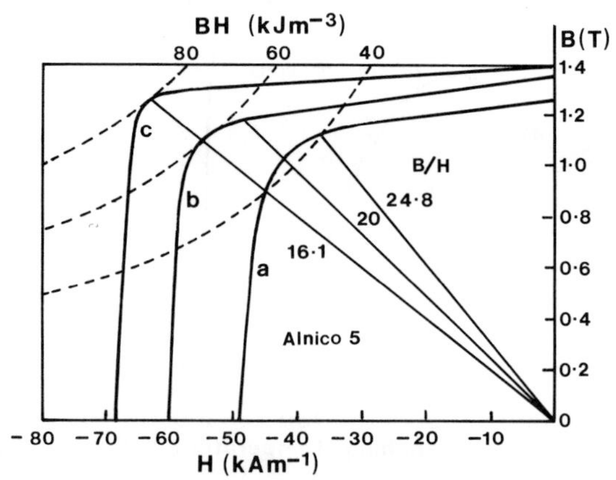

Fig. 5.16 Typical demagnetization curves for various forms of Alnico 5: (a) randomly oriented grains; (b) semi-grain-oriented; (c) 'fully' grain-oriented.

Table 5.2 Compositions and magnetic properties of anisotropic thermomagnetically treated commercial Alnicos.

Characteristics	Alnico	Composition (wt%, balance is Fe)					Magnetic properties		
		Al	Ni	Co	Cu	Others	B_r (T)	$_BH_c$ (kA m^{-1})	$(BH)_{max}$ (kJ m^{-3})
Field-treated, random grain	5	7.8–8.5	12–15	23–25	2–4	0–0.5 Ti 0–1 Nb	1.2–1.3	52–46	40–44
Field-treated, random grain, higher-H_c	5	7.8–8.5	13–16	23–25	2–4	1–1.5 Ti 1–2 Nb	1.05–1.15	62–56	30–36
Field-treated, random grain, high-H_c	8	7–8	14–16	32–36	4	4–6 Ti O-1Nb	0.8–0.9	140–110	40–45
Field-treated, random grain, extra-high-H_c	8	7–8	14–15	37–40	3	7–8 Ti	0.74–0.78	170–150	44–48
Field-treated, directed-grain form of Alnico 5	5DG 5–7	7.8–8.5	13–15	24–25	2–4	0–1 Nb	1.3–1.4	62–56	56–64
Field-treated, directed-grain form of Alnico 8	9	7–8	14–16	32–26	4	4–6 Ti 0–1 Nb 0.3 S	1.0–1.1	140–110	60–75

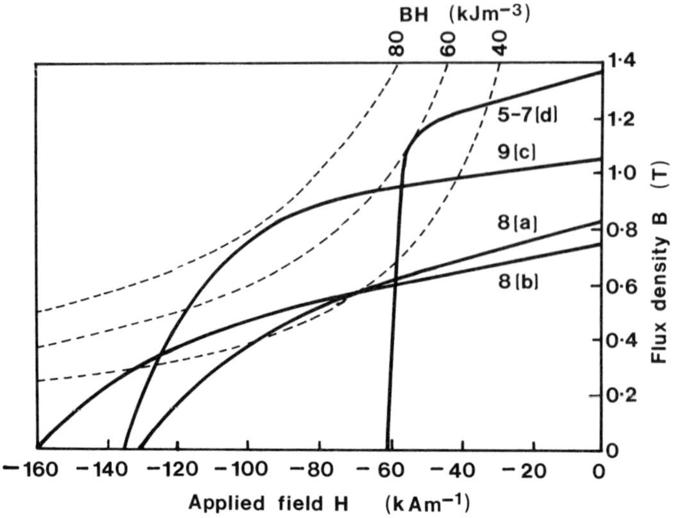

Fig. 5.17 Typical demagnetization curves for Alnicos 8 and 9. Curve 8(a) is standard Alnico 8, while curve 8(b) is extra-high-coercivity Alnico 8. Curve 5–7(d), which is representative of Alnicos 5–7, is shown for comparison.

Alnico 8 alloys are homogenized at 1250°C to produce the high-temperature bcc phase, and then cooled at a controlled rate to about 820°C, i.e. at a temperature just below the Curie temperature of the Fe–Co-rich α_1 phase, at which temperature they are heat-treated isothermally in a saturating field of about 300 kA m^{-1} for 5–10 min, after which they are tempered for about 6 h at 650°C, followed by about 24 h at 550°C. The isothermal heat treatment in the magnetic field is necessary because the relatively large Ti^{2+} ions lower the rates of diffusion, so that sufficient time at 820°C must be allowed for the preferred growth of the α_1-phase particles to occur parallel to the applied field. This thermomagnetic heat treatment, combined with the increased cobalt and titanium contents, increase the elongation, alignment and saturation magnetization of the α_1 particles and hence the coercivity of the magnet.

5.5.5 Columnar Alnico 9

Columnar Alnico 9 is produced by columnar crystallization of Alnico 8 (see Fig. 5.17). Unfortunately the high titanium content of Alnico 8 decreases its grain size and thus makes columnar crystallization difficult. However, this problem can be overcome by small additions of sulfur, selenium or tellurium, which increase the grain size. The fully columnar grain orientation is obtained by casting the alloy in a heated or exothermic mould with a chilled steel or copper base plate. The columnar grains grow preferentially perpendicular to the plane of the chill plate parallel to the [100] direction. The alloy is then given the same thermomagnetic and two-stage heat treatment as Alnico 8. The improvements in the magnetic properties parallel to the columnar axis are obtained at the expense of those perpendicular to the columnar axis. The highest energy products are of course obtained in single crystals, the highest yet reported (Naastepad, 1966) for any Alnico is 107 kJ m^{-3}, with a coercivity $_BH_c$ of 122 kA m^{-1}. The composition of the single crystal was 35 wt% Fe–34.8% Co–14.9% Ni–7.5% Al–5.4% Ti–2.4% Cu.

It appears that the Alnicos, which have been available since 1931, have reached the zenith of their development, and there is little prospect that any further substantial improvement in their magnetic properties can be made.

5.5.6. Sintered Alnicos

All the Alnicos can also be made by sintering the powdered alloys at 1250–1350°C. After sintering, they are given a heat treatment appropriate to their composition. The resulting magnets have properties inferior to those of the cast alloys, but they have the important advantage over the latter in that small magnets with intricate shapes can be manufactured that would be very difficult or very expensive to produce from cast Alnicos, which have to be ground to their final dimensions. Typical demagnetization curves for sintered Alnicos, 2, 5, 6 and 8 are shown in Fig. 5.18.

5.5.7 Bonded or Moulded Alnico Magnets

Very small magnets can be produced by bonding Alnico powders into a thermoplastic binder so that they can be shaped as required. Since the volume fraction of the alloy is less than 1, the magnetic properties are inferior to those of the sintered magnets.

5.5.8 Microstructure and Coercivity Mechanism in the Alnicos

The electron micrographs of Alnico 8 shown in Fig. 5.19 clearly show the rodlike shape of the α_1-phase particles and their regular spacing, which is a characteristic of phase separation by spinodal decomposition. Since the average length and diameter of the

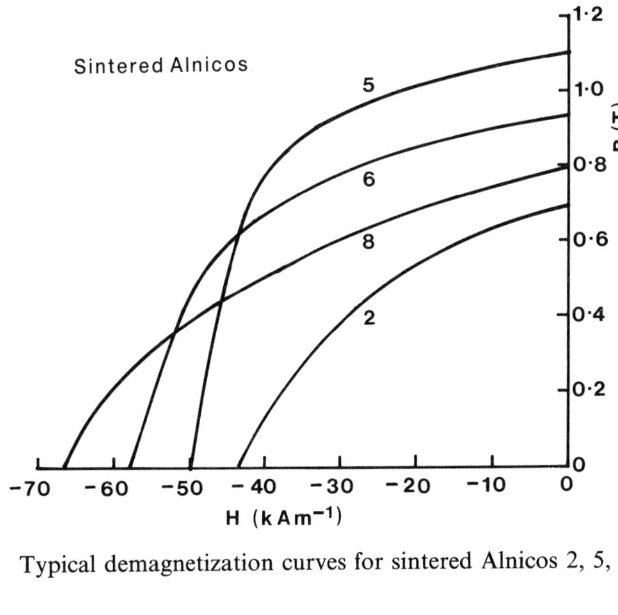

Fig. 5.18 Typical demagnetization curves for sintered Alnicos 2, 5, 6 and 8.

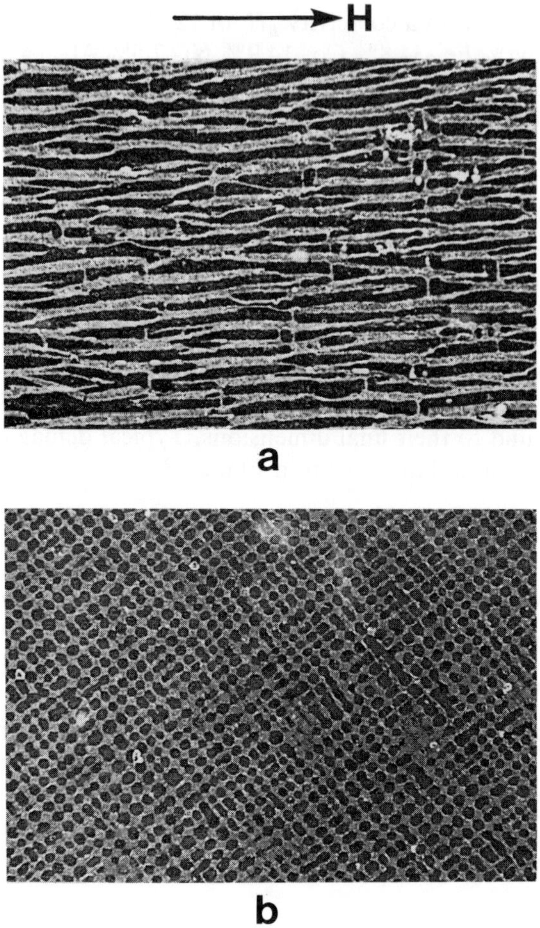

Fig. 5.19 Highly oriented microstructure of Alnico 8 or 9: (a) in a plane parallel to the direction of the applied-field during thermomagnetic treatment; (b) in a plane perpendicular to the direction of the applied field during thermomagnetic treatment ($\times$45,000) (de Vos, 1969).

particles are about 600 and about 30 nm respectively, it may be assumed that they are small enough to be single domains with an elongation $l/d \approx 20$, which corresponds to a difference in the demagnetizing factors $D_z - D_x \approx 0.5$ (see Chapter 1). Assuming that the saturation magnetization of the Fe–Co-rich α_1-phase particles $M_s \approx 1.7 \times 10^6$ A m^{-1}, the theoretical intrinsic coercivity ${}_iH_c$ due to the uniaxial shape anisotropy is given by

$$ {}_iH_c = (D_z - D_x)M_s = 850 \text{ kA m}^{-1}. $$

Since this value is very much larger than the observed maximum intrinsic coercivity of about 190 kA m^{-1}, it is clear that, in spite of the high degree of elongation of the single-domain particles, magnetization reversal does not occur by coherent rotation of the magnetization vector according to the Stoner–Wohlfarth (1948) theory (see Chapter 1). McCurrie and Jackson (1980) have measured the rotational hysteresis energy

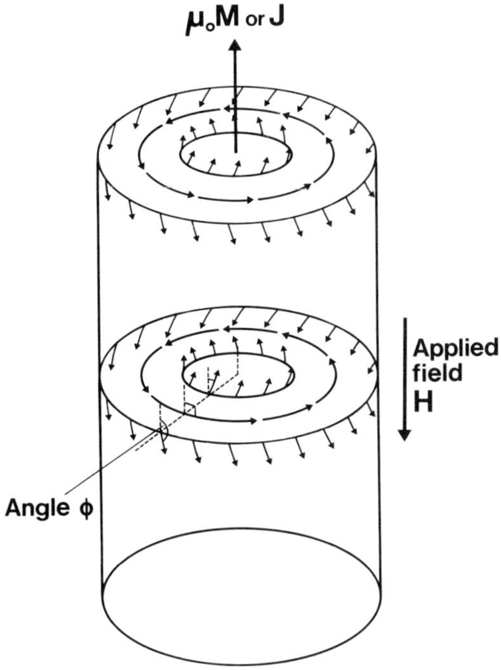

Fig. 5.20 Curling mechanism of magnetization reversal. Magnetic moments further from the axis of the cylindrical particle rotate first, but the whole reversal process occurs very rapidly and in much lower fields than are predicted by the Stoner–Wohlfarth (1948) coherent rotation process.

and angular variation of the coercivity in Alnicos 8 and 9, and have concluded that magnetization reversal occurs by the curing mechanism as shown in Fig. 5.20, in which concentric tubes of ion magnetic moments undergo progressively larger rotations from the centre outwards to the surface of the cylindrical particle. Complete reversal occurs very rapidly, and takes place in a much lower reverse field than is predicted by the Stoner–Wohlfarth (1948) coherent reversal model. A detailed mathematical theory of the curling mechanism has been given by Kondorsky (1952) and Frei *et al.* (1957), and has been reviewed by Aharoni (1966).

The microstructures of Alnico 5 and 5 DG show significant elongation and alignment of the α_1-phase particles, but they are not as well shaped and distinct as those in Alnicos 8 and 9.

A detailed review of the Alnicos and a summary has been published by McCurrie (1982, 1992).

5.6 Fe–Cr–Co ALLOYS

The permanent magnet properties of Fe–Cr–Co alloys were first reported by Fahlenbrach (1960, 1962) and later developed by Kaneko *et al.* (1971), who showed that magnets with properties comparable to those of Alnico 5 can be obtained relatively easily, with the advantage that they are ductile and can be cold-formed at room temperature. They have the additional advantage that magnetic properties comparable to those of many of the Alnicos can be obtained with lower cobalt contents and a consequent reduction in cost.

The magnetic hardness is due to shape anisotropy and domain wall pinning in a strongly ferromagnetic bcc Fe-rich α_1 phase together with a less strongly ferromagnetic bcc Cr-rich α_2 phase. These phases are formed by spinodal decomposition from a high-temperature bcc α phase. In order to avoid the formation of the non-ferromagnetic fcc γ phase, the alloys must be solution-treated above about 1250°C and then rapidly quenched. The suppression of the undesirable $\alpha \rightarrow \gamma$ transformation is assisted by the addition of small quantitites of bcc α-phase-forming elements such as Nb, Al, V, Ti or Zr. It is important to limit impurities such as C, N, O and Mn, since these elements stimulate the formation of the non-ferromagnetic γ-phase. The alloys have a particularly strong tendency to absorb atmospheric nitrogen, so that melting must be carried out in an argon-assisted vacuum and the heat treatment in argon.

The isotropic permanent magnet properties can be developed by a high-temperature solution treatment and an isothermal spinodal decomposition followed by step cooling. The heat-treatment schedule is very sensitive to the composition of the alloys, especially the cobalt content. The spinodal decomposition temperature increases as the latter increases, and thus increases the rate at which the spinodal decomposition proceeds because of the increased atomic diffusion rate at higher temperatures.

Anisotropic permanent magnets can be produced either by (a) thermomagnetic treatment, i.e. by applying a magnetic field during the spinodal decomposition followed by slow cooling, or (b) by deformation of the alloy after spinodal decomposition followed by controlled cooling.

5.6.1 Isotropic Magnets

These may be produced as described above, except that no magnetic field is applied during the spinodal decomposition. Alternatively, the alloys may be cooled rapidly from above the spinodal decomposition temperature, held at an intermediate temperature and then cooled very slowly to room temperature. The initial rapid cool produces the optimum 'particle' sizes for the α_1 and α_2 phases, while the slow cool results in the optimum composition difference and the maximum difference in the saturation magnetization between two phases. Alloys with lower cobalt contents require much lower cooling rates during the second cooling stage. Typical properties of the isotropic magnets are given in Table 5.3.

An important application of Fe–Cr–Co magnets is in ring armature telephone receivers, which require a cup-shaped magnet to provide a DC-bias magnetic field. The ductility of the isotropic Fe–28% Cr–10% Co alloy (known as Chromindur II) gives it an advantage over Remalloy (Fe–20% Mo–12% Co), because it can be rapidly cold-formed at room temperature whereas Remalloy requires blanking and drawing at 1250°C. Other applications include speedometers, hysteresis and timing motors, and security labels.

5.6.2 Anisotropic Magnets Produced by Thermomagnetic Treatment

If a magnetic field of about 160 kA m^{-1} is applied during the spinodal decomposition, the resulting α_1- and α_2-phase particles tend to elongate in the direction of the field, although, unlike the Alnicos, there is no strong preferential growth or alignment parallel

Table 5.3 Magnetic properties of Fe–Cr–Co alloys.

Alloy (composition in wt%)	B_r (T)	${}_BH_c$(kA m^{-1})	$(BH)_{max}$ (kJ m^{-3})
Isotropic magnets			
Fe–31 Cr–23 Co	0.84	50	18
Fe–28 Cr–10.5 Co (Chromindur II)	0.98	32	14
Fe–28 Cr–7 Co	0.97	26	11
Anisotropic magnets (thermomagnetic treatment and ageing)			
Fe–30 Cr–25 Co–3 Mo	1.15	62	40
Fe–24 Cr–15 Co–3 Mo (⟨100⟩ columnar)	1.54	67	76
Fe–27 Cr–16 Co–1 Si	1.34	43	38
Fe–23 Cr–15 Co–5 V	1.32	43	40
Fe–22 Cr–18.5 Co–3 Mo (single crystal)	1.58	73	91
Fe–22 Cr–15 Co	1.56	51	66
Fe–25 Cr–12 Co	1.45	50	61
Fe–27 Cr–9 Co	1.30	46	50
Fe–28 Cr–7 Co	1.20	41	42
Fe–30 Cr–5 Co	1.34	42	42
Fe–33 Cr–2 Co	1.24	38	34
Anisotropic magnets (deformation and ageing treatment)			
Fe–33 Cr–23 Co–2 Cu	1.3	86	78
Fe–33 Cr–16 Co–2 Cu	1.25	76	55
Fe–33 Cr–11.5 Co (Chromindur III)	1.2	61	44
Fe–33 Cr–5 Co	1.15	25	19
Anisotropic magnets (thermomagnetic, deformation and ageing treatment)			
Fe–30 Cr–23 Co–1 Si (single crystal)	1.3	78	64
Sintered magnets			
Fe–29.5 Cr–25 Co–3 Mo	0.95	69	27
Fe–25 Cr–12 Co	1.40	44	42
Fe–31 Cr–5 Co	1.23	40	35

to any particular crystallographic directions such as ⟨100⟩. The thermomagnetic treatment results in a squaring of the hysteresis loop, and hence a significant increase in the remanence, coercivity and energy product parallel to the heat-treatment field direction.

Although crystallographic texture has little or no effect on the growth direction of the α_1- and α_2-phase 'particles' in the ternary Fe–Cr–Co alloys, the addition of Mo favours elongation in one of the ⟨100⟩ directions, so that if the heat-treatment field is applied as close as possible to the [100] direction then there will be an enhancement of particle growth in the direction of the field. If the alloy is cast with a columnar crystallographic texture, a very high degree of particle elongation and alignment can be achieved if the heat-treatment field is applied parallel to the [100] columnar axis. This results in a very significant improvement in magnetic properties. The best properties are of course obtained for single crystals. The properties of field-treated anisotropic Fe–Cr–Co magnets are given in Table 5.3, from which it can be seen that the best properties for the thermomagnetically treated magnets are those for the single-crystal and columnar magnets containing 3% Mo. Although very good magnetic properties can be obtained for alloys with very low Co contents (e.g. Fe–33% Cr–2% Co), very low cooling rates of order 0.1°C h^{-1}, are necessary to achieve these properties. Thus the cost saving on the low cobalt content may be offset by the increased cost of running the furnaces for very long periods.

5.6.3 Anisotropic Magnets Produced by Deformation and Heat Treatment

The solution-treated quenched alloy is first cooled through the spinodal decomposition temperature in the absence of a magnetic field, thus developing almost spherical particles. These are then elongated and aligned by a mechanical deformation process such as extrusion, drawing or rolling. The alloy is then heat-treated isothermally followed by controlled cooling to develop the maximum difference in the saturation magnetizations of the α_1 and α_2 particles. As can be seen from Table 5.3, the properties of these magnets compare very favourably with those of thermomagnetically treated magnets. It is also clear that adequate magnetic properties can be obtained for alloys with intermediate cobalt contents i.e. 10–15%, and it is for this reason that considerable attention has focused on these materials, since they offer considerable cost savings. The alloy known as Chromindur II (Table 5.3) was produced by cooling from 680°C to 610°C at 40°C h^{-1}, wire-drawing to 67% reduction in area, reheating to 600° and then controlled cooling to 500°C. The resulting properties of $B_r = 1.2$ T, ${}_BH_c = 61$ kA m^{-1} and $(BH)_{max} = 44$ kJ m^{-3} are comparable to those of Alnico 5, and could therefore provide a cheaper alternative in many applications. The best properties for the deformation-aged magnets were obtained for the alloy Fe–33% Cr–23% Co–2% Cu, for which $(BH)_{max} = 78$ kJ m^{-3}. The improvement resulting from the addition of 2% Cu is due to the increased coercivity.

5.6.4 Anisotropic Magnets Produced by Both Thermomagnetic and Deformation Ageing

Kaneko *et al.* (1976) have produced a permament magnet with composition Fe–30% Cr–23% Co–1% Si that has the following properties: $B_r = 1.3$ T, ${}_BH_c = 7.8 \times 10^4$ A m^{-1} and $(BH)_{max} = 64$ kJ m^{-3}. These properties were obtained with a single crystal that had been given the following treatment. The alloy was first solution-treated at 1300°C, then quenched in water and heat-treated at 640°C for 40 min in a magnetic field of 160 kA m^{-1}. After the thermomagnetic treatment, the crystal was step-tempered at 600, 580 and 560°C for 1 h each and annealed at 540°C for 5 h. Finally, the crystal was swaged parallel to the [100] direction to give a reduction of 70%. Measurements were also made on crystals with the (110) and (111) orientations, but Kaneko *et al.* (1976) showed that the magnetic properties did not depend on the crystallographic orientation. However, they found that the properties of all the crystals were significantly improved if they were swaged after the thermomagnetic treatment, and that the best properties were obtained for the (100) crystal. The magnetic properties of the polycrystalline alloy were also significantly improved by swaging.

Typical *B–H* demagnetization curves for various Fe–Cr–Co permanent magnets are shown in Fig. 5.21.

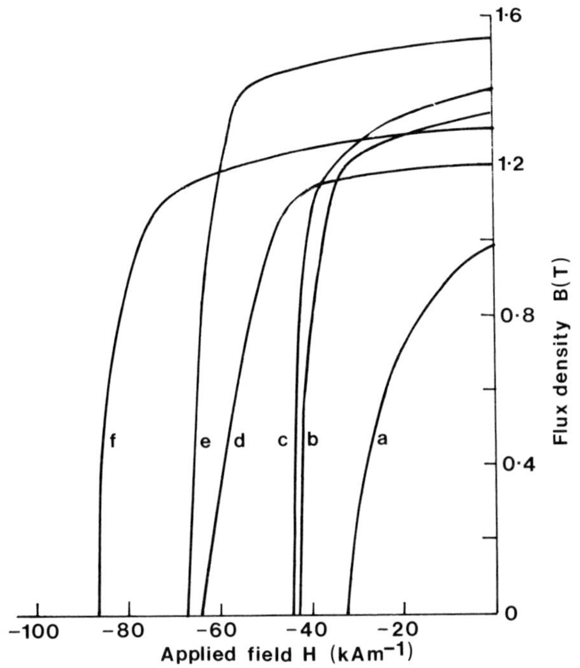

Fig. 5.21 Typical demagnetization curves for Fe–Cr–Co alloys: (a) Fe–28 wt% Cr–10.5% Co (isotropic); (b) Fe–30% Cr–5% Co; (c) Fe–25% Cr–12% Co (sintered); (d) Fe–33% Cr–11.5% Co–2% Cu (deformation-aged); (e) Fe–24% Cr–15% Co–3% Cu (⟨100⟩ columnar); (f) Fe–32% Cr–23% Co–2% Cu (deformation-aged) (Wernick and Chin, 1992).

5.6.5 Sintered Alloys

Since the Fe–Cr–Co alloys are ductile and machinable, unlike the Alnicos, there is probably little need to produce the magnets by standard powder sintering. Fe–Cr–Co magnets can be produced by sintering at 1300–1400°C in dry hydrogen followed by controlled slow cooling. However, it can be seen from Table 5.3 that the magnetic properties are superior to the isotropic magnets, but inferior to the anisotropic magnets.

5.6.6 Origin of Coercivity in Fe–Cr–Co Magnets

Since both the α_1 and α_2 phases are ferromagnetic, it is clear that the coercivity cannot be attributed entirely to the shape anisotropy of isolated Fe–Co-rich α_1-phase particles as has been accepted for the Alnico alloys. The microstructure of Fe–Cr–Co alloys has been studied extensively by transmission electron microscopy, and it has been observed that in some specimens the domain walls extend throughout the two-phase microstructure. This has led to the suggestion that the coercivity is controlled by domain wall pinning. However, in some optimally processed specimens there is an obvious shape anisotropy in the α_1 and α_2 phases and a high degree of alignment parallel to a ⟨100⟩ direction. There is some degree of interconnection between the phases, so that the origin of the coercivity is still uncertain. It should be borne in mind that even a small degree of elongation (e.g. 5%) can have a significant effect on the shape anisotropy and hence the coercivity. The difference $D_z - D_x$ in the demagnetizing factors changes by about 30% for a 5% change in the length-to-diameter ratio for a single-domain particle.

Mössbauer spectroscopy studies indicate that at the maximum coercivity there is only a relatively small difference in the saturation magnetizations of the α_1 and α_2 phases, so it would appear that domain wall pinning is a more appropriate model of the coercivity than a Stoner–Wohlfarth model based on coherent rotation (or curling) in single-domain particles with uniaxial shape anisotropy.

The shape anisotropy is most pronounced in the swaged alloys, as can be seen from Fig. 5.22, which shows the microstructure of the alloy Fe–30% Cr–23% Co–1% Si parallel and perpendicular to the swaging direction (Kaneko *et al.*, 1975). However, it

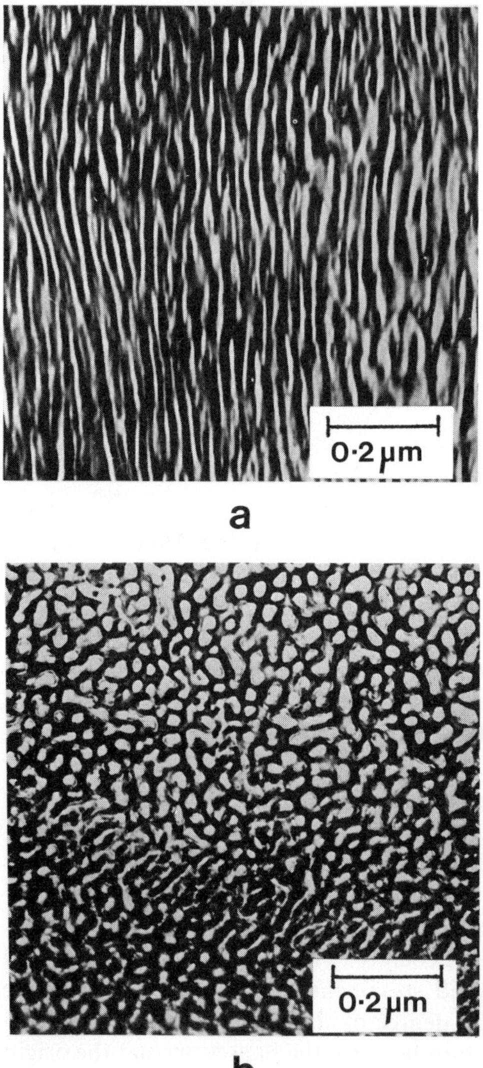

Fig. 5.22 Highly oriented microstructure of Fe–30 wt% Cr–23% Co–1% Si ⟨100⟩ single crystal that has been both thermomagnetically treated and swaged to 70% reduction followed by step tempering (a) parallel to the swaging direction (b) perpendicular to the swaging direction (Kaneko *et al.*, 1975).

may be that the most appropriate model for the coercivity in Fe–Cr–Co alloys is a combination of domain wall pinning and magnetization rotation, and that the dominant mechanism will depend on the composition and precise processing of the alloy. Detailed studies of the microstructures of Fe–Cr–Co alloys have also been made by Okada *et al.* (1978). In any case, domain wall movement involves rotation of the magnetization vectors, so there may not be any clear distinction between the two models.

A review of the structure and properties of Fe–Cr–Co magnets has been published by Rossiter and Houghton (1984). A short review has also been published by Wernick and Chin (1992).

5.7 Cu–Ni–Fe AND Cu–Ni–Co ALLOYS

5.7.1 Cu–Ni–Fe Alloys

Although the magnetic properties of the Cu–Ni–Fe alloys are much inferior to those of the Alnico series, they have found extensive practical application because they combine reasonably high coercivity and remanence with ductility. They can therefore be drawn into wires or more complex shapes without the necessity for grinding, which is an expensive process. Cu–Ni–Fe magnets are widely used in measuring instruments, e.g. speedometers. Furthermore, they can also be cold-formed and machined in the magnetically hard condition and can be precision-stamped to the required shape at high speed. Neumann *et al.* (1937) have shown that for alloys with compositions close to 60% Cu–30% Ni–10% Fe coercivities ($_BH_c$) as high as 3.2×10^4 A m^{-1} can be obtained after suitable heat treatment. This usually involves quenching the alloy from about 1000°C, followed by annealing for several hours at a temperature in the range 600–750°C. Neumann *et al.* (1937) have also shown that the magnetic properties can be improved by severe cold working of the alloy after solution treatment and quenching. This produces anisotropic alloys in which the axis of elongation in the worked material becomes the preferred axis of magnetization with the best permanent magnet properties.

The cold-working treatment usually involves a reduction in area of about 90% in order to obtain the best properties. The demagnetization curves for an alloy containing 60% Cu–20% Ni–20% Fe before and after cold working are shown in Fig. 5.23. Note the marked deterioration of properties perpendicular to the preferred axis of magnetization (curve c). The Cu–Ni–Fe alloys are known commercially as the Cunife alloys.

5.7.2 Cu–Ni–Co Alloys

These alloys, first developed by Dannöhl and Neumann (1938), have very similar magnetic properties to the Cu–Ni–Fe alloys, although they generally have higher coercivities and remanences, typically in the range 35–64 kA m^{-1} and 0.32–0.5 T respectively. The Cu–Ni–Co alloys are also ductile and malleable in the magnetically hard state and, like the Cu–Ni–Fe alloys, their best magnetic properties are obtained by severe cold working. The usual commercial composition of Cunico I is 50% Cu–21% Ni–29% Co. In order to avoid problems associated with segregation during casting, the alloy is first cast into small bars and then given a prolonged homogenization treatment at 1100°C, followed by cold rolling. After punching and

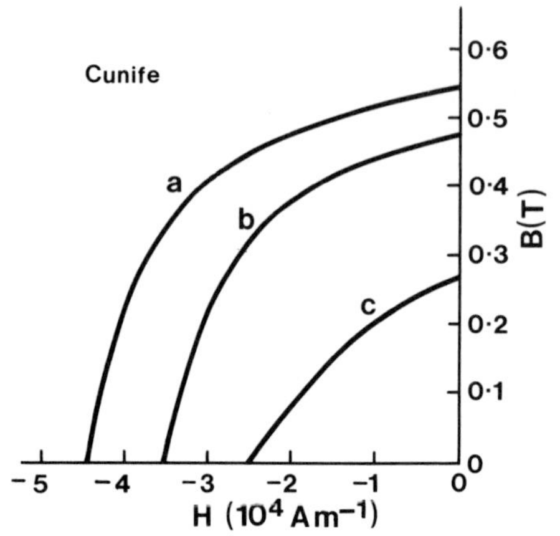

Fig. 5.23 Typical demagnetization curves for Cunife, 60 wt% Cu–20% Ni–20% Fe: (a) isotropic, as cast; (b) after 90% reduction in area by cold working; (c) perpendicular to preferred axis of magnetization.

forming into the required shape, the alloy is reheated to 1100°C, quenched and then tempered for several hours at 650–750°C.

5.7.3 The Mechanism of Magnetic Hardening in Cu–Ni–Fe and Cu–Ni–Co Alloys

The high coercivities of the Cu–Ni–Fe and Cu–Ni–Co alloys are due to the shape anisotropy of precipitated single-domain particles of an Ni–Fe- or Ni–Co-rich phase in a non-ferromagnetic or weakly ferromagnetic matrix. At high temperatures the alloys are fcc solid solutions, but at low temperatures the separated phases are both slightly tetragonal. The ferromagnetic phase has a ratio $c/a<1$, while the Cu-rich non-ferromagnetic phase has $c/a>1$. X-ray studies show that there is a periodic variation in the lattice parameters and compositions, which suggests that the phase separation occurs by spinodal decomposition. From replication electron micrographs, Biedermann and Kneller (1956b) have concluded that in the alloy $Cu_{0.65}Ni_{0.20}Fe_{0.15}$ the segregated particles are thin Ni–Fe-rich lamellae of thickness about 14 nm and diameter about 110 nm, and are therefore sufficiently small to be single domains. Biedermann and Kneller (1956) attribute the coercivity to the shape anisotropy of these thin lamellae, in which the two long axes are assumed to be of unequal length. Thus a lamella of thickness 14 nm, length 140 nm and width 84 nm is expected to have a coercivity of about 7.2×10^{3} A m^{-1}. This is certainly of the correct order of magnitude, but there is unfortunately no direct electron-microscopic evidence for this. Thus it seems probable that the magnetic hardness in both Cu–Ni–Fe and Cu–Ni–Co alloys is due to the precipitation of Ni–Fe-rich or Ni–Co-rich single-domain particles with uniaxial shape anisotropy. Since the permanent magnet properties of both alloys can be improved by cold working, it is reasonable to assume that this is due to preferential growth and orientation of the axes of elongation in the single-domain lamellae with $a\neq b\neq c$. The above conclusions are supported by the detailed magnetic measurements made by Joffe and Gaunt (1971). A study of the phase separation process and the morphology of the particles by Butler and Thomas (1970), using transmission electron microscopy, is also consistent with the above conclusions.

5.8 Fe–Co–V ALLOYS OR VICALLOYS

The permanent magnet properties of the Fe–Co–V alloys were first reported by Nesbit and Kelsall (1940) and Nesbit (1946). There are two main classes of these alloys, known as Vicalloy 1 and Vicalloy 2, though the class 1 alloys have three subdivisions. The composition ranges of interest are as follows

Vicalloy 1	Fe–40 wt% Co–4% V	magnetically soft
Vicalloy 1-α	Fe–(45–55%)Co–(4–9%)V	magnetically hard
Vicalloy 1-γ	Fe–(45–55%)Co–(9–11%)V	magnetically hard
Vicalloy 2	Fe–52% Co–(11–20%)V	magnetically hard.

All of the above alloys are mechanically soft, so they can be rolled and drawn. The coercivities $_BH_c$ are typically in the range 8×10^3–32×10^3 A m^{-1}, with a maximum energy product of about 22 kJ m^{-3}. The best permanent magnet properties are obtained only after cold working the alloys by rolling and drawing. The alloys are then aged for 1–2 h at about 500–600°C. The properties improve as the degree of cold working and the vanadium content increase, provided that the latter is kept below about 14%. The effects of composition and cold working have been discussed in detail by Nesbit (1946). The dependence of the magnetic properties on the vanadium content is summarized in Fig. 5.24. Typical demagnetization curves for Vicalloys 1 and 2 are shown in Figs 5.25 and 5.26. An extensive study of the effects of various thermomagnetic treatments has been made by Shur *et al.* (1957a, b), who have shown that exceptionally good permanent magnet properties can be obtained for a Vicalloy with 36 wt% Fe–52% Co–12% V if it is heat-treated at 575°C while under tensile stress. The resulting demagnetization curves are shown in Fig. 5.27, from which it can be seen that the best properties are obtained for an applied stress of 2992 MN m^{-2}, namely $_BH_c = 7\times10^4$ A m^{-1}, $B_r = 1.25$ T

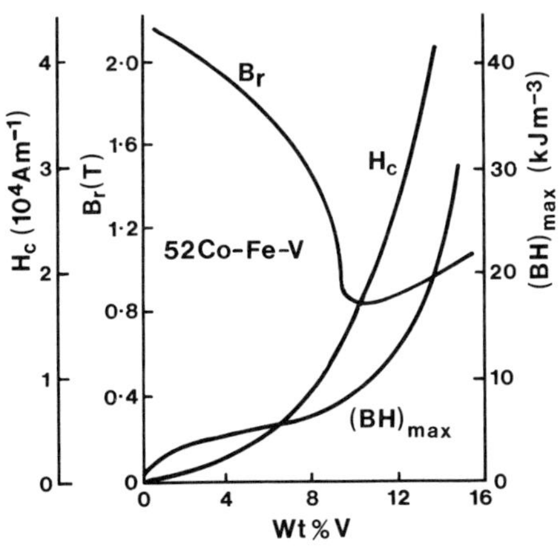

Fig. 5.24 Dependence of permanent magnet properties of 52 wt% Co–Fe–V alloys (Vicalloys) on vanadium content.

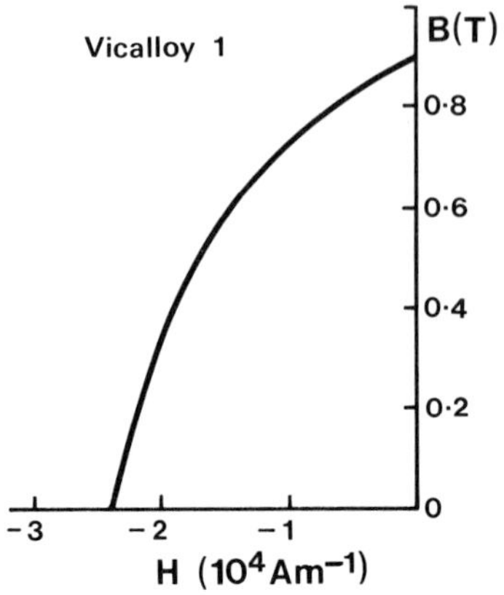

Fig. 5.25 Typical demagnetization curve for Vicalloy $1-\alpha$.

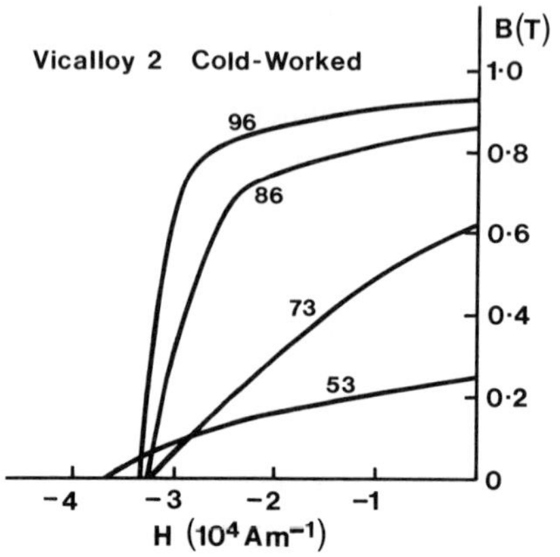

Fig. 5.26 Typical demagnetization curves for Vicalloy 2 after various degrees of reduction (%).

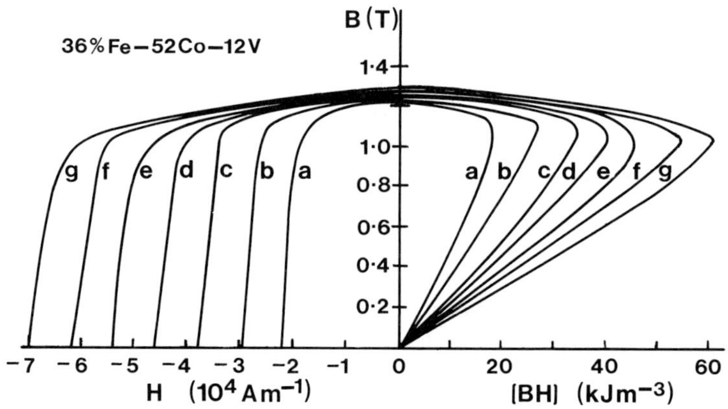

Fig. 5.27 Demagnetization curves for Vicalloy 2 (52 wt% Co–36% Fe–12% V) after heat treatment at 575°C while under tensile stress σ: (a) $\sigma=0$; (b) 500 MN m^{-2}; (c) 990 MN m^{-2}; (d) 1491 MN m^{-2}; (e) 1991 MN m^{-2}; (f) 2492 MN m^{-2}; (g) 2992 MN m^{-2} (after Shur *et al.*, 1957a).

Table 5.4 Composition and properties of some Fe–Co–V alloys.

Composition (wt%)			B_r (T)	$_BH_c$ (kA m^{-1})	$(BH)_{max}$ (kJ m^{-3})
Co	V	Others			
52	10	—	0.9	24	8
53	12	—	1.0	24	14.3
52	4	—	1.8	4.8	4.8
52	14	—	1.0	41.5	28
53	8	4 Cr	1.15	29.5	25
53	8	4 Cr	1.05	24	15
53	5	4 Cr	1.3	10.4	8.8
45	4	6 Ni	1.5	4.8	4.8
52	12	—	1.02	33.4	24
52	12	—	0.99	44	29.5

and $(BH)_{max}=61.6$ kJ m^{-3}. These properties are a considerable improvement on those of the conventional Vicalloys given above.

The magnetic properties of various Vicalloys are summarized in Table 5.4.

5.8.1 Mechanism of Magnetic Hardening in the Vicalloys

The magnetic properties of the Vicalloys are very complex, and as yet there is no completely satisfactory theory of the coercivity. However, as a result of both magnetic and microstructural investigations, it has been concluded that the principal stages in the development of the permanent magnet properties of the type 2 Vicalloys can be summarized as follows:

1. Heat treatment of the homogenized alloy at temperatures greater than about 450–500°C causes a decomposition into a ferromagnetic bcc α phase that is Fe–Co-rich and a non-ferromagnetic fcc γ phase. The transformation is accompanied by a contraction in volume of the specimen, and is thought to be martensitic.
2. During heat treatment, the bcc Fe–Co lattice becomes progressively more ordered.
3. Since the magnetocrystalline anisotropy of the bcc Fe–Co phase is very small and vanishes at 45% Co, it is possible that the coercivity is due principally to shape anisotropy, though some contribution to this from stress anisotropy cannot be excluded.
4. Cold rolling of the Vicalloys gives rise to a magnetic anisotropy in which the preferred or easy direction of magnetization is perpendicular to the rolling plane. Since the roll-induced anisotropy vanishes at about 300°C, it clearly is not related to the rolling texture that can be observed by X-ray diffraction.
5. When the cold-rolled alloys are tempered, the high-temperature fcc phase transforms martensitically into the ferromagnetic bcc α phase and a non-ferromagnetic fcc γ phase. The bcc α phase becomes progressively more ordered, as before.

Although the detailed mechanism of magnetic hardening is unknown, it seems to be generally agreed that the coercivity is due to impeded magnetization reversal resulting from the fact that the ferromagnetic bcc α phase is distributed as fine particles in the non-ferromagnetic fcc γ matrix. Since the particles or regions of the bcc Fe–Co phase are too small to be observable by replication electron microscopy, it seems probable that they are sufficiently small to make domain wall formation difficult or unfavourable. Those walls that do form will then have to negotiate their movement through a rather tangled interlocking aggregate of Fe–Co particles with variable shape anisotropy. This conclusion is further supported by the observation that the coercivity is not strongly dependent on the temperature; i.e. if the coercivity is proportional to the saturation magnetization then ${}_iH_c(T)\propto M_s(T)$. For the alloy 36 wt% Fe–52% Co–8% V–4% Cr the coercivity is almost constant from 673 K down to 73 K. An increase of about 10% in ${}_iH_c$ would be expected if the alloy behaved according to the above rather simplified mechanism.

However, it should be pointed out that Henkel (1966a, b) has proposed an entirely different mechanism for the magnetic hardening process. Henkel has suggested that on quenching, needles of the ferromagnetic α phase are formed by a diffusionless martensitic transformation from the γ phase and that the transformation can also be produced by cold working. In Vicalloy 2, Henkel suggests that martensite is formed by the cold working and that the subsequent heat treatment causes precipitation and dispersion of single-domain α particles in a non-ferromagnetic matrix. The coercivity is thought to be due to shape anisotropy. Unfortunately, there is no convincing experimental evidence to confirm or refute these suggestions.

5.8.2 Applications of Vicalloys and Related Alloys

Since the Vicalloys are malleable and ductile, they are particularly suitable for applications requiring small magnets, since these can easily be manufactured in the form of wire or strip and shaped as required. The Vicalloys are widely used for the rotors of hysteresis motors, in ammeters, measuring instruments, clock motors, magnetic escapements, speedometers and compass needles. The properties and applications of the Vicalloys have been discussed by Fahlenbrach (1964) and Bretthauer (1968).

5.8.3 The Sixtus and Tonks Magnetic Switching Phenomenon

Sixtus and Tonks (1931) showed that when hard-drawn 10–20% Ni–Fe wires with positive magnetostriction and low magnetocrystalline anisotropy were magnetized to saturation under tension they formed uniaxial single domains with an easy axis of magnetization parallel to the axes of the wires. It was shown that magnetization reversal occurred only when the reverse field reached a critical value and that the magnetization reversal corresponded to the rapid propagation of a domain wall i.e. a large Barkhausen discontinuity. Sixtus and Tonks (1931) showed that the deformation and stress converted the original smoothly varying hysteresis loop into a rectangular hysteresis loop with a coercivity corresponding to the critical field. A similar effect can be created using Vicalloy wires and has been patented by Wiegand (1981) who produced what are now known as Wiegand sensors. For a review of Wiegand magnetoelastic sensors see Rauscher and Radeloff (1989).

5.8.4 Vicalloys and the Wiegand Effect

The ductility of Vicalloy permanent magnets is of great value in the production of various security devices, e.g. Vicalloy wire segments can be embedded in plastic key cards. Strips of Vicalloy can also be used as antitheft devices by tagging library books, clothing or other goods. When the item is passed through a suitable detection gate, the tag emits a signal unless it is properly desensitized or removed. Currently, about 10 000 kg of Vicalloy strips approximately 50 μm thick are produced for these purposes, but they have to compete with microwave and radiofrequency security systems. The principal advantage of the Vicalloy strips is that they are cheap (approximately 5 pence per tag), but the security area is only about 1 m^2. Microwave and radiofrequency systems can operate over a distance of 1–5 m.

The Vicalloy wire segments used in plastic key cards are also Wiegand sensors. The magnetic properties of Vicalloys are very sensitive to plastic deformation, as has already been described. Wiegand found that the magnetic properties of a wire could be selectively altered by twising the wire clockwise and anticlockwise about the axis. After heat treatment at about 300°C, a helical pattern of strain could be incorporated close to the surface of the wire. This produced a wire in which the core had a relatively low coercivity surrounded by a cylindrical shell of higher coercivity. In the presence of a critical field the wire behaves as a bistable device in which the direction of magnetization parallel to the wire axis can be switched very rapidly from one direction to the other. The change in the direction of the magnetization results in the generation of relatively large emfs, approximately 5–10 V. The advantage of Wiegand array codes is that, unlike particulate recorded information, they cannot easily be duplicated. Wiegand-sensor Vicalloy wires have also been used for position sensors and voltage pulse generators, e.g. in automobile ignition triggers, speed and position sensors, and antilock systems. Improved alloys with a wider range of coercivities such as Fe–8% Mo–1% Ni and Fe–12% Cr–5% Mo are also suitable for these applications, and have the further advantage that they are cobalt-free, hence resulting in a reduction in cost.

5.9 Fe–Co–V–Cr AND Fe–Co–Cr ALLOYS

Fahlenbrach and Baran (1964) have shown that the replacement of some or all of the vanadium in the Vicalloys by chromium results in an increase in the remanence, although the coercivity is not changed significantly. These alloys also respond to cold working, and the resulting increased remanences give squarer hysteresis loops, which are advantageous in applications such as rotors for small hysteresis motors.

Since chromium is cheaper than vanadium, the replacement of some of the vanadium by chromium is desirable on economic grounds. Indeed, many of the investigations of the Vicalloys cited earlier were carried out on alloys containing about 4 wt% Cr. Vicalloy 2 usually has a composition close to 36 wt% Fe–52% Co–8% V–4% Cr.

The above conclusions concerning the principal stages in the development of the permanent magnet properties of the Vicalloys are also expected to apply to those containing small additions of chromium. However, in alloys that contain larger amounts of chromium, such as 55 wt% Fe–30% Co–15% Cr (Fahlenbrach and Baran, 1964), which have magnetic properties similar to those of the Vicalloys, the mechanism of magnetic hardening is probably quite different (see Section 5.6).

5.10 Fe–Co–Mo ALLOYS

The useful permanent magnet properties of Fe–Co–Mo alloys were first investigated by Seljesater and Rogers (1932) and Köster (1932). Although their applications are limited, they are still available commercially. The alloys are variously known as Comalloy, Remalloy, Indalloy or Comol, usually have a composition close to 71% Fe–12% Co–17% Mo and should contain no carbon. The permanent magnet properties are obtained by quenching from about 1200–1300°C followed by tempering at 675–725°C for 1–2 h. The coercivity in the quenched state is about 800 A m^{-1}, but

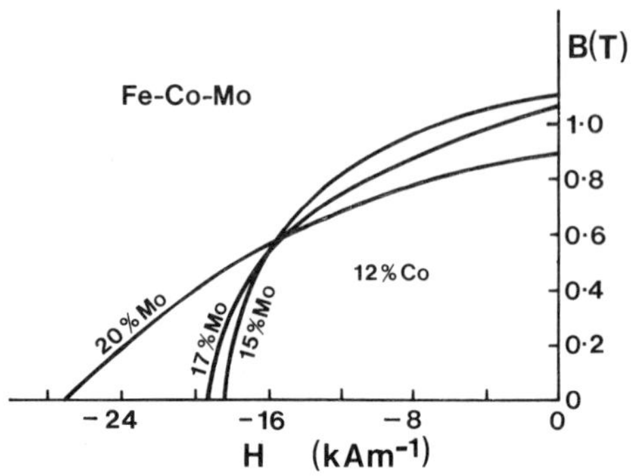

Fig. 5.28 Typical demagnetization curves for Fe–Co–Mo alloys containing 15, 17 and 20 wt% Mo.

after tempering it rises to 2×10^4 A m^{-1}, depending on the Mo content. After tempering, the alloys become rather brittle, so any forming processes are carried out at high temperatures where the alloys are single-phase. The alloys have only one major application, namely, as bias magnets in telephone receivers. Typical demagnetization curves are shown in Fig. 5.28.

The coercivity is thought to be due to the precipitation of non-ferromagnetic lamellae of a complex ε-Fe_3Mo_2 phase and a Co–Mo phase precipitated in a bcc iron solid solution, which hinder domain wall movement and rotation of the magnetization vectors in or through the ferromagnetic regions. Unfortunately, the origin of the coercivity in these alloys does not appear to have been investigated in detail. Since the coercivities of Fe–Co–Mo alloys are generally rather low, it seems very unlikely that single-domain particle anisotropy is involved.

The magnetic properties can be significantly improved by cold working; energy products of about 8–10 kJ m^{-3} have been obtained.

The alloys can also be readily hot-worked, even to very thin sheets suitable for punching to any shape and size. This property makes them particularly appropriate for use in telephone receivers, an important application for which many millions are required, although the material is being replaced by Fe–Co–Cr alloys, see e.g. Chin *et al.* (1978) and Wonsiewicz *et al.* (1979).

5.11 SEMIHARD PERMANENT MAGNET ALLOYS

There are some applications in which mechanically workable materials with a high degree of squareness of the demagnetization curve and high remanence but moderate coercivity are required; these include pulse-switched devices, e.g. fereed switches and latching relays (Chin *et al.*, 1969; Tokuyoshi, 1971).

In these devices the connection is required to be broken by an applied field of 0.4–8 A m^{-1}. Another important application of semihard magnetic materials is for the rotors of hysteresis motors, where high coercivities are undesirable. The torque on the rotor is due to the fact that the magnetization vector lags behind the rotating magnetizing field. The magnitude of the torque is proportional to the area enclosed by the B–H loop (or the area of the M–H or J–H loop) of the rotor material, which is also proportional to the coercivity. However, if the coercivity is too high, the available magnetizing field may be insufficient to produce a high enough magnetization. Thus the rotor material must be chosen very carefully so as to maximize the torque available for the magnetizing field. Materials with high saturation magnetic polarizations of about 1.8 T and coercivities in the range 15–20 kA m^{-1} are suitable for the rotors of hysteresis motors.

The use of modified Vicalloys (Fe–Co–V) has been mentioned above, but many other alloys have also been considered for such applications.

Since the assembly of devices that use semihard permanent magnet alloys may involve the accidental application of stress, it is important that this should not affect the magnetic properties significantly. Nesbitt *et al.* (1968) have investigated the alloy 82% Co–12% Fe–6% Au, in which the Co/Fe ratio has been chosen to minimize the saturation magnetostriction. This alloy has a high demagnetization curve squareness ratio, a coercivity ${}_BH_c \approx 0.8$–1.6 kA m^{-1} and a remanence $B_r = 1.8$ T. Sasaki *et al.* (1970) described a similar alloy with a composition 85% Co–12% Fe–3% Nb, which can be quenched and drawn into thin wires, and after tempering gives ${}_BH_c = 1.6$ kA m^{-1} and $B_r = 1.56$ T.

The magnetic properties of Co–Fe–Nb alloys have also been discussed by Okada *et al.* (1973) and those of Co–Fe–Nb–Mo by Suzuki *et al.* (1978), who have shown that the addition of 2–3% Mo to the alloys increases the coercivity without reducing their workability. They found ${}_BH_c = 35$ kA m^{-1}, $B_r = 1.32$ T and $J_s = 1.7$ T.

For a discussion of alloys suitable for latching reed relays the reader is referred to the paper by Tokuyoshi (1971), who has discussed the magnetic properties of Fe–Ni–Al–Ti alloys, which have a high degree of demagnetization-curve 'squareness', coercivities in the range 0.8–5.6 kA m^{-1} and remanences up to a maximum of 1.5 T.

Chin *et al.* (1969) found that the alloy 80% Co–11% Fe–9% Ti was age-hardenable and ductile and has a high J_s, a low remanence and coercivities in the range 0.8–3.6 kA m^{-1}. The magnetic properties of Co–Fe–Ti alloys, in particular 82% Co–12% Fe–6% Ti, have been studied by Chin *et al.* (1969). The origin of the coercivity and the relationship between the microstructure and the magnetic properties have been investigated by Shilling and Soffa (1973) and Prokop and Soffa (1978). The former authors showed that the microstructure of the solution-treated, quenched and aged alloy consisted of a period array of non-ferromagnetic precipitates and that the coercivity was directly proportional to the average precipitate size.

Kawaguchi and Ogawa (1971) investigated cold workable Fe–Cu alloys and obtained coercivities and remanences in the ranges 1.6–20 kA m^{-1} and 0.6–1.3 T respectively.

Olsen and Stoffers (1971) found that the alloy Fe–3% Co–15% Cr was ductile (provided that the carbon content was in the range 0.04–0.08%), and also had useful semihard magnetic properties. The Fe–Co–Cr system can also be used to produce workable permanent magnets with relatively high coercivities (about 80 kA m^{-1}), as described earlier.

High remanence and square hysteresis loops have been observed in Fe–Ni and Fe–Mn alloys by Jin and Tiefel (1980).

The origin of the coercivity in Co–Fe–Ni alloys with Al and Ti additions has been investigated by Radeloff and Pfeiffer (1975).

The magnetic properties of square-hysteresis-loop semihard permanent magnet materials have been reviewed by Pinnel (1976, 1992), and are summarized in Table 5.5.

Table 5.5 Alloys proposed as having possible application in reed contacts and their nominal properties (Pinnel, 1992; courtesy of Pergamon Press).

Alloy composition (%)	Coercive force ${}_BH_c$ (kA m^{-1})	Remanent flux density B_r	Squareness ratio B_r/B_s	Magnetostrictive coefficient (10^{-6})	Thermal expansion coefficient (10^{-6}) 20–400°C	Formability	Plateability	Solderability	Sealability	Ageing temperature (°C)
49 Co–48 Fe–3 V (Remendur)	2.4	1.8	0.90	50	10.2	Good	Good	Good	Good	610
38 Co–39 Fe–20 Ni–3 Nb	2.4	1.6	0.95	~45	9.8	Good				600
Co–Fe–Ni–Al–Ti (Vacozet 655)	3.2	1.4	0.90	48	11	Good	Good	Poor	Good	550
85 Co–12 Fe–3 Nb (Nibcolloy)	1.6	1.5	0.95	2	12.1	Good	Good	Good	Good	700
82 Co–13 Fe–5 Mo	2.4	1.3	0.92	Slightly positive	~12	Good				650
84 Co–12 Fe–4 Ti	1.8	1.4	>0.90	Slightly positive	~12	Good				650
82 Co–12 Fe–6 Au (Coballoy)	1.1	1.6	0.85	Small	12.5	Good	Good	Good	Good	550
89 Co–10 Fe–1 Be	2.4		0.90	Small	~12	Good				800
54 Fe–28 Ni–18 Co (Kovar)	<0.8	1.8	<0.70		5	Good	Good	Good	Good	
80 Fe–16 Ni–3 Al–1 Ti	2.4	1.25–1.4	0.8–0.95	0.4	12.5	Good	Good	Poor		600
78 Fe–16 Ni–5 Cu–1 W	2.8	1.4	0.95		13	Good	Good	Good		550
80 Fe–18 Cu–2 Mn	2.4	1.5	0.90		11.5	Good	Good			500
C-steel	1.6–3.2	1.0	<0.70	Slightly negative	13	Good	Good	Good	Fair	
Cr–Co–C steel	1.6–4	1.0–1.25		Small	13				Fair	300–500
92 Fe–8 Mn	2.2	1.8	0.96	1.4		Good			Poor	400–450
Fe–7 Mo–5 Ni	2.4	1.9	0.97	Small		Good			Good	600

5.12 Mn–Al–C MAGNETS

15.12.1 Mn–Al Alloys

The potentially useful permanent magnet alloys of Mn–Al have compositions in the range 52–58 at% Mn. The ferromagnetism in the system is confined to the metastable tetragonal τ-phase which can be produced by controlled cooling from the disordered hcp high temperature ε-phase through the transition temperature$\simeq$750°C, or by heat treating the rapidly quenched hcp phase at temperatures in the range 380–600°C (Kono, 1958; Köster and Wachtel, 1960; Koch *et al.*, 1960; McCurrie *et al.*, 1978). If the heat treatment is prolonged, the τ-phase decomposes into two non-ferromagnetic phases β-Mn and η. The τ-phase is actually ferrimagnetic rather than ferromagnetic. Since the τ-phase is produced by cooling the alloy from the ε-phase, it is necessary to have a slight excess of Mn ions in the alloy to ensure that the necessary parent hexagonal ε-phase is present at high temperatures. It is now known that the τ-phase is formed by a martensitic transformation from the hexagonal ε-phase via an intermediate ordered orthorhombic ε^1-phase to form the face-centred tetragonal τ-phase i.e.

$$\varepsilon(\text{hcp}) \rightarrow \varepsilon^1(\text{orthorhombic}) \rightarrow \tau(\text{fct}).$$

The saturation magnetic polarization of the alloy depends on the degree of long-range order and hence on the thermal history but is usually in the range 0.4–0.85 T.

The magnetic properties of Mn–Al alloys close to the composition $Mn_{55}Al_{45}$ have been investigated by Kono (1958) and Koch *et al.* (1960). The latter showed that if the cast alloy is swaged, permanent magnets can be produced with $B_r = 0.428$ T, ${}_iH_c = 368$ kA m^{-1}, ${}_BH_c = 220$ kA m^{-1} and $(BH)_{max} = 28$ kJ m^{-3}.

According to Kamino *et al.* (1966) the metastable τ-phase of $Mn_{55}Al_{45}$ has a uniaxial magnetocrystalline anisotropy energy of about 10^5 J m^{-3} with the easy axis of magnetization along the *c*-axis of the face-centred tetragonal unit cell. This corresponds to an anisotropy field of about 2.5×10^6 A m^{-1} so that if single domain particles of the alloy could be made which reverse their magnetization by coherent rotation coercivities $\sim 2.5 \times 10^6$ A m^{-1} are theoretically possible. However, the observed coercivities are only about one-fifth of the theoretical value even for particles a few microns in diameter. Zijlstra and Haanstra (1966) have suggested that this may be due to the nucleation and pinning of domain walls at antiferromagnetic stacking faults or antiphase boundaries (APB) which divide two oppositely magnetized domains.

By means of electron microscope studies Jakubovics and Jolly (1977) concluded that the antiphase boundaries acted as domain wall nucleation sites and thus lowered the coercivity whereas the stacking faults, resulting from the phase transformations, acted as domain wall pinning sites and thus raised the coercivity.

Van Landuyt *et al.* (1978) have suggested that the high remanences and coercivities of the extruded or swaged magnets is due to the absence of antiphase boundaries and to the high density of other defects such as stacking faults and possibly precipitation. The Mn–Al alloy studied contained an addition of 0.5 wt% C but the microstructures were very similar to those observed by Jakubovics and Jolly (1977).

5.12.2 Isotropic Mn–Al–C Magnets

It was discovered that the addition of carbon improved the stability of the ferromagnetic τ phase in Mn–Al alloys and at the same time increased the coercivity and remanence (see Yamamoto, 1972). This led to the production of the first successful isotropic Mn–Al–C magnet, with $(BH)_{max} = 9.6$ kJ m^{-3}, $B_r = 0.28$ T and $_BH_c = 120$ kA m^{-1}. This alloy had a composition 72 wt% Mn–27% Al–1% C. The permanent magnet material was prepared by induction melting at 1400°C. The alloy was then solution-treated at 1000°C and cooled to below 600°C, after which it was annealed for one hour at 600°C. A secondary property of these magnets that is of considerable importance is that they are machinable and have high mechanical strength.

5.12.3 Anisotropic Mn–Al–C Magnets

Although Mn–Al–C alloys are hard and brittle at room temperature, they can be plastically deformed at temperatures between 600 and 800°C. Anisotropic magnets are made by solution-treating the cast alloy (70 wt% Mn–29.5% Al–0.5% C) at 1000°C for one hour, after which it is cooled to below 600°C. The billet is then coated with a lubricant and extruded by a hydraulic press at about 700°C, followed by a short anneal at the same temperature. The resulting magnet is anisotropic, with easy direction of magnetization parallel to the axis of the extruded bar. The magnetic properties of these Mn–Al–C magnets are summarized in Table 5.6.

Typical demagnetization curves for the isotropic and anisotropic Mn–Al–C alloys are shown in Fig. 5.29.

The production of these anisotropic Mn–Al–C magnets by hot extrusion of the cast alloy has the important advantage that it avoids the expensive thermomagnetic treatment necessary to obtain comparable properties in the Alnicos. Also, since the alloys can be cut and machined relatively easily, they can be mass-produced by a continuous hot extrusion process. Mn–Al–C magnets are very resistant to corrosion by the atmosphere, even at relatively high temperatures and high relative humidity. The thin oxide film that is formed protects the magnets from further attack and remains unchanged for very long periods. The alloys are of course corroded by acids and alkalis. Since Mn and Al are relatively abundant and cheap compared with Co and Ni, which are required for the manufacture of Alnico magnets, they have a considerable cost advantage – particularly since magnetic properties comparable to those of the Alnicos can be obtained by a much simpler and cheaper process. The metallurgical structure of Mn–Al–C magnets is stable up to about 500°C, so that the magnets can be completely thermally demagnetized by heating above the Curie temperature of 300°C. The theoretical maxima

Table 5.6 Magnetic properties of anisotropic Mn–Al–C permanent magnets.

Maximum energy product $(BH)_{max}$	40–56 kJ m^{-3}
Remanence B_r	0.52–0.62 T
Coercivity $_BH_c$	160–208 kA m^{-1}
Average recoil permeability μ_{rec}	1.0–1.2
Curie temperature T_c	300°C

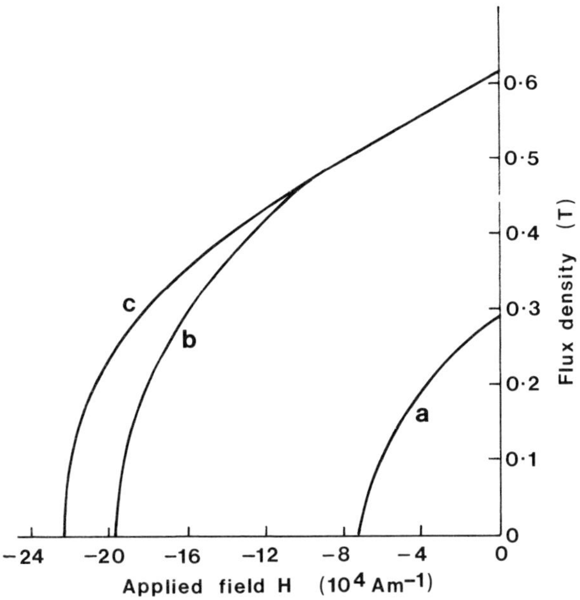

Fig. 5.29 Demagnetization curves for 70 wt% Mn–29.5% Al–0.5% C: (a) before extrusion; (b) after extrusion; (c) after ageing (after Ohtani *et al.*, 1977).

for the energy product and coercivity $_BH_c$ are 104 kJ m^{-3} and 3.2 MA m^{-1}, so it is clear that there is considerable potential for improvement.

Sakamoto *et al.* (1980) have developed a technique for producing radially or circumferentially magnetized anisotropic nickel-doped Mn–Al–C magnets (69.5 wt% Mn–29.3 wt% Al–0.5 wt% C–0.7 wt% Ni). Bars of the alloy are first hot-extruded at 700°C and then after cooling, they are plastically deformed parallel to the extrusion axis at 660°C. From magnetic torque measurements it was concluded that the bars were uniformly magnetized in the plane perpendicular to the axis of extrusion and plastic deformation. The optimum magnetic properties in this easy plane of magnetization were:

$$B_r = 0.46\ \text{T}$$

$$_BH_c = 232\ \text{kA m}^{-1}$$

$$_iH_c = 328\ \text{kA m}^{-1}$$

$$(BH)_{max} = 32.8\ \text{kJ m}^{-3}.$$

The magnetic properties parallel to the extrusion axis are significantly inferior viz.

$$B_r = 0.27\ \text{T}$$

$$_BH_c = 168\ \text{kA m}^{-1}$$

$$(BH)_{max} = 12\ \text{kJ m}^{-3}.$$

These radially magnetized magnets are therefore particularly suitable for applications which require multipole magnetization such as in motors and generators. They are also suitable for small loudspeakers.

Yamaguchi *et al.* (1989) have shown that high density oriented magnets with $(BH)_{max} \sim 50$ kJ m^{-3} can be produced by the extrusion of gas atomized powders.

5.12.4 Applications of Mn–Al–C Permanent Magnets

Very small Mn–Al–C magnets find applications in many devices and machines, e.g. in control equipment, reed switches, watches, sewing machines, loudspeaker magnets, business machines and Hall-effect integrated circuits. Since the magnets have a high energy product per unit weight, good machinability and high mechanical strength, they are leading to the design of more compact, lighter and more efficient magnetic circuits.

5.13 Pt–Co ALLOYS

The permanent magnet properties of the platinum–cobalt alloys were discovered by Jellinghaus (1936) and subsequently investigated by Newkirk *et al.* (1950). Near the equiatomic composition PtCo, the alloys may have intrinsic coercivities in the range 0–5.3 × 10^5 A m^{-1} and remanences up to a maximum of 0.72 T. The high coercivity is due to a transformation from a disordered fcc phase to an ordered tetragonal phase with $c/a = 0.98$. From the phase diagram, given by Newkirk *et al.* (1951a, b), it can be seen that the ordering transformation occurs at about 825°C.

According to Martin (1958), the optimum permanent magnet properties are obtained when the alloy is cooled from 1000°C at 150°C min^{-1} and heat-treated at 600°C for 5 h, although excellent properties can also be obtained by heat treatment at temperatures up to about 730°C. The magnetic properties are therefore not strongly dependent on the heat-treatment temperature for the ordering reaction.

Since the alloy is ductile and very resistant to corrosion (even to mineral acids or caustic alkalis), it is still used in many specialized applications, e.g. in surgery, various high-quality instruments, hearing aids and synchronous motors for watches. The

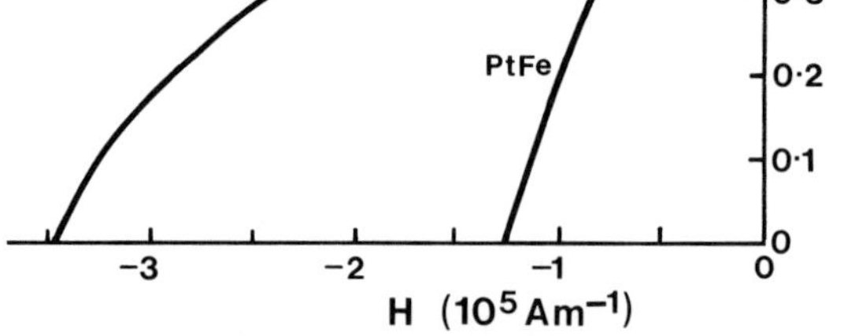

Fig. 5.30 Typical demagnetization curve for 77 wt% Pt–Co. A demagnetization curve for 78% Pt–Fe is shown for comparison.

corrosion resistance of the alloy is especially useful in pumping or metering devices, where the magnet may be in direct contact with a corrosive liquid. The main disadvantage of the material is its very high cost (it contains 77 wt% platinum). Partial substitution of the platinum by Pd, Rh, Ir or Ru has been explored, but all the resulting alloys had inferior properties. The material is marketed under the name Platinax II. The properties and applications of Platinax II have been discussed by Mintern (1961). A typical demagnetization curve is shown in Fig. 5.30, for which the $(BH)_{max}$ product is 76 kJ m^{-3}.

The highest energy product for Pt–Co, of order 100 kJ m^{-3}, and an intrinsic coercivity $({}_iH_c)$ of 400 kA m^{-1}, has been obtained by Kaneko *et al.* (1968). These values were obtained by quenching the alloy from 1000°C into a salt bath at 680–720°C, holding for 3–10 min, after which the alloy was water-quenched and finally heat-treated at 600°C for 30–100 min. No significant improvements in the permanent magnet properties of Pt–Co alloys have been reported in the last 10 years, and it therefore appears that they have reached the zenith of their development.

5.13.1 Mechanism of Magnetic Hardening in Pt–Co

McCurrie and Gaunt (1965, 1966) have shown that both the disordered fcc and ordered fct phases in Pt–Co are ferromagnetic, with saturation magnetic polarizations of approximately 0.87 and 0.73 T respectively. There is a significant variation in the values of J_s quoted by different authors, but this may be attributed to composition variations and variations in the degree of long-range order depending on the heat treatment. Brissonneau *et al.* (1966) found that $J_s = 0.94$ and 0.966 T for the disordered and ordered Pt–Co respectively.

Although the ratio c/a $(=0.98)$ is very close to unity, McCurrie and Gaunt (1965, 1966, 1969) and Brissonneau *et al.* (1966) have shown that the ordered tetragonal phase has an extremely high uniaxial magnetocrystalline anisotropy energy, with $K_1 \approx 3 \times 10^6$ J m^{-3} and with the c-axis as the easy direction of magnetization. The disordered cubic phase, however, has normal cubic anisotropy, with $K_1 \approx -6.0 \times 10^4$ J m^{-3} and $\langle 111 \rangle$ as the easy directions of magnetization (McCurrie and Gaunt, 1966).

The crystallography of the phase transformation has been investigated by Newkirk *et al.* (1950, 1951a, b), who have shown that the ordered platelets form on the $\{110\}$ planes of the parent matrix and that they are related by $\{110\}_c \parallel (101)_t$, i.e. the (101) tetragonal plane is parallel to one of the $\{110\}$ planes of the parent cubic matrix as shown in Fig. 5.31. This orientation relationship presumably minimizes the strain energy,

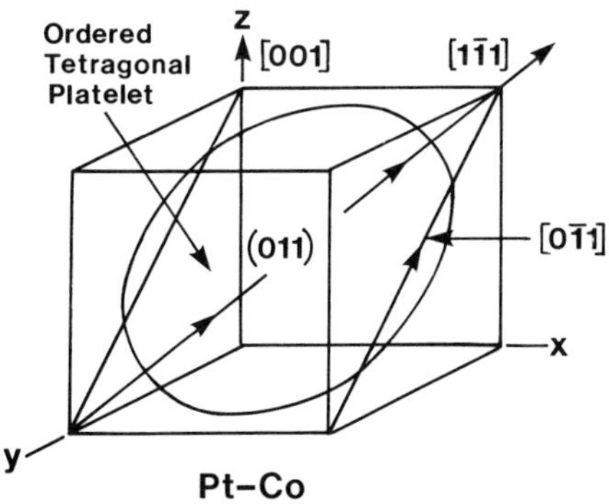

Fig. 5.31 Orientation relationship of ordered tetragonal platelets to parent disordered fcc matrix in partially transformed Pt–Co.

and since the principal axes of the cubic phase become the principal axes of the tetragonal phase, this implies that the easy *c*-axis is at 45° to the plane of the platelets. The occurrence of these small platelets on $\{110\}$ planes is also consistent with the presence of the much larger surface striations that are observed on the surfaces of the heat-treated alloys in the permanent magnet state. The appearance of these striations is probably due to the cooperative growth of the platelets in parallel sheets. The fact that the striations are parallel to $\{110\}$ planes can easily be confirmed by standard stereographic analysis.

Measurements of the magnetic properties of single crystals at PtCo and their temperature dependence have been made by McCurrie and Gaunt (1969), who have shown that the best properties are usually obtained when the material is saturated parallel to one of the $\langle 111 \rangle$ directions in the parent matrix. (Note that the $\langle 111 \rangle$ directions are the easy directions of magnetization in the disordered cubic phase.)

High coercivities can be obtained after partial transformation of the disordered phase into small particles of the ordered phase with dimensions of about 20 nm. It is believed that the optimum permanent magnet state is obtained when about 50% of the disordered fcc phase is transformed to the ordered tetragonal phase. McCurrie and Gaunt (1965, 1966) and Gaunt (1966) have suggested that the high coercivity is due to impeded domain wall motion rather than magnetization reversal by coherent rotation. A similar model has also been proposed by Mishin (1967). Since the small tetragonal platelets have very high uniaxial magnetocrystalline anisotropy energy compared with that of the cubic phase, they act as barriers to domain wall motion. The energies and widths of the domain walls in the tetragonal and cubic phases may be estimated from the following relations (see Chapter 1):

$$\gamma = 4(AK_1)^{1/2},$$

$$\delta = \pi(\mathrm{A}/\mathrm{K}_1)^{1/2},$$

with $A = kT_c/a$. Since the Curie temperature $T_c = 880$ K and the lattice parameter $a = 0.38$ nm, we obtain the following approximate values

cubic (disordered)	tetragonal (ordered)
$A = 3.2 \times 10^{-11}\ \mathrm{J\,m^{-1}}$	$A = 3.2 \times 10^{-11}\ \mathrm{J\,m^{-1}}$
$\gamma = 5.65 \times 10^{-3}\ \mathrm{J\,m^{-2}}$	$\gamma = 3.9 \times 10^{-2}\ \mathrm{J\,m^{-2}}$
$\delta = 72$ nm	$\delta = 10$ nm

Thus in the tetragonal phase the domain walls are very narrow, with high energy, whereas in the cubic phase the reverse is the case. During magnetization reversal, we may assume that the domain walls are nucleated in the low-anisotropy fcc phase or at some other point where nucleation is relatively easy, after which its movement is impeded by the tetragonal particles. When the domain wall intersects such a particle, its energy must increase so that a relatively high external field is required to force the domain wall out of the cubic phase into the tetragonal particles. The critical diameter for single-domain particles of Pt–Co can be estimated from the relation (see Chapter 1)

$$d_c = \frac{18\gamma}{\mu_0 M_s^2}.$$

With the values 6.92×10^5 and 5.81×10^5 A m^{-1} for M_s, the critical diameters for the cubic and tetragonal phases are 0.17 and 1.7 μm respectively.

Ralph and Brandon (1965) have shown by field-ion microscopy that the partially ordered tetragonal regions have dimensions of about 15–20 nm, so that they are certainly small enough to be single domains. The ordering process has also been studied by Southworth and Ralph (1966) and by Tsong and Müller (1967). The latter authors have shown that the platelets are 2–8 nm thick. However, since the surrounding disordered cubic matrix is also ferromagnetic, with a relatively low magnetocrystalline anisotropy energy, magnetization reversal must occur by domain wall displacement. During the magnetization reversal process, the domain walls must at some time intersect the tetragonal platelets, so their energy is increased due to the incorporation of a large number of particles with high uniaxial magnetocrystalline anisotropy energy. The high coercivity is therefore determined by the field required to move the cubic domain walls through the mixed cubic and tetragonal phases.

Shur *et al.* (1968) and Hadjipanayis and Gaunt (1979) have found that the optimum magnetic hardness can also be obtained when the alloy is in the fully tetragonal state, so that the above mechanism of magnetic hardening is not valid. Gaunt (1977a) has suggested that the difference in behaviour may be due to differences in the homogenization treatments. If the alloy is not homogenized at a high enough temperature for a long enough time, the nucleation of the tetragonal phase will probably occur at heterogeneous sites, so resulting in a large number of nuclei, all of which attempt to grow into each other, so that wall pinning may still occur at the boundaries between the tetragonal lamellae. However, even if the tetragonal phase forms an array of equiaxial uniaxial crystallites, a high coercivity is to be expected, because domain wall displacement will also be difficult in this case. In well-homogenized alloys the tetragonal nuclei form and grow more slowly, so that the alloy consists of both the cubic and tetragonal phases.

The structure and mechanism of magnetic hardening in a Pt–Co alloy containing 50.5 at% Co have been investigated in detail by Shur *et al.* (1968). Since the alloy studied was in the fully ordered tetragonal form (with coercivities still in the range 240–320 kA m^{-1}), they suggest that the high coercivities can be explained in terms of domain wall impedance by antiphase domains of ordering.

An electron-microscope study of the structure and morphology of a magnetically hard Pt–Co alloy containing 52 at% Co has been made by Hadjipanayis and Gaunt (1979), who showed that in the high-coercivity state (${}_iH_c = 320$ kA m^{-1}) the alloy is fully tetragonal and consists of finely divided ordered particles whose c-axes are randomly distributed and parallel to the a axes of the original disordered cubic phase. According to Hadjipanayis and Gaunt (1979), no simple model of domain wall interaction is consistent with such a structure, and they have therefore suggested that further experimental and theoretical research is necessary in order to construct a domain wall impedance model that is consistent with the structural and magnetic data presented in their paper.

Mishin *et al.* (1965) have shown that a macroscopic uniaxial anisotropy can be induced by cold-rolling the disordered alloy. Shimizu and Hashimoto (1968) and Shimizu and Horiuchi (1970) have found that an additional uniaxial anisotropy can also be induced by heat treatment under compressive stress. Eurin and Pauleve (1969) have shown that the alloy 50 at% Pt–Co responds to heat treatment in a magnetic field and develops a uniaxial anisotropy parallel to the direction of the applied field, provided that this is parallel to one of the fourfold axes in the initial disordered cubic phase. All of these induced anisotropies are small relative to the magnetocrystalline anisotropy of ordered Pt–Co, and so far no significant improvement in the permanent magnet properties has been reported. A study of the texture and domain structure of ordered equiatomic Pt–Co has been made by Brissonneau and Schlenker (1967) and Brissonneau *et al.* (1968).

5.14 HEXAGONAL FERRITES, $BaFe_{12}O_{19}$ AND $SrFe_{12}O_{19}$

5.14.1 Crystal and Magnetic Structure

The compounds $BaFe_{12}O_{19}$ and $SrFe_{12}O_{19}$ are the most important hexagonal ferrites, and are used on a very large scale as permanent magnets. The crystal structure, known as the magnetoplumbite structure, is shown in Fig. 5.32. The unit cell contains two formula units, and can be divided into four blocks, designated as SRS*R*, arranged as shown in Fig. 5.33. The asterisk signifies that the block has been rotated by 180° about the c-axis. The S blocks have the spinel structure, but are orientated so that the [111] direction is parallel to the c-axis. Since the R blocks have a rhombohedral structure, the overall symmetry of the unit cell is rhombohedral (since this has a lower symmetry), with $a=0.588$ nm and $c=2.32$ nm. This rhombohedral unit cell has one-third of the volume of the hexagonal unit cell and has the same symmetry. The relationship between the two unit cells is shown in Fig. 5.34. For crystallographic studies it is usually more convenient to use the smaller rhombohedral cell, though $BaFe_{12}O_{19}$ and $SrFe_{12}O_{19}$ and similar materials are often referred to as the hexagonal ferrites. The ferrimagnetism is of course due entirely to the 24 Fe^{3+} ions, each of which has a magnetic moment of $5\mu_B$. These ions occupy three kinds of crystallographic sites: tetrahedral, octahedral and fivefold coordination sites. The directions of the individual magnetic moments are shown in Fig. 5.33, and the magnetic structure is summarized in Table 5.7, from which the magnetic moment per unit cell (which contains two formula units, i.e. $2BaFe_{12}O_{19}$) is

$$\mu=(16-8)\times 5\mu_B=40\mu_B.$$

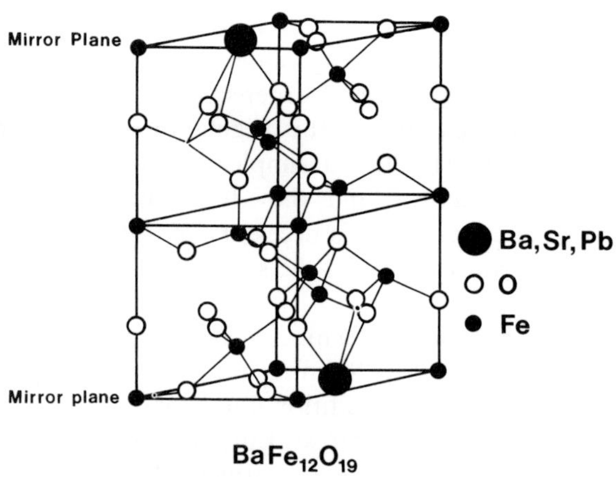

Fig. 5.32 Rhombohedral cell of hexagonal unit cell of $BaFe_{12}O_{19}$ or $SrFe_{12}O_{19}$.

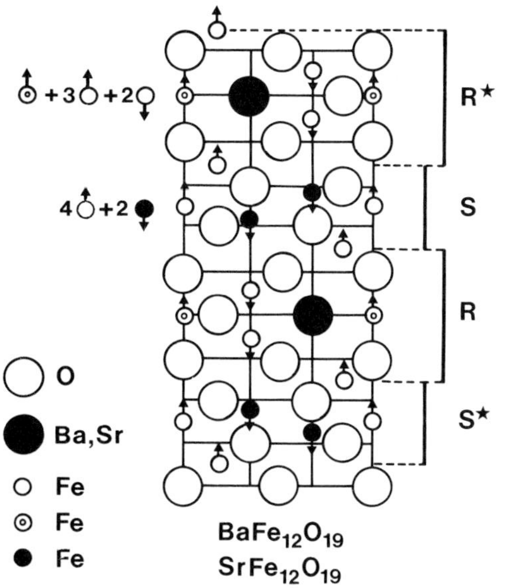

Fig. 5.33 Projection of the unit cell of barium or strontium ferrite, showing that it can be divided into four blocks designated as R*SRS*. The directions of the magnetic moments of the Fe^{3+} ions are also shown. The asterisk indicates rotation of a block by 180° about the c-axis (after Smit and Wijn, 1959).

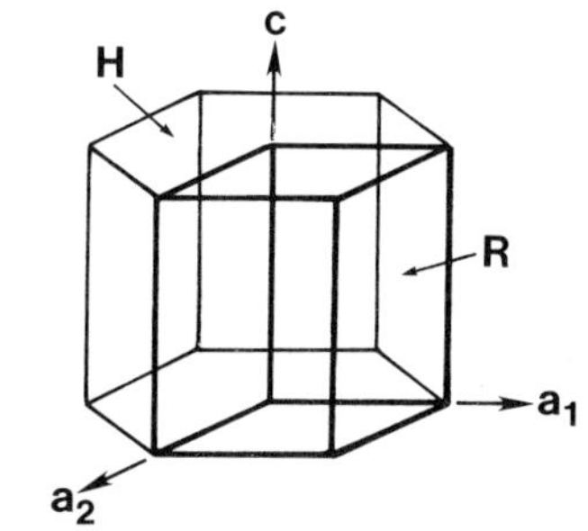

Fig. 5.34 Relationship between rhombohedral (R) and hexagonal (H) unit cells in $BaFe_{12}O_{19}$ or $SrFe_{12}O_{19}$.

Since the dimensions of the unit cell are $a = 0.588$ nm and $c = 2.32$ nm, the expected saturation magnetic polarization at 0 K is

$$J_s = \mu_0 M_s = \frac{40\mu_B\mu_0}{\text{volume of unit cell}} = \frac{40\mu_B\mu_0}{a^2 c \sin 60°}.$$

Hence

$$J_s = \frac{40 \times 9.27 \times 10^{-24} \times 4\pi \times 10^{-7}}{(0.588)^2 \times 2.32 \times 10^{-27} \times 0.866} = 0.6 \text{ T},$$

Table 5.7 Magnetic moment distribution $BaFe_{12}O_{19}$ and $SrFe_{12}O_{19}$ for unit cell containing two formula units (after Smit and Wijn, 1959).

Block	Tetrahedral sites	Octahedral sites		Fivefold coordination sites
R*	—	3 ↑	2 ↓	1 ↑
S	2 ↓	4 ↑	—	—
R	—	3 ↑	2 ↓	1 ↑
S*	2 ↓	4 ↑	—	—

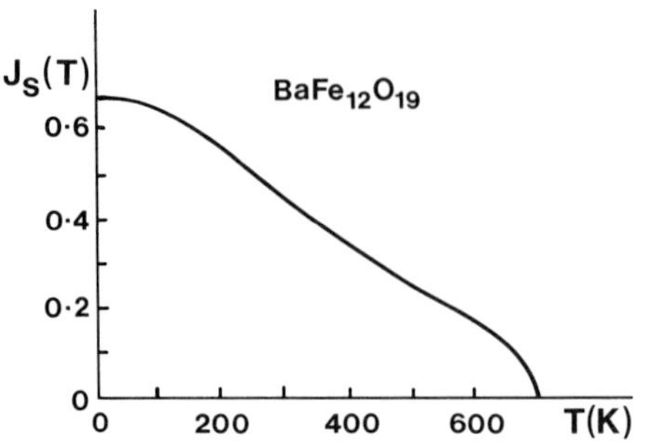

Fig. 5.35 Variation of saturation magnetic polarization with temperature for $BaFe_{12}O_{19}$.

which is in excellent agreement with the experimental value of 0.663 T at 0 K. The variation of the saturation magnetic polarization with temperature is shown in Fig. 5.35.

One of the most important properties of $BaFe_{12}O_{19}$ (and $SrFe_{12}O_{19}$) is the very high uniaxial magnetocrystalline anisotropy. The easy directions of magnetization are of course parallel and antiparallel to the hexagonal *c*-axis, so that a single crystal can be easily saturated along this axis. However, if increasing fields are applied perpendicular to the *c* axis in a single crystal, the magnetization increases linearly until the crystal is saturated as shown in Fig. 5.36. This linear magnetization curve shows that the uniaxial magnetocrystalline anisotropy energy can be accurately described by the relation

$$E = K_0 + K_1 \sin^2 \theta,$$

where θ is the angle between the magnetization vector and the *c* axis. Since K_1 is 3.22×10^5 J m^{-3}, the anisotropy field is

$$H_A = \frac{2K_1}{\mu_0 M_s} = 1.35 \times 10^6 \text{ A m}^{-1},$$

so the crystal can be saturated in a direction perpendicular to the *c* axis by this field (see Fig. 5.36). Thus if spherical single-domain particles of the material can be

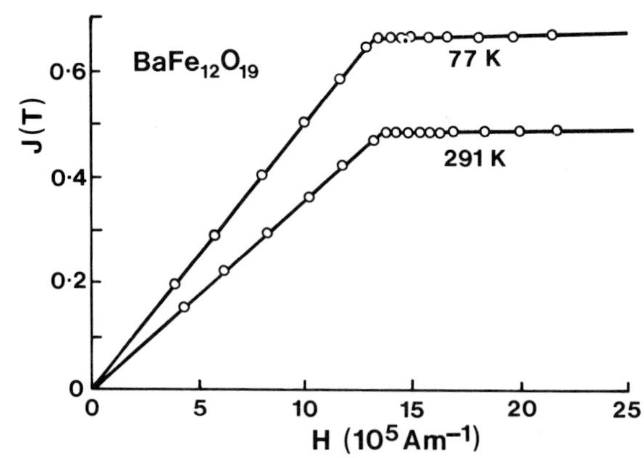

Fig. 5.36 Magnetization curves perpendicular to the easy hexagonal *c*-axis in $BaFe_{12}O_{19}$ at 77 and 291 K.

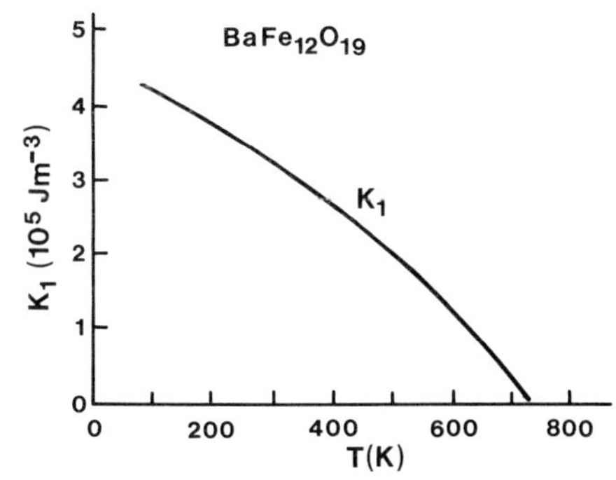

Fig. 5.37 Temperature variation of the uniaxial magnetocrystalline anisotropy coefficient K_1 for $BaFe_{12}O_{19}$.

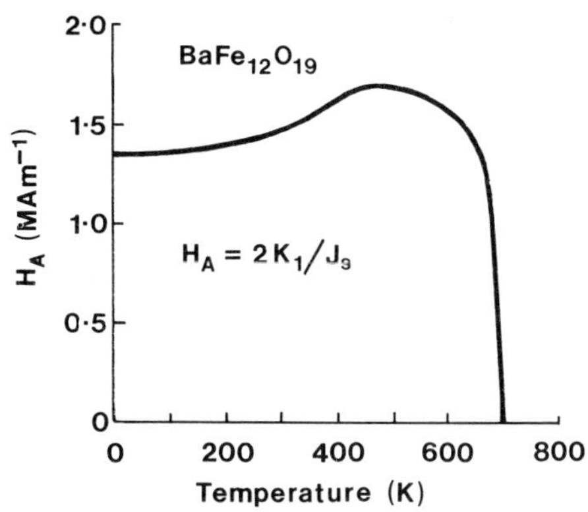

Fig. 5.38 Temperature variation of the anisotropy field H_A for $BaFe_{12}O_{19}$.

prepared, a reverse field of 1.35×10^6 A m^{-1} should be required to reverse the direction of magnetization by coherent rotation, i.e. the theoretical intrinsic coercivity is 1.35×10^5 A m^{-1}. The temperature variations of K_1 and the anisotropy field $H_A = 2K_1/J_s$ for $BaFe_{12}O_{19}$ are shown in Figs 5.37 and 5.38. The basic magnetic properties of the hexagonal ferrites have been reviewed by Kojima (1982).

5.14.2 Permanent Magnet Properties of $BaFe_{12}O_{19}$ and $SrFe_{12}O_{19}$

The permanent magnet properties of $BaFe_{12}O_{19}$ were discovered by Went *et al.* (1951/2), who showed that when fine particles of the material with diameters of order 1 μm are sintered into a solid compact of high density, it is possible to obtain intrinsic coercivities of about 380 kA m^{-1} in combination with a remanence of about 0.2 T and an energy product of 6.8 kJ m^{-3}. Barium ferrite magnets are usually prepared as follows. Fine powders of $BaCO_3$ and Fe_2O_3 are first carefully mixed in the ratio 1:6; they are then calcined in a furnace at about 1200°C, after which the resulting material $BaFe_{12}O_{19}$ is ground to a fine powder with a particle size of order 1 μm. This material is then pressed with or without a magnetic field and sintered at 1200°C in an oxidizing atmosphere. After the calcining process platelets of the hexagonal $BaFe_{12}O_{19}$ are formed in which the easy direction of magnetization is perpendicular to the plane of the platelets but parallel to the easy hexagonal axis of magnetization as shown in Fig. 5.39. This fortuitous

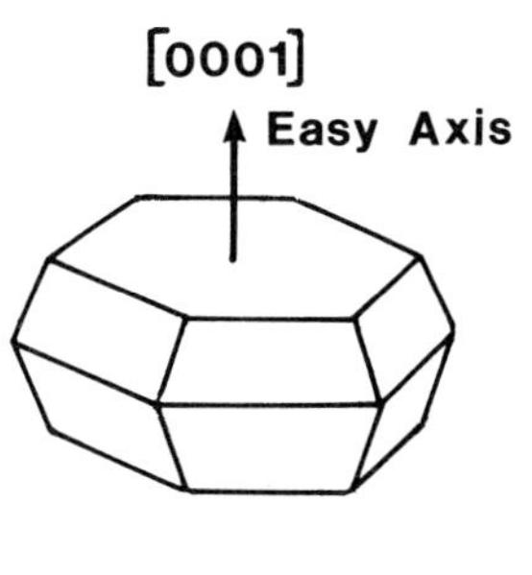

Fig. 5.39 Hexagonal crystal platelet of $BaFe_{12}O_{19}$, showing easy axis of magnetization.

growth habit is very advantageous, since during the pressing process, the planes of the hexagonal platelets tend to lie parallel to each other so as to minimize the total volume of the compact in the most energetically favourable way. However, since the preferred axis of magnetization due to the magnetocrystalline anisotropy is parallel to a hard direction because of the shape anisotropy, the demagnetizing field due to the latter assists the magnetization reversal process.

The magnetic properties of barium and strontium ferrite permanent magnets are very strongly influenced by composition, additives, particle size, the partial pressure of oxygen, and the sintering temperature and time. The addition of small quantities of SiO_2 or Al_2O_3 (0.5–1%) assists in the formation of the ferrites, and in particular it has been shown that the addition of SiO_2 results in the formation of low-melting-point eutectic phases or in some cases vitreous eutectics. Thus the sintering of the magnets takes place in the presence of a liquid phase. The effects of adding 0.5% SiO_2 to strontium hexaferrite magnets on the remanence and coercivity compared with those of the silica-free magnets are shown in Fig. 5.40.

The magnetic properties of fine-milled barium and strontium ferrites have been investigated by Richter and Dietrich (1968). These authors have shown that the magnetic properties deteriorate after excessive milling, a result that they attribute to the introduction of lattice defects during the milling process. However, they also showed that the original magnetic properties could be almost completely restored by annealing the powders at 1200°C.

Strongly anisotropic $BaFe_{12}O_{19}$ magnets (first prepared by Stuijts *et al.*, 1954) are made by wet-pressing the powder in a magnetic field before sintering. This treatment improves the alignment of the hexagonal axes. The microstructure of an anisotropic $BaFe_{12}O_{19}$ magnet is shown in Fig. 5.41. Note that in (b) the easy axis of magnetization

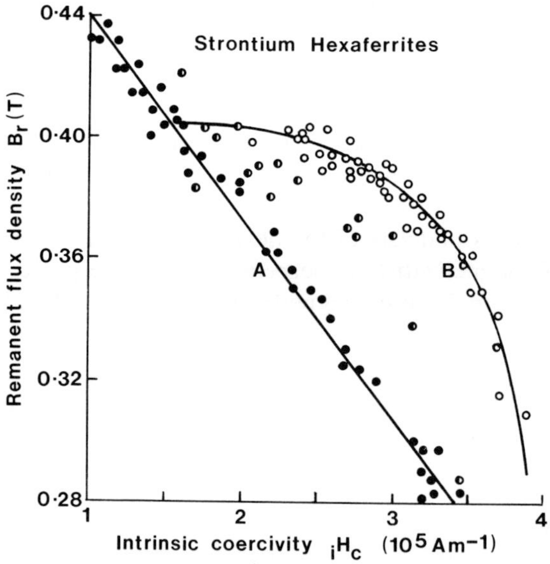

Fig. 5.40 Relationship between remanence B_r and coercivity $_BH_c$ for various strontium hexaferrites: (A) silica-free; (B) 0.5% silica (SiO_2) (after Krijtenberg, 1965; Stuijts, 1968).

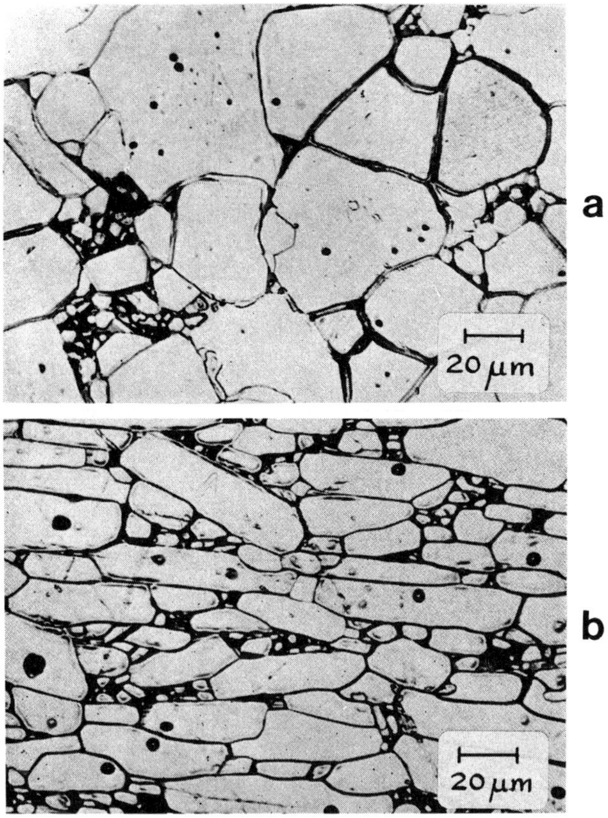

Fig. 5.41 Microstructure of an anisotropic sintered $BaFe_{12}O_{19}$ magnet: (a) in a plane perpendicular to the preferred direction of magnetization; (b) in a plane parallel to the preferred direction of magnetization (Smit and Wijn, 1959).

is perpendicular to the long axes of the particles or grains. The improvement of the properties parallel to the preferred axis is of course obtained at the expense of those perpendicular to this axis. There is a gradual decrease in remanence and energy product at intermediate angles.

Strontium ferrite $SrFe_{12}O_{19}$ is almost identical with $BaFe_{12}O_{19}$ in its crystallographic structure and magnetic properties, except that the sintered magnets generally have higher coercivities. Cochardt (1963) investigated a large number of anisotropic magnetoplumbite magnets in which the principal phase was selected to have a composition $(M_1O)_{1-x}(M_2O)_x kFe_2O_3$, where M_1 and M_2 represent Ba, Sr, Pb and Ca, x was varied between 0 and 1, and k was varied between 3 and 8, with the exception that in the expression $(CaO)_x$, x was never greater than 0.5. A summary of the most important results is shown in Fig. 5.42, from which it can be seen that the $SrFe_{12}O_{19}$-based magnets have the best properties. The highest $(BH)_{max}$ for the $SrFe_{12}O_{19}$-based magnets was 38.4 kJ m^{-3}. Typical intrinsic and flux demagnetization curves for isotropic and anisotropic $BaFe_{12}O_{19}$ and $SrFe_{12}O_{19}$ magnets are shown in Fig. 5.43. The permanent magnet properties are strongly dependent on the sintering conditions and other

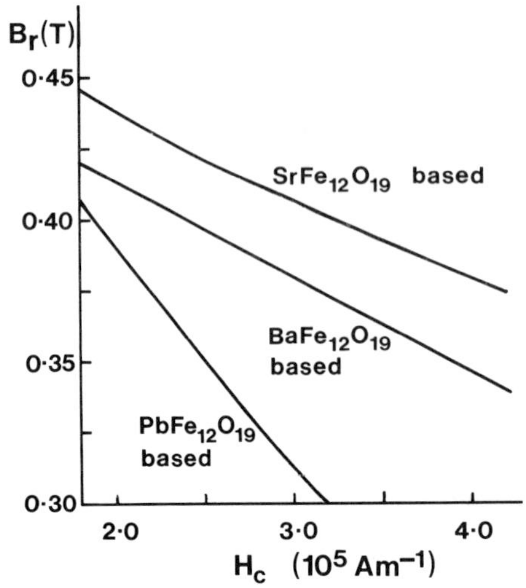

Fig. 5.42 Relationship between remanence B_r and coercivity H_c for various $BaFe_{12}O_{19}$-, $SrFe_{12}O_{19}$- and $PbFe_{12}O_{19}$-based magnets (after Cochardt, 1966).

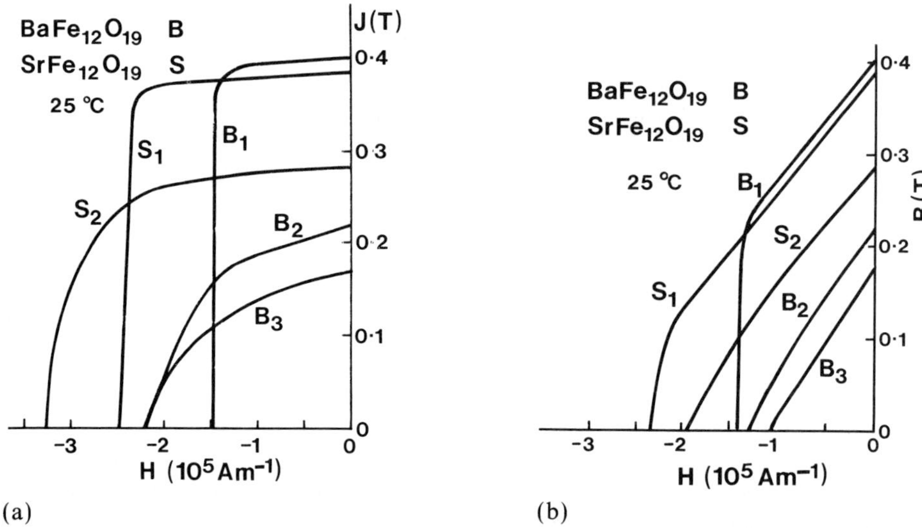

Fig. 5.43 (a) Typical intrinsic (J) demagnetization curves for various isotropic and anisotropic $BaFe_{12}O_{19}$- and $SrFe_{12}O_{19}$-based magnets. (b) The associated flux density (B) demagnetization curves for the magnets shown in (a).

subtleties, which are usually classified by the manufacturers. Generally speaking, the magnets with the higher coercivities have the lower remanences.

Barium and strontium ferrites can be heated in air to temperatures well above the Curie temperature (about 450°C) without structural damage, and hence no irreversible loss of their permanent magnet properties, which reappear on cooling. Indeed, the structure remains stable up to about 1400°C in air before oxygen is released and phase transformations occur.

Barium and strontium ferrite magnets also have the great advantage that the raw materials (natural haematite, α-Fe_2O_3, and barium carbonate, $BaCO_3$) are cheap and widely available. Both materials are easy to handle, and impurities up to about 1% by weight can be tolerated, so that extreme purity of the starting materials is not required. They are now the most widely used of all permanent magnets. A very detailed review of their properties has been published by Stäblein (1982).

5.14.3 Coercivity Mechanism in Barium and Strontium Ferrites

The high coercivities that can be observed in barium and strontium ferrites are due fundamentally to the high uniaxial magnetocrystalline anisotropy energies. The coercivities of bulk specimens of either $BaFe_{12}O_{19}$ or $SrFe_{12}O_{19}$ are initially low, but comminution by milling generally results in an increase in the intrinsic coercivities H_c, although after prolonged milling their coercivities approach limiting values and sometimes decrease thereafter. If the milled particles are annealed, there is a further substantial increase in the coercivities. It is now generally agreed that the low coercivities of bulk samples are due to the fact that they contain a large number of domains. The milling process reduces the size of the particles to the extent that they are either close to being single domains or that the nucleation of domain walls within them is difficult.

Since the highest coercivities of $SrFe_{12}O_{19}$ and $BaFe_{12}O_{19}$ are observed when the materials are ground to fine powders or are in the form of sintered compacts of fine powders, it is clear that the particle size is of fundamental importance in determining the coercivities. The critical diameter d_c for the existence of spherical single-domain particles (Zijlstra, 1982) can be estimated from

$$d_c = \frac{18\gamma}{\mu_0 M_s^2}, \tag{5.8}$$

$$\gamma = 4(AK_1)^{1/2}, \tag{5.9}$$

where γ is the domain wall energy of the material, M_s is the saturation magnetization, A is the exchange energy and, K_1 is the uniaxial magnetocrystalline anisotropy coefficient. An estimate of the exchange energy A can be obtained from the Curie temperature T_c and the distance a between ferromagnetically coupled magnetic moments, since at the Curie temperature the thermal energy kT_c is just capable of destroying the exchange coupling, i.e. $A \approx kT_c/a$.

We use the representative values of the parameters for $BaFe_{12}O_{19}$ (there is a range of published values, depending on the details of preparation and composition), $K_1 = 3.1 \times 10^5\ \mathrm{J\,m^{-3}}$, $T_c = 723$ K and $M_s = 3.80 \times 10^5\ \mathrm{A\,m^{-1}}$, and take a equal to the lattice parameter $a_0 = 0.58$ nm. Then, with $k = 1.38 \times 10^{-23}\ \mathrm{J\,K^{-1}}$ and $\mu_0 = 4\pi \times 10^{-7}\ \mathrm{H\,m^{-1}}$, we have $A = kT_c/a \approx 1.7 \times 10^{-11}\ \mathrm{J\,m^{-1}}$, and the domain wall energy per unit a rea $\gamma = 4(AK_1)^{1/2} = 10^{-2}\ \mathrm{J\,m^{-2}}$, and the domain wall width $\delta = \pi(A/K_1)^{1/2} = 0.023\ \mu$m or 23 nm, and $d_c = 1\ \mu$m.

Since the diameters of the grains in the sintered permanent magnets are about 5 μm and the domain wall width $\delta \approx 23$ nm, it is clear that the grains are large enough to support domain walls, so that magnetization reversal is very unlikely to occur by coherent rotation of the magnetization vector. Magnetization reversal is most likely to occur by domain wall nucleation followed by rapid reversal of the magnetization of the whole grain. The temperature dependences of the theoretical and experimental coercivities of barium and strontium ferrites are shown in Figs 5.44 and 5.45 respectively, from which it is clear that, even after allowance has been made for the demagnetizing fields, there is still a very large difference between the theoretical and experimental values. Since the easy axis of magnetization in the grains is parallel to the short axis, it is necessary to subtract the demagnetizing fields due to shape anisotropy. The demagnetizing fields are due to the flux leakage at the sinter grain boundaries, where there is a weaker exchange interaction. For a specimen with randomly oriented grain platelets, the intrinsic coercivity is given by (Stoner and Wohlfarth, 1948)

$$ {}_iH_c = 0.479\left(\frac{2K_1}{J_s} - DM_s\right), $$

where D is the demagnetizing factor. Suppose for simplicity that the platelets are infinitely thin discs, so that $D=1$. At 77 K the demagnetizing field DM_s is therefore 5.25×10^5 A m^{-1}, which reduces the expected or theoretical intrinsic coercivity from 6.5×10^5 A m^{-1} to 3.95×10^5 A m^{-1}, but the observed coercivities are still considerably

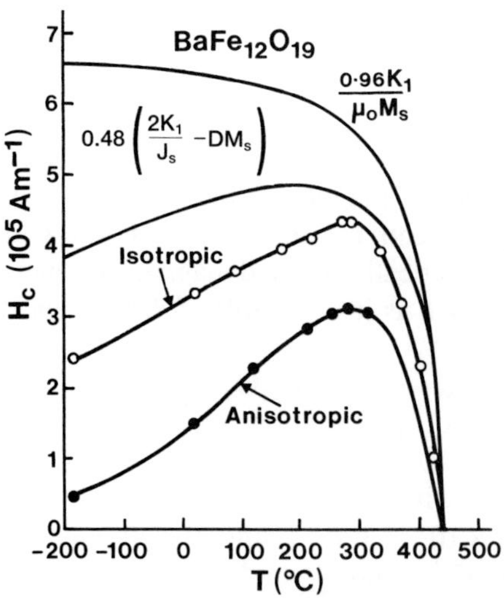

Fig. 5.44 Temperature dependence of the coercivities of $BaFe_{12}O_{19}$ compared with various theoretical values (after Sixtus et al., 1956).

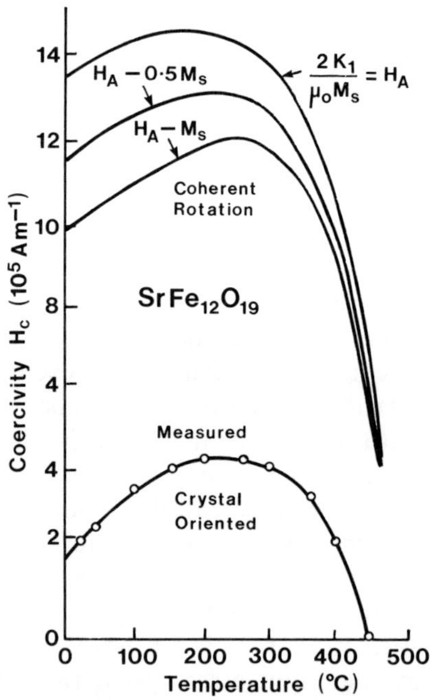

Fig. 5.45 Temperature dependency of the coercivities of $SrFe_{12}O_{19}$ compared with various theoretical values (after Jahn and Müller, 1969).

lower than this. It is generally agreed that in bulk-sintered barium and strontium magnets, magnetization reversal takes place by domain wall nucleation, followed by rapid reversal of the magnetization within the whole grain. Very few domain walls can be observed within the grains, even in the fully field-demagnetized state (McCurrie and Jackson, 1987, Hadjipanayis, 1988, 1990). The failure of the intrinsic coercivity to approach the theorctical value of ${}_iH_c = 2K_1/J_s - DM_s$ can be attributed to the presence of crystallographic defects of various kinds, including the sinter grain boundaries, although these act as domain wall pinning centres, which prevent magnetization reversal from spreading from one reversed grain to contiguous grains unless a critical field is reached. Thus we may regard the nucleation process as occurring from within the sinter grain boundaries. All of these factors of course depend fundamentally on the high magnetocrystalline anisotropy energy, the high wall energy and the narrowness of the domain walls, which enables them to interact strongly with the sinter grain boundaries.

Particle coercivities of about 90% of the theoretical values for $BaFe_{12}O_{19}$ have been observed by Mee and Jeschke (1963), Haneda and Kojima (1973) and Haneda *et al.* (1974), who used chemical co-precipitation techniques. Presumably these particles have a high degree of crystallographic perfection and are therefore free of stacking faults and other defects that act as relatively easy domain wall nucleation sites. Roos *et al.* (1980), Roos and Hempel (1980) and Schippan and Hempel (1983) have concluded from microwave absorption experiments and hysteresis measurements on single particles of $BaFe_{12}O_{19}$ that magnetization reversal occurs by wall nucleation.

5.14.4 Applications of Barium and Strontium Magnets

There are a very large number of applications of barium and strontium ferrite magnets for which their properties and low cost make them very suitable. They are now the most important of all permanent magnets in terms of both tonnage and monetary value of sales. About 10^8 kg are produced in Western countries at present (1992). Unfortunately, their relatively low Curie temperatures (about 450°C) and poor temperature stability preclude their use in measuring instruments and other applications. Barium and strontium ferrites have the unusual property that their coercivities rise with increasing temperature, but unfortunately their remanences decrease more rapidly than those of of nearly all other magnets. Another disadvantage is that the coercivity decreases as the temperature is lowered, so that in some applications, such as starter motors for cars, the lowest environmental temperature likely to be experienced must be taken into account. Typical demagnetization curves at −45, 25 and 85°C are shown in Fig. 5.46.

Since barium and strontium ferrite magnets are insulators, they are very useful in dynamic applications such as in electric motors, because the eddy currents are very small. Furthermore, their insulation property enables them to be magnetized more easily by pulsed fields.

Barium and strontium ferrite particles are currently being investigated for use in perpendicular recording, where their high coercivities, chemical stability and low cost make them potentially very suitable. The most important properties and applications of some commercial ferrite magnets are summarized in Table 5.8.

The magnetic properties of barium and strontium ferrite magnets have been reviewed by van den Broek and Stuijts (1977) and Stäblein (1982). The reader is also referred to the books by Heimke (1976) and by Heck (1974).

Table 5.8 Commercial barium and strontium ferrite permanent magnets (typical values) (after Kools, 1992; data from McCaig and Clegg, 1987; *Philips Components Data Handbook*, 1991).

		Me[a]	B_r (mT)	$_BH_c$ (kA m^{-1})	$_iH_c$ (kA m^{-1})	K (mT)[b]	$(BH)_{max}$ (kJ m^{-3})	μ^c_{rec}	H_s (kA m^{-1})	$\alpha(B_r)$ (% K^{-1})[d]	$\alpha(H_c)$ (% K^{-1})[d]	ρ (Ω m)	Density (kg m^{-3})	Typical application	Market Share (%)
Sintered															
Isotropic	F_1	Ba	215	145	>240	>311	7.4	1.2		−0.2	0.4	10^4		Rotors for cycle dynamos	~10
Anisotropic															
low $_iH_c$	F_2	Ba	400	160	165	466	29.5	1.1	560	−0.2	0.48	10^4	4900	Flat rings for loudspeakers	~70
medium $_iH_c$	F_3	Sr	370	245	255	472	25.5	1.1	875	−0.2	0.37	10^4	4650	Segments for windscreen wiper motors	
			425	250	260	529	33.6		955		0.37	10^4	4900		
high $_iH_c$	F_4	Sr	380	280	320	508	26.8	1.1	1115	−0.2	0.30	10^4	4700	Segments for blower motors	
			400	295	330	532	30.5		1100		0.29	10^4	4850		
very high $_iH_c$	F_5	Sr	385	285	360	529	27.6	1.1	1200	−0.2	0.26	10^4	4850	Segments for starter motors	
Plastic-bonded															
Flexible isotropic	F_6	Ba	125	88	190	201	2.8	1.15	800	−0.2	0.4	10^7	3100	Door catches (e.g. for refrigerators), small DC motors (e.g. for fuel pumps)	~10
			145	96	190	221	3.6	1.15	800			10^6	3700		
anisotropic	F^7	Ba	200	140	200	280	8.0	1.05		−0.2	0.4				
			250	176	240	346	12.0	1.05							
Rigid isotropic	F_8	Ba	80	58	190	156	0.9	1.15	800	−0.2	0.4	10^8	2500	Correction magnets for TV	~10
			155	104	190	231	4.4	1.15	800	−0.2	0.4	10^4	3900		
anisotropic	F_9	Sr	245	100	260	349	12.0	1.05	800	−0.2	0.4	10^5	3500	Small DC motors	
			270	196	260	374	14.0		800			10^5	3900		

[a] Normally applied M ferrites.
[b] K (mT) $= B_r$ (mT) $+ 0.4\,_iH_c$ (kA m^{-1}); the quality factor is independent of the aluminium content.
[c] Recoil permeability.
[d] α = temperature coefficient (0.8 kA m^{-1} K^{-1} for BaM and 0.9 kA m^{-1} K^{-1} for SrM).

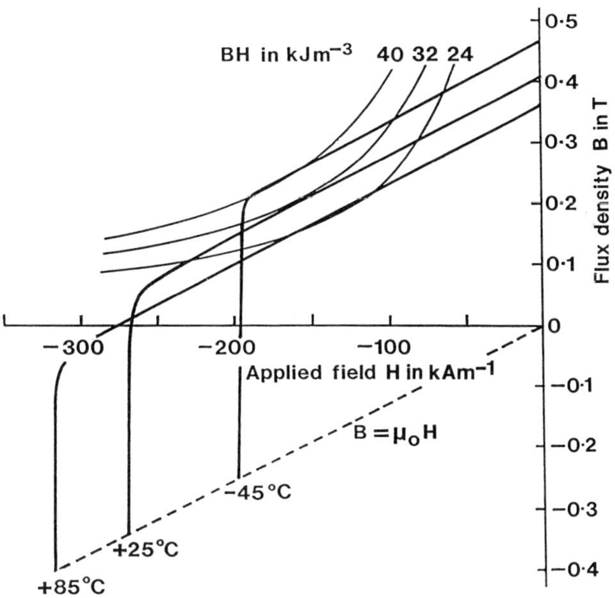

Fig. 5.46 Typical demagnetization curves for sintered $BaFe_{12}O_{19}$-based magnets at 85, 25 and 45°C (after Stäblein, 1982).

5.14.5 Bonded Barium and Strontium Ferrites, or Plastoferrite Magnets

The bonded ferrites are composite materials in which the finely powdered ferrite magnet material is embedded in a non-ferromagnetic matrix. The magnetic properties of the bonded magnets (Stäblein, 1982) are very much inferior to the solid sintered magnets because of the reduced volume fraction of the permanent magnet material. However, this disadvantage is offset by the very low cost, ease of formability and the improvement in other non-magnetic properties such as strength, pliability, elasticity and resistance to fracture. The bonded magnets can be formed by extrusion, calendering, injection moulding or die pressing into a very wide range of shapes and magnetization patterns.

The most commonly used materials for bonding are rubber, polyvinylchloride (PVC), polyamides (PA), polyethylene, polypropylene, and phenol and polyester resins. Both isotropic and anisotropic bonded magnets can be made, but in the latter case particles small enough to be aligned (i.e. with an overall easy axis of magnetization) in a magnetic field must be used. Usually a volume fraction of about 50–65% of ferrite particles is used for anisotropic magnets, whereas for injection moulded or die-pressed magnets the volume fraction is typically 70–80%.

The strength of the rubber-bonded magnets is increased by vulcanization, while magnets bonded into thermosetting plastics are irreversibly cured by heat treatment. Any extra components such as axles or pole sheets can of course be bonded into the magnets during production.

Typical B–H demagnetization curves for isotropic and anisotropic bonded ferrites are shown in Fig. 5.47. Note that the magnetic properties of the anisotropic magnet

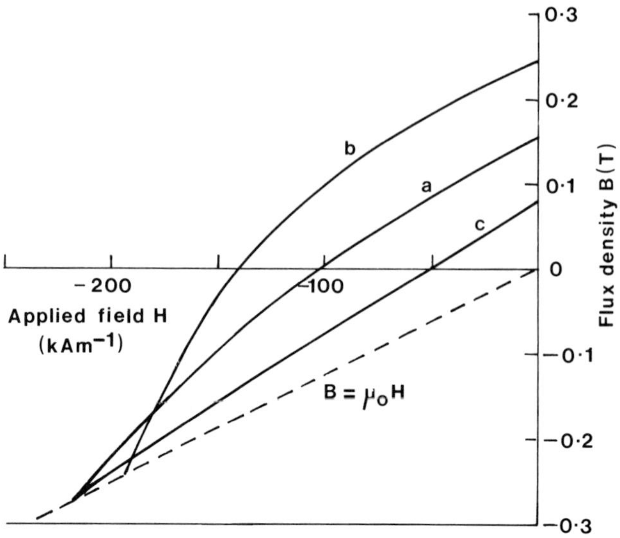

Fig. 5.47 Typical demagnetization curves for plastoferrites: (a) isotropic; (b) anisotropic; (c) perpendicular to preferred axis of anisotropic material (after Ebeling and Krause, 1978).

perpendicular to the preferred axis of magnetization are significantly inferior to those of the isotropic magnet. Typical magnetic properties for three grades of bonded ferrite magnets are shown in Table 5.9.

Applications of plastoferrites

The plastoferrites are used in a very wide range of applications, principally as holding magnets – especially for refrigerator door seals, where the contact between the rubber magnet and the mid steel frame produces a reliable, almost-airtight seal, thus improving the efficiency and safety of the refrigerator. They are also used in toys, planning boards, notices, correction magnets for television tubes, fractional-horsepower motors and dynamos, actuating magnets, electronic ignition switches and separators.

Bonded barium and strontium ferrites are now being considered for application in information storage, in particular for perpendicular magnetic recording (Fayling, 1979) (see Chapter 4).

Table 5.9 Typical magnetic properties of plastoferrites.

	B_r (T)	${}_BH_c$ (kA m^{-1})	${}_iH_c$ (kA m^{-1})	Density (10^3 kg m^{-3})
Isotropic	0.60	50	175	2.3
Isotropic	0.135	85	175	3.8
Anisotropic	0.22	145	190	3.4

5.15 PERMANENT MAGNET PROPERTIES OF LANTHANIDE–COBALT INTERMETALLIC COMPOUNDS AND RELATED ALLOYS

5.15.1 The Intermetallic Compound $SmCo_5$

The potential permanent magnet properties of lanthanide–cobalt intermetallic compounds were first recognized by Hubbard *et al.* (1960) and later developed by Hoffer and Strnat (1966). Sintered $SmCo_5$ magnets are now available commercially with energy products of about 180 kJ m^{-3} combined with flux coercivities of about 700 kA m^{-1} and remanent flux densities of about 0.95 T. Significantly higher values can be obtained under laboratory conditions.

The crystal structure of $SmCo_5$, which has the hexagonal space group p6/mmm, is shown in Fig. 5.48. The lattice parameters are $a=0.5004$ nm, $c=0.3971$ nm and $c/a=0.7936$, which correspond to a density of 8.58×10^3 kg m^{-3}. Since the electronic configuration of the samarium atom is $\{Xe\}4f^5 5d^1 6s^2$, where $\{Xe\}$ is the full shell xenon core, the samarium ion may be either doubly or triply ionized, but in either case the magnetic moment is due to the electrons in the unfilled 4f shell. The orbital and spin magnetic moments of the Sm ion are approximately antiparallel, and very nearly compensate each other. Neutron diffraction experiments show that the magnetic structure of $SmCo_5$ is that of a simple ferromagnetic alignment of the magnetic moments of the cobalt ions.

The permanent magnet properties of $SmCo_5$ are ultimately attributable to its exceptionally high uniaxial magnetocrystalline anisotropy energy. The easy axis of magnetization is, of course, the hexagonal axis, along which a single crystal is readily saturable by domain wall movement. If an attempt is made to magnetize the crystal in any direction parallel to the basal plane, i.e. perpendicular to the easy c axis, it is found that a field of 5 MA m^{-1} produces a magnetic polarization only about 20% the saturation value ($J_s \approx 1.14$ T). The magnetization curves parallel and perpendicular to the c-axis of a single crystal of $SmCo_5$ are shown in Fig. 5.49. Complete saturation perpendicular to the c axis requires a field of about 24 MA m^{-1}, although this varies with the preparation and composition of the crystal. The fact that the magnetization curve in any direction perpendicular to the c axis is linear suggests that the uniaxial magnetocrystalline anisotropy energy can be expressed by the relation

$$E=K_0+K_1\sin^2\theta,$$

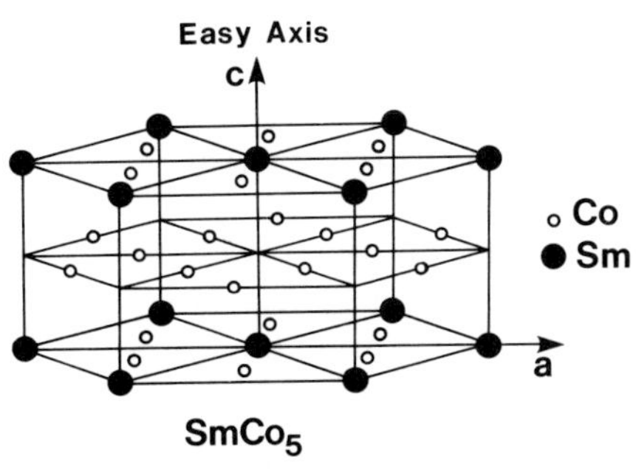

Fig. 5.48 Hexagonal crystallographic structure of $SmCo_5$.

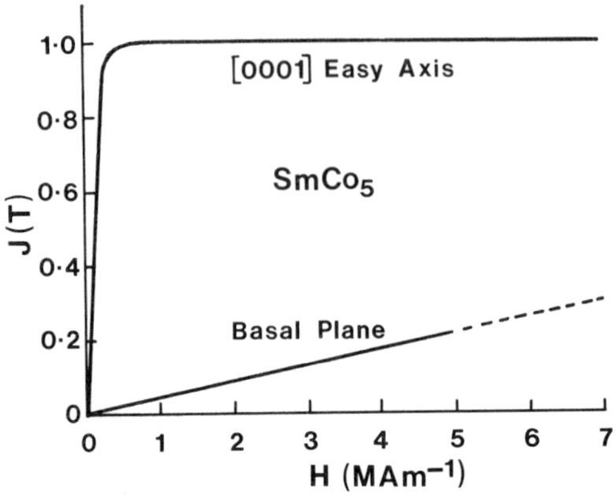

Fig. 5.49 Magnetization curves for single-crystal $SmCo_5$ parallel and perpendicular to the easy axis of magnetization (i.e. the *c* axis) (after Buschow and Velge, 1969).

with the same notation as in Chapter 1. The slope of the magnetization curve perpendicular to the *c* axis is $\mu_0 M_s^2/2K_1$, so that the anisotropy field $H_A = 2K_1/\mu_0 M_s$, or $H_A = 2K_1/J_s$, which for $SmCo_5$ is about 24 MA m^{-1}. This field is of course equal to the maximum possible intrinsic coercivity for a single-domain particle, assuming coherent rotation of the magnetization vector.

The coercivity of cast polycrystalline $SmCo_5$ is usually very low, about 40 kA m^{-1}, i.e. only about 0.15% of the theoretical maximum. However, if the alloy is ground to a fine powder, e.g. by ball milling in a protective atmosphere, the coercivity increases very rapidly as the particle size decreases. Particles with intrinsic coercivities as high as 3 MA m^{-1} have been obtained in this way. Prolonged comminution, however, results in a gradual decrease in the intrinsic coercivity, because of the disruption and damage to the crystal structure caused by plastic deformation.

Kumar *et al.* (1978) have obtained a room-temperature intrinsic coercivity of 5.14 MA m^{-1} for a $SmCo_5$ film prepared by plasma spraying, but even this is only about 20% of the anisotropy field.

Although $SmCo_5$ is often referred to as an intermetallic compound, this classification is an over-simplification, since the magnets are generally non-equilibrium multiphase metallurgical systems with complex microstructures often containing other elements apart from Sm and Co. Since the phases present in the alloys are all derivatives of the binary intermetallic compounds, the equilibrium phase diagram is a very useful guide to the phase content of the alloys. The Sm_2Co_{17} intermetallic compound also forms the basis for another very important group of permanent magnets, and will be discussed later. Substitution of other lanthanide metals such as Pr, Ce or Ce-rich mischmetal is also of importance in extending the range of magnetic properties and in lowering the cost of the alloys.

5.15.2 Preparation of Sintered $SmCo_5$ Permanent Magnets

The intermetallic compound $SmCo_5$, or rather an alloy close in composition to it, may be formed by melting the elements in vacuo or in a protective atmosphere such as argon or helium. A slight excess of Sm is included in order to compensate for losses due to evaporation or oxidation. Cech (1974) has shown that many lanthanide–cobalt alloys can be prepared by direct reduction of the oxides in the presence of calcium, together with the appropriate quantity of cobalt. Since the reduction is followed by diffusion, the process is known as the R–D process. A very similar process, known as co-reduction (KOR) has also been developed by Domazer (1974). Full details of the commercial processes are given by Strnat (1988).

The cast or 'reduced' $SmCo_5$ alloy is then ground in liquid hexane (or other protective liquid) to a fine powder (3–10 μm particles) and blended with another alloy powder close in composition to Sm_2Co_7, i.e. richer in Sm. This alloy has two functions: (1) it allows the Sm content of the alloy to be adjusted, and (2) it produces a small amount of a phase that temporarily melts during the sintering process and hence accelerates the densification of the alloy close to 100%, as is required for a high-quality magnet. If a magnet with temperature compensation is required, a third alloy powder such as Gd–Co may be added at this stage.

The fine particles, which are uniaxial single crystals, are then pressed in a die in the presence of a magnetic field of order 1 MA m^{-1} so as to align and compact them to a 'green' density sufficient to allow handling. Alternatively, the particles may be isostatically pressed in the presence of the aligning field. The latter method gives the best results, and has the advantage that a pulsed field 3–5 times larger can be used. The 'green' compacts are then sintered at about 1150°C for 3 h in helium, argon or hydrogen—which must be free of oxygen and nitrogen. The sintering temperature must be carefully controlled in order to obtain a homogeneous high-density magnet with a uniform fine-grain texture; excessive grain growth must be prevented so as to avoid the presence of large grains, in which domain wall nucleation will occur in low fields and hence produce a poorer-quality magnet. The best permanent magnets are obtained after a post-sintering heat treatment above the eutectoid temperature, ideally in the range 850–900°C, for about 4 h, followed by rapid cooling through the critical range 800–500°C. This avoids the eutectoid decomposition of $SmCo_5$ into the Sm_2Co_7 and Sm_2Co_{17} phases, which act as easy reverse field nucleation sites for domain walls and hence result in low-coercivity magnets (Kronmüller and Hilzinger, 1976).

5.15.3 Properties of Sintered $SmCo_5$ Magnets

Demagnetization curves parallel to the preferred axis of magnetization (i.e. the alignment axis) for sintered $SmCo_5$ magnets are shown in Fig. 5.50, from which it can be seen that the intrinsic magnetic polarization does not decrease significantly until the reverse field approaches the coercivity. The flux density B within the magnet therefore decreases linearly as the reverse field increases. If the particles were perfectly aligned with identical individual coercivities, the intrinsic demagnetization curve would be rectangular – but this is of course impossible to obtain. However, this ideal is approached by many magnets prepared under laboratory conditions. The relative recoil permeability or relative dynamic permeability for these magnets $\mu_{rec} \approx 1$, and has very important implications for the design of devices containing $SmCo_5$ magnets.

Although much of the very high intrinsic coercivity might appear to be redundant, since it is greater than the flux coercivity, this is certainly not the case. The very high intrinsic coercivity gives the magnet a very much greater stability, especially in dynamic applications such as repulsion and torque devices, in which the magnet may have nearly zero flux. Even in this state, the intrinsic magnetization M (or J) will still be high, thus enabling the magnet to restore the flux in the magnetic circuit after the strong demagnetizing field is removed. Also, since the intrinsic coercivity and the flux coercivities are much higher than the maximum possible demagnetizing fields, the magnets can be

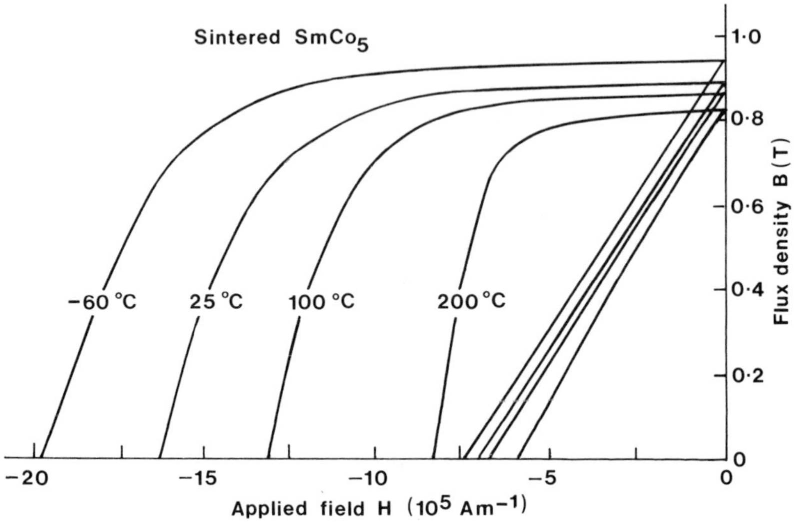

Fig. 5.50 Intrinsic and flux demagnetization curves for a sintered $SmCo_5$ magnet at four temperatures (after Mildrum *et al.*, 1981).

used in the form of thin slabs magnetized perpendicular to the plane of the slab. For a magnet with $B_r = J_r = 0.8$ T, the remanent magnetization $M_r = J_r/\mu_0 = 6.36 \times 10^5$ A m^{-1}. Thus if we take the demagnetizating factors of the slab to be $D_x = D_y = 0$ and $D_z = 1$ (approximately), the demagnetizing field $H_D = D_z M_r = 6.36 \times 10^5$ A m^{-1} for magnetization parallel to the z direction. The intrinsic coercivities of most $SmCo_5$ magnets are greater than this, so it is clear that this maximum possible demagnetizing field will not produce any significant intrinsic demagnetization. If the magnet is required as a source of magnetomotive force to magnetize a soft magnetic material or produce a field in an air gap, this can be achieved by using a very short or thin magnet; indeed, $SmCo_5$ magnets work most efficiently in this form because the load line must intersect the demagnetization quadrant at the $(BH)_{max}$ point (if the volume of the magnet is to be a minimum), and in order to achieve this, the demagnetizing field of the magnet must be high because its coercivity is high. For a linear demagnetization B–H 'curve', this working point occurs at $H = \frac{1}{2}M_r$ and $B = \frac{1}{2}B_r$.

In general, therefore, if high-coercivity and high-energy product magnets are to be used in any magnetic device or circuit, careful thought must be given to the design criteria, which are quite different from those used for materials with ${}_BH_c < \frac{1}{2}M_r$, and this usually means that the device must be completely redesigned in order to make full use of the magnetic properties and to minimize the volume of the magnets, which are still relatively expensive. If this is done, further saving can often be made, because the device itself can be smaller. Thus the use of a more expensive magnet may result in a more efficient and cheaper device.

The temperature dependence of the coercivity of sintered $SmCo_5$ magnets given a variety of heat treatments has been investigated in detail by Kütterer *et al.* (1977), who have shown that the coercivity decreases rapidly with increasing temperature and that the decrease is approximately linear over a wide temperature range.

Detailed studies of the temperature dependence of the magnetic properties of $SmCo_5$ and other lanthanide–cobalt alloys have been published by Li *et al.* (1988).

5.15.4 Origin of the High Coercivity and Mechanism of Magnetization Reversal in $SmCo_5$

Although it is clear that the high coercivity of $SmCo_5$ is attributable to the exceptionally high uniaxial magnetocrystalline anisotropy, this is not a sufficient condition, because there are many lanthanide–cobalt alloys that also have very high uniaxial magnetocrystalline anisotropy yet their coercivities are relatively low. The critical size for single domain particles of $SmCo_5$ can be estimated from the relation

$$d_c = \frac{18\gamma}{\mu_0 M_s^2},$$

where M_s is the saturation magnetization, $\gamma = 4(AK_1)^{1/2}$ is the domain wall energy per unit area, A is the exchange energy and K_1 is the uniaxial magnetocrystalline energy coefficient. An estimate of the exchange energy can be obtained from $A \approx kT_c/a$, where T_c is the Curie temperature and a is the distance between ferromagnetically aligned spins, which for simplicity can be taken as half the lattice parameter. For $SmCo_5$, with $M_s = 8.58 \times 10^5$ A m^{-1}, $T_c = 1000$ K, $K_1 = 1.3 \times 10^7$ J m^{-3} and $a = 0.25$ nm, we find that $d_c = 2$ μm, $\gamma = 0.1$ J m^{-2} and the domain wall width $\delta = \pi(A/K_1)^{1/2} = 8$ nm. Thus the critical diameter for single-domain particles is somewhat smaller than the grain sizes in sintered $SmCo_5$ magnets, which are typically in the range 3–10 μm and are therefore big enough to be multidomain particles. Magnetization reversal is therefore most likely to occur by domain wall nucleation.

When the magnet is finally cooled after post-sintering heat treatment, it is in a demagnetized state, with the individual grains containing many domains with magnetization vectors parallel to the hexagonal axes of the grains. After the application of the magnetizing field parallel to the preferred axis of the magnet, the domain walls are either removed from the magnet, or (more probably) they are forced into the sinter grain boundaries, where they remain pinned until they are subjected to a sufficiently large reverse field to 'renucleate' them into the main body of the grains and thus reverse the grain magnetization by domain wall displacement. However, in the field-demagnetized state very few domain walls are visible (Kerr magneto-optical effect), which suggests that the magnetization of the grain is reversed during demagnetization, so at the coercivity point there are equal volumes of grains magnetized to saturation parallel and antiparallel to the preferred axis of the magnet. Multidomain structures are usually observed in oversize grains (>20 μm). Thus the demagnetization process occurs by a switching or reversal of the individual grains by sudden nucleation and displacement of domain walls. The domain structure observed using the polar Kerr effect of a sintered $SmCo_5$ magnet in the partially demagnetized state is shown in Fig. 5.51. The plane of the photograph is perpendicular to the preferred c axis of magnetization. The dark and light regions are grains (i.e. whole domains) magnetized in opposite directions parallel to the c axis. A few multidomain grains are visible. The above calculation for the critical size d_c for single-domain particles is somewhat oversimplified. More rigorous calculations by Brown (1969) suggest that the particle size ranges used for the manufacture of $SmCo_5$ magnets are certainly larger than the critical size for single domains but smaller than the critical size d_m where they will certainly be multidomains. Even for those calculations, it is assumed that the particles are isolated; however, it seems clear that the grain sizes

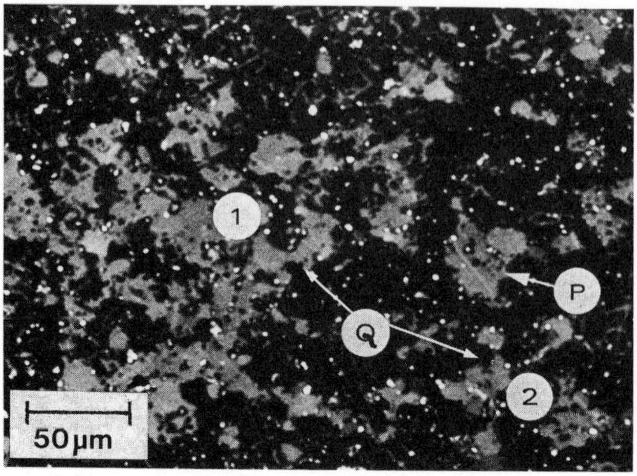

a

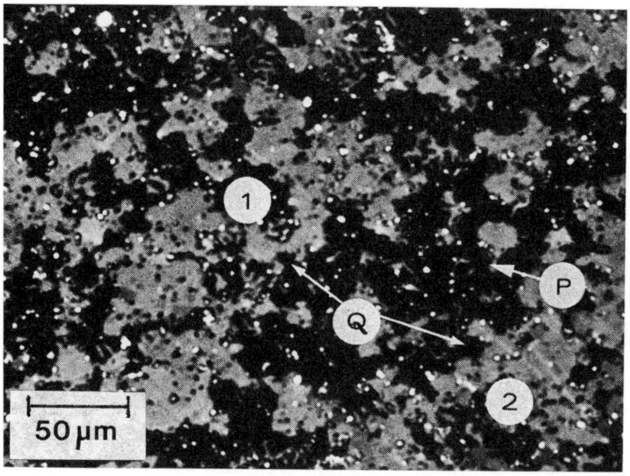

b

Fig. 5.51 Kerr polar magneto-optical effect photograph showing oppositely magnetized domains in a sintered $SmCo_5$ magnet parallel (dark contrast) and antiparallel (light contrast) to the applied field, which is itself parallel to the preferred hexagonal axis of magnetization: (a) in a reverse field of 1280 kA m^{-1}; (b) in a reverse field of 1520 kA m^{-1}. Note the increase in the area (and volume) of the light-contrast reverse domains. In the higher field some of the domains have been remagnetized in the original direction owing to magnetostatic interactions.

in high-coercivity $SmCo_5$ magnets lie between d_c and d_m. Although there is ample confirmation of the fact that domain walls can be nucleated in sintered $SmCo_5$ magnets, micromagnetic theory (Brown, 1963) predicts that the field required to nucleate a domain wall is equal to the anisotropy field minus the demagnetizing field, i.e.

$$H_n = \frac{2K_1}{\mu_0 M_s} - DM_s.$$

However, it is well established that the nucleation fields are very much lower than this. In any case, the precise definition of the nucleation field is complicated by the fact that even after 'saturation', the magnet may contain very small vestigial domains that cannot be resolved by the Kerr magneto-optic effect or even by a very sensitive magnetometer. The term 'nucleation' may refer to either (a) the abrupt change in magnetization due to the nucleation of a domain wall folllwed by rapid movement throught the grain, or (b) the unpinning of a domain wall from within a sinter grain boundary followed by rapid movement through the grain. The distinction between these two types of nucleation is not clear. Both mechanisms may occur during magnetization reversal. There is no evidence that domain walls are strongly pinned at precipitates, second phases or other crystallographic defects with the grains. Electron micrographs of $SmCo_5$ are relatively featureless (Fidler, 1982). The changes in the values of A and K_1 at the sinter grain boundaries result in a lowering of the domain wall energy $\gamma = 4(AK_1)^{1/2}$, and thus gives rise to domain wall pinning according to the theory proposed by Hilzinger and Kronmüller (1975) and by Friedberg and Paul (1975). Gaunt and Mylvaganum (1977) have shown that the width of the pinning region compared with the domain wall width $\delta = \pi(A/K_1)^{1/2}$ must reach a critical value, and hence explains why ordinary metallurgical grain boundaries in the cast alloy do not give rise to pinning and hence to high coercivities. The sinter grains are not in strong magnetic exchange contact with each other, and hence behave independently to some extent – thus domain walls nucleated in one grain do not necessarily invade contiguous grains. This explains why Kerr magneto-optical observations during magnetization reversal show that the magnetization of individual grains reverse independently of the other grains: often isolated grains or groups of grains are reversed while the surrounding grains remain magnetized in the original direction as shown in Fig. 5.51. Since $SmCo_5$ is mechanically very hard it is possible to observe the domain structures after mechanical polishing only (McCurrie and Caswell, 1970). This makes it relatively easy to observe the domain structures during magnetization either by means of the Kerr magneto-optic effect or by the Bitter colloid technique. Domain wall pinning by planar defects has been reviewed by Hilzinger (1977), Zijlstra (1982) and Kronmüller (1991).

5.15.5 Cerium Mischmetal–Cobalt Magnets

In view of the high cost of Sm, attempts have been made to substitute some or all of the Sm with cerium mischmetal (MM), which is very much cheaper and more abundant. Cerium mischmetal has an approximate composition 53 wt% Ce, 24 wt% La, 16 wt% Nd, 5 wt% Pr and 2 wt% of other lanthanide metals, although the composition may vary considerably, depending on the source of the ores.

Magnets with compositions $(MM_{1-x}Sm_x)Co_5$ have energy products in the range 110–130 kJ m^{-3} and intrinsic coercivities ${}_iH_c$ in the range 0.73–1.9 MA m^{-1}. The best magnetic properties for sintered $MMCo_5$ corresponding to $x=0$ are $B_r = 0.8$ T, ${}_BH_c =$ 624 kA m^{-1}, ${}_iH_c = 720$ kA m^{-1} and $(BH)_{max} = 124$ kJ m^{-3}.

Alloys in the system $(MM_{1-x}L_x)_{0.85}Sm_{0.5}Co_5$ where L is Ce, La, Nd, or Pr with $0<x<0.5$, have, with suitable choices of L and x been shown to have good permanent magnet properties, some comparable to those of $SmCo_5$. The alloy $(MM_{0.9}Nd_{0.1})_{0.85}Sm_{0.15}Co_5$ has an energy product of 144 kJ m^{-3} and an intrinsic coercivity of 1.28 MA m^{-1}.

For further details see Menth *et al.* (1978), Buschow (1986, 1988, 1991), Strnat (1988) and Kumar (1988).

5.15.6 $Pr_{5.6}Sm_{0.4}Co_5$

Although $PrCo_5$ has potentially even better permanent magnet properties than $SmCo_5$, this potential has not yet been realized. $PrCo_5$ has a very high uniaxial magnetocrystalline anisotropy and a high saturation magnetic polarization $J_s = 1.2$ T. If Pr is substituted for some of the Sm, as in $Pr_{0.6}Sm_{0.4}Co_5$, a significantly better permanent magnet can be obtained, with an energy product of about 200 kJ m^{-3}.

5.15.7 $Sm_{0.58}Gd_{0.42}Co_5$ and $Sm_{0.8}Ho_{0.2}Co_5$

These alloys have very low temperature coefficients of flux density change, which are very close to zero at room temperature. Although the remanent flux densities are only about 0.63 and 0.71 T respectively, their stability is very useful in applications such as travelling wave tubes, watt-hours meters and gyroscopes.

5.15.8 $Sm_{0.5}La_{0.5}Co_5$

The intrinsic and flux demagnetization curves for this alloy show almost ideal characteristics. The exceptionally high intrinsic coercivity and loop squareness give a linear B–H relationship extending into the third quadrant, thus giving the magnet outstanding recoil characteristics.

5.15.9 Sm(Co, Fe, Cu)$_{5-6.8}$ and (Sm, Ce)(Co, Fe, Cu)$_{5-6.8}$ Alloys

The permanent magnet properties of these alloys are based on the high uniaxial magnetocrystalline anisotropy of $SmCo_5$; the expensive cobalt is partly replaced by Fe and a small amount of Cu. The alloys have compositions extending beyond the stoichiometric ratio 1:5. Their properties are inferior to sintered $SmCo_5$, though they are easier to magnetize. Even cheaper magnets can be produced by replacing some or all of the Sm with Ce, e.g. Ce(Co, Fe, Cu)$_5$. Since Sm and Ce can be alloyed in any ratio, a whole range of properties can be obtained from the (Sm, Ce)(Co, Fe, Cu)$_{5-7}$ system, enabling a choice of magnets with cost–performance characteristics to be made for particular applications.

In all of these magnets the high coercivities are obtained by heat treatments that produce second-phase precipitation. The addition of copper is essential in order to stabilize and increase the homogeneity range of the $SmCo_5$ phase structure and hence enable significant quantities of Fe to be added, with a resulting increase in remanence. Metallography of the alloys in the optimum high-coercivity state has revealed that the microstructure consists of a fine precipitate of Sm_2(Co, Fe)$_{17}$ platelets in a matrix of Sm(Co, Cu)$_5$, and similarly in the (Sm, Ce)(Co, Fe, Cu)$_{5-7}$ system. Observations of the domain structures confirm that the coercivity is controlled by the strength of the domain wall pinning at the second-phase precipitates.

15.16 PERMANENT MAGNETS BASED ON THE INTERMETALLIC COMPOUND Sm_2Co_{17}

5.16.1 Sm(Co, Fe, Cu, M)$_{7.2-8.5}$

These magnets are based on the intermetallic compound Sm_2Co_{17}, which has a high uniaxial magnetocrystalline anisotropy and forms the basis of a whole series of multiphase alloys in which some of the cobalt is replaced by a number of transition elements. M is usually Cr, Mn, Ti, Hf or Zr. The crystallographic structures of Sm_2Co_{17}, which may be either hexagonal or rhombohedral, are shown in Figs 5.52(a) and (b). The $SmCo_5$ structure can be converted to the Sm_2Co_{17} structure by substituting two Co ions for the Sm ions in alternate layers in the $SmCo_5$ structure. For the Sm(Co, Fe, Cu, M)$_z$ alloys the value of z is usually in the range 7.2–8.5. The addition of Cr and Mn have been found to increase the magnetocrystalline anisotropy, and results in single-phase magnets with high coercivities and energy products, e.g. $Sm(Co_{0.8}Fe_{0.14}Mn_{0.04}Cr_{0.02})_{8.5}$, as shown in Fig. 5.53 curve (a). In this and similar alloys the coercivity depends on domain wall pinning within the grain boundaries. Once the domains walls are re-nucleated within the grains, they move relatively easily. The coercivity mechanism is therefore very similar to that in $SmCo_5$. By contrast, the coercivity of the alloy $Sm(Co_{0.87}Cu_{0.13})_{7.8}$ (Fig. 5.33 curve (b)) is determined principally by domain

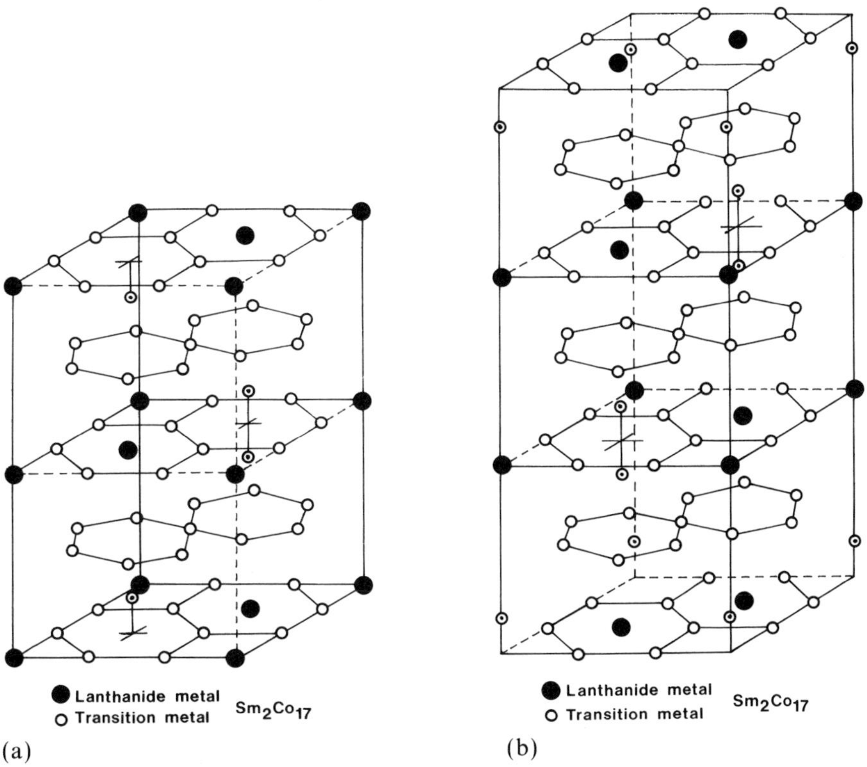

Fig. 5.52 (a) Crystal structures of Sm_2Co_{17}: (a) hexagonal; (b) rhombohedral (after Strnat, 1988).

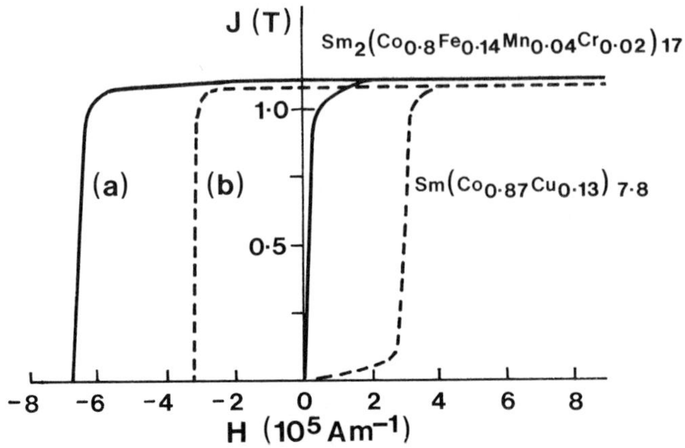

Fig. 5.53 Comparison of intrinsic magnetization and demagnetization curves: (a) for $Sm_2(Co_{0.8}Fe_{0.14}Mn_{0.04}Cr_{0.02})_{17}$, whose coercivity is controlled by domain wall nucleation; (b) for $Sm(Co_{0.87}Cu_{0.13})_{7.8}$, whose coercivity is controlled by domain wall pinning (after Menth *et al.*, 1978).

wall pinning within the grains themselves. The initial magnetization and demagnetization curves for the two alloys are also shown in Fig. 5.53. The very different initial magnetization curves are characteristic of magnets for which the coercivities are controlled by domain wall nucleation (or grain boundary pinning only) as shown by $Sm(Co_{0.8}Fe_{0.14}Mn_{0.04}Cr_{0.02})_{8.5}$ or by domain wall pinning as shown by $Sm(Co_{0.87}Cu_{0.13})_{7.8}$. These distinctions will be discussed more fully below.

Further improvements can be made by the addition of Zr, Ti or Hf. Alloys containing Zr, such as $Sm(Co_{0.65}Fe_{0.28}Co_{0.05}Zr_{0.02})_{7.67}$, are particularly important, and have energy products up to about 264 kJ m^{-3}. They have high coercivities and are gradually replacing $SmCo_5$, especially in high-temperature applications. Unfortunately, they are more difficult to produce and to magnetize. Attempts to increase the magnetic polarization by increasing z, either by increasing the iron content or decreasing the copper content, result in a lowering of the coercivity. The presence of Zr in the alloys has the effect of preventing the precipitation of the low-coercivity Fe–Co phase, thus enabling a higher remanence to be obtained

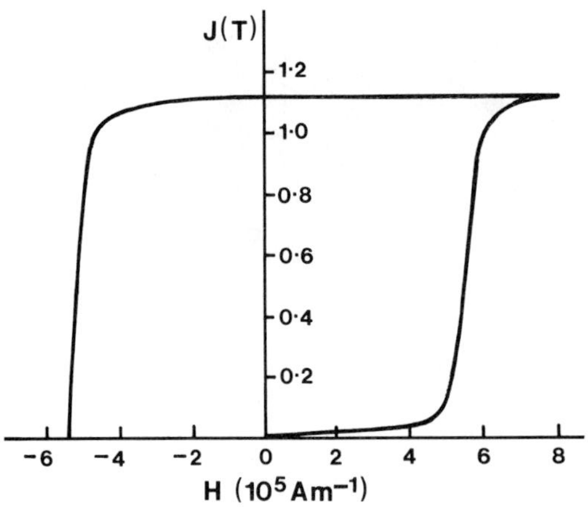

Fig. 5.54 Intrinsic initial and demagnetization curves for $Sm(Co_{0.675}Fe_{0.213}Cu_{0.1}Zr_{0.013})_{7.42}$ with composition 25.5 wt% Sm–50% Co–15% Fe–8% Cu–1.5% Zr. The coercivity is due to domain wall pinning, as indicated by the high threshold field on the initial magnetization curve, after which the net magnetization increases rapidly (after Ojima *et al.* (1977).

by the increased iron content. Intrinsic magnetization and demagnetization curves for the alloy 25.5 wt% Sm–50% Co–15% Fe–8% Cu–1.5% Zr, with the approximate formula $Sm(C_{0.675}Fe_{0.213}Cu_{0.1}Zr_{0.013})_{7.42}$, are shown in Fig. 5.54.

The high coercivities are controlled by domain wall pinning at precipitates that are produced by the heat treatment following sintering. The microstructures of the $Sm(Co, Cu, Fe, Zr)_{7.2-8.5}$ alloys at peak coercivity have been investigated by Livingston and Martin (1977) and Fidler and Skalicky (1982). The cellular microstructure is shown in Fig. 5.55, and consists of a twinned rhombohedral (R) $Sm_2(Co, Fe)_{17}$-rich phase containing platelets of a hexagonal (H) $Sm_2(Co, Fe, Cu, Zr)_{17}$-rich phase normal to the

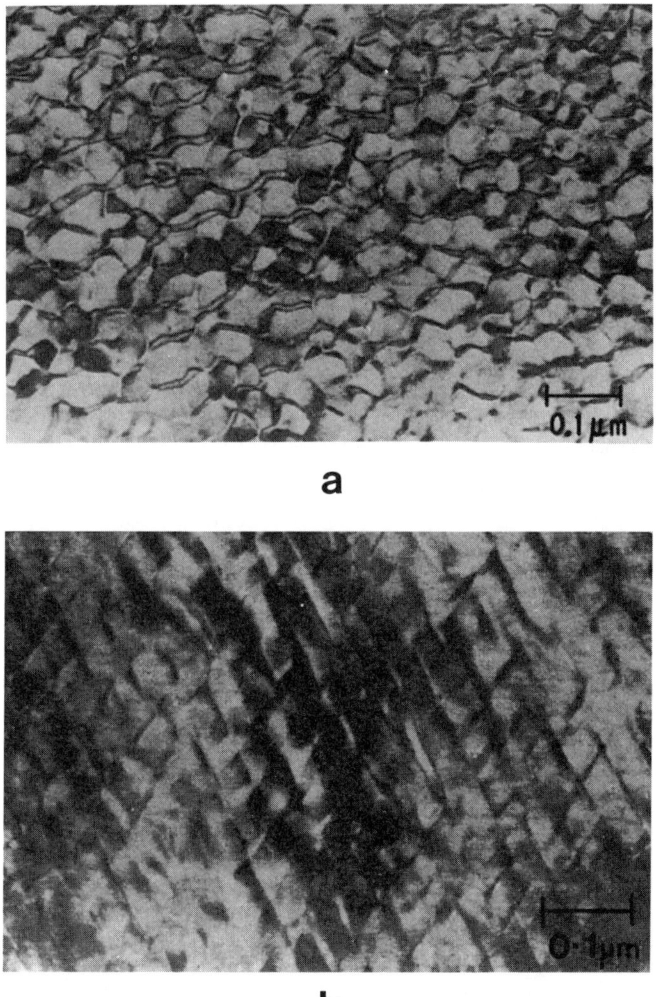

Fig. 5.55 Cellular precipitation microstructure in a single grain of a sintered $Sm(Co_{0.8}Cu_{0.15}Fe_{0.05})_7$ magnet. The cell interiors have a rhombohedral Sm_2TM_{17} structure, while the fully coherent cell boundaries have the $SmTM_5$ structure (TM = transition metals). (a) A section perpendicular to the easy axis of magnetization (i.e. the hexagonal [0001] axis). (b) A section parallel to the hexagonal [0001] axis (Livingston and Martin, 1977).

[0001] axis. The rhombic cells (see Fig. 5.55b) are separated by an $Sm(Co, Cu)_5$-rich coherent boundary phase as shown in Fig. 5.56. The pinning of a domain wall AB against the $Sm(Co, Cu)_5$-rich boundary phase is illustrated in Fig. 5.56. The same domain wall pinning in planes parallel and perpendicular to the [0001] direction is illustrated in Figs 5.57(a) and (b), which should be compared with the microstructure shown in Figs 5.55(a) and (b), of which it is a schematic model. The hexagonal platelets (shaded phase) are believed to provide diffusion paths for the rapid segregation of Cu

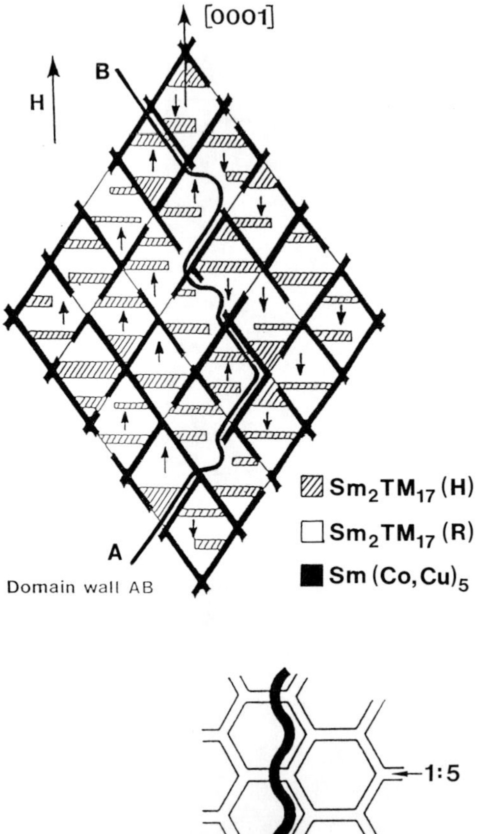

Fig. 5.56 Detailed microstructure of the rhombic cells shown in Fig. 5.55(b). The cell interiors consist of a rhombohedral (R) Sm_2TM_{17} phase (where TM = transition metals) containing platelets of a hexagonal (H) Sm_2TM_{17} phase. These are separated by a $Sm(Co, Cu)_5$-rich phase. AB represents a domain wall pinned or blocked by the $Sm(Co, Cu)_5$ boundary phase (after Fidler and Skalicky, 1982).

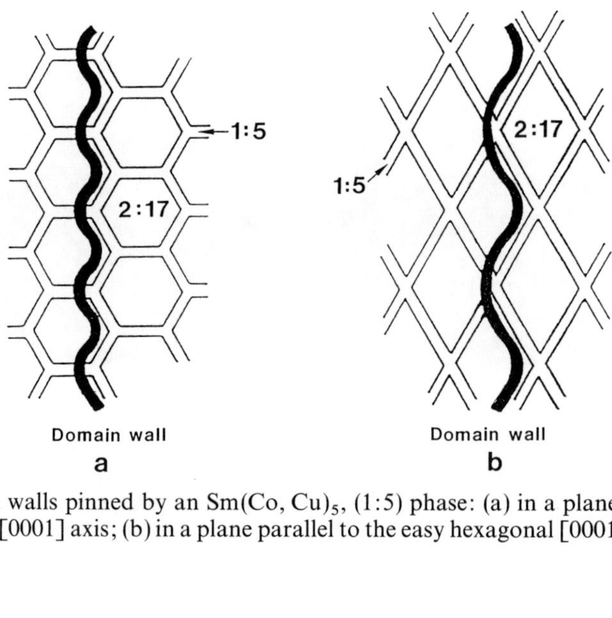

Fig. 5.57 Domain walls pinned by an $Sm(Co, Cu)_5$, (1:5) phase: (a) in a plane perpendicular to the easy hexagonal [0001] axis; (b) in a plane parallel to the easy hexagonal [0001] axis (Livingston and Martin, 1977).

Table 5.10 Properties of lanthanide–transition metal permanent magnets based on $SmCo_5$ and Sm_2Co_{17} compounds or their structures.

Material[a]	B_r (T)	$_BH_c$ (10^5 A m^{-1})	$_iH_c$ (10^5 A m^{-1})	$(BH)_{max}$ (kJ m^{-3})
$SmCo_5$	1.00	7.9	30	196
$Sm_{0.2}MM_{0.8}Co_5$	0.82	6.4	13	128
$MMCo_5$	0.80	6.2	7	124
$Sm_{0.4}Pr_{0.6}Co_5$	1.03	8.0	13	208
$Sm_{0.58}Gd_{0.42}Co_5$	0.64	4.5	30	74
$Sm_{0.75}Ce_{0.25}Co_{3.3}Cu_{1.2}Fe_{0.5}$	0.70	4.0	5	98
$Sm(Co_{0.76}Fe_{0.10}Cu_{0.14})_{6.8}$	1.04	4.8	5	212
$Sm(Co_{0.87}Cu_{0.13})_{7.8}$	1.08	3.0	3	204
$Sm(Co_{0.68}Fe_{0.21}Cu_{0.10})_{7.4}$	1.10	5.1	5	236
$Sm_2(Co_{0.8}Fe_{0.14}Mn_{0.04}Cr_{0.02})_{17}$	1.13	8.8	10	248
$Sm(Co_{0.63}Fe_{0.28}Cu_{0.05}Zr_{0.02})_{7.7}$	1.2	16	10	264
Sm_2TM_{17} (polymer-bonded)	0.68	4	7.6	84
Sm_2TM_{17} (epoxy-resin-bonded)	0.89	5.4	12	136

[a] MM, mischmetal, principally Ce; TM, transition metals Fe, Co, Cu, Mn, Cr.

to the boundary phase. The presence of Zr in these alloys has two important effects: (1) it increases the magnetocrystalline anisotropy, and (2) it assists in the formation of the hexagonal platelets and hence the segregation of Cu to form the rhombic cell boundaries, with a resulting increase in domain wall pinning because of differences in the wall energy in the various phases.

The cost of these alloys can be reduced by substituting Ce for some of the Sm, so that alloys with compositions in the range (Sm, Ce)(Co, Fe, Cu, Zr or $Ti)_{6.5-8.5}$ are now of commercial importance.

5.16.2 (Sm, Gd)(Co, Fe, Cu, $Zr)_{7.2-8.5}$ and (Sm, Er)(Co, Fe, Cu, $Zr)_{7.2-8.5}$ Alloys

These alloys have compositions close to those discussed in the previous section. The partial substitution of Gd or Er for Sm improves the temperature stability, and they may ultimately replace the (Sm, Gd)Co_5 and (Sm, Ho)Co_5 alloys, although, like the latter, they are relatively expensive compared with the alloys containing Sm only.

The properties of a selection of lanthanide–transition metal permanent magnets based on $SmCo_5$ and Sm_2Co_{17} compounds or their structures are given in Table 5.10. Reviews of the lanthanide-transition metal permanent magnet alloys have been published by Menth *et al.* (1978), Buschow (1986, 1988, 1991), Strnat (1988), Kumar (1988) and Ray (1985). See also the book by Nesbitt and Wernick (1973) for earlier work.

5.17 Nd–Fe–B ALLOYS

Neodymium–iron–boron permanent magnets were first produced on a commercial scale in the USA and Japan in 1983, although high coercivities in an Nd–Fe alloy were first reported by Drozzhina and Janus in 1935. The high coercivities are due to the very

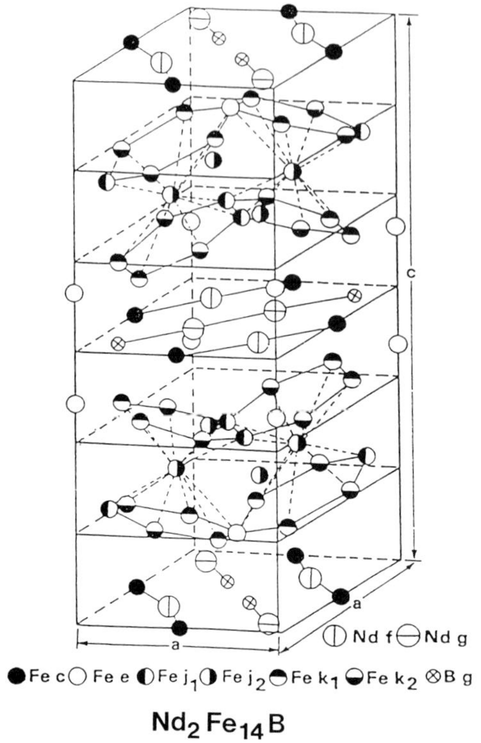

Fig. 5.58 Tetragonal unit cell of $Nd_2Fe_{14}B$. The ratio c/a is exaggerated to emphasize the puckering of the hexagonal iron nets (Herbst *et al.*, 1984).

high uniaxial magnetocrystalline anisotropy of the tetragonal $Nd_2Fe_{14}B$ phase, whose crystal structure of which is shown in Fig. 5.58. The addition of boron to the Nd–Fe alloy was shown to be essential in order to stabilize the tetragonal structure. The alloys can be prepared either by vacuum induction melting or by calciothermic reduction. Energy products up to about $400\,kJ\,m^{-3}$ have already been obtained, see e.g. Narasimhan (1985), Croat *et al.* (1984), Lee (1985) and Sagawa *et al.* (1987a).

5.17.1 Sintered Magnets

The cast alloy is pre-milled to a coarse powder either in a high-energy hammer mill under nitrogen or by hydrogen decrepitation, after which it is milled to a fine single-crystal powder with a narrow size distribution. Milling may be carried out in an attritor mill in cyclohexane (to prevent oxidation) or by jet milling. The latter produces a narrower particle size distribution. An aligned powder compact is obtained by die pressing or isostatic pressing in a magnetic field, which may be parallel or perpendicular to the direction of pressing. The green compact is then sintered at 1150°C. In order to obtain a high-coercivity magnet, a post-sintering heat treatment at 630°C for about 1 h is necessary. Energy products up to about $400\,kJ\,m^{-3}$ can be obtained by this method. Sintered Nd–Fe–B magnets are somewhat tougher than $SmCo_5$ magnets, and much less susceptible to chipping and fracture. In view of the susceptibility of Nd–Fe–B magnets to corrosion, they are coated with aluminium by ion vapour deposition or by

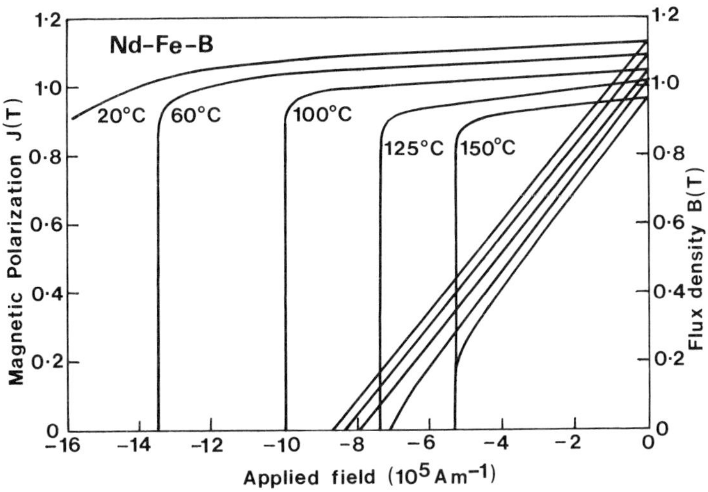

Fig. 5.59 Intrinsic and flux density demagnetization curves for a sintered Nd–Fe–B magnet at various temperatures (after Ward and Taylor, 1989).

a cathodic epoxy electropaint. Typical intrinsic and flux demagnetization curves are shown in Fig. 5.59.

(*i*) *Magnequench I*

In this method the molten alloy is very rapidly quenched (at about 10^6 °C s^{-1}) onto a cooled rotating metallic drum. The quench rate has to be carefully controlled because for the optimum high coercivity the resulting ribbons or flakes must consist of a very finely divided microstructure of amorphous and crystalline regions. The latter have very high uniaxial magnetocrystalline anisotropy and are probably single domains.

After grinding, the flakes are then bonded into an epoxy resin, die-pressed and cured in an oven. This process results in an isotropic magnet known as Magnequench I or MQI. After magnetization along any axis in a field of 3 T, an energy product of 56–72 kJ m^{-3} can be obtained. Since the magnets can very easily be shaped, cut or ground, they are finding an ever-increasing number of applications.

(*ii*) *Magnequench II*

If the crushed flakes are themselves formed into a magnet by hot pressing, the resulting magnet is known as Magnequench II or MQII. These magnets are also isotropic, but they have energy products in the range 100–120 kJ m^{-3}. The maximum operating temperature is about 150°C.

(*iii*) *Magnequench III*

These magnets are produced by a two-stage hot pressing procedure. Hot pressing at 700°C develops sufficient plasticity to achieve 100% density, which is followed by hot

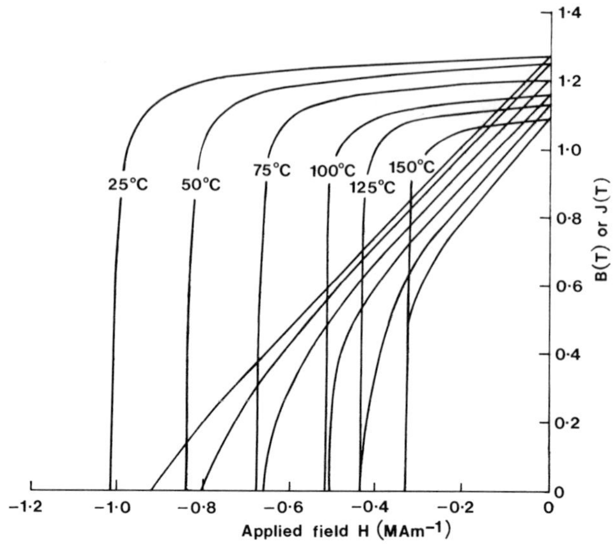

Fig. 5.60 Intrinsic and flux density demagnetization curves for a Magnequench III magnet (Delco Remy, General Motors).

pressing to a thinner dimension, which develops a magnetic alignment and hence gives an anisotropic magnet. The magnetic properties parallel to the pressing direction are a significant improvement on the isotropic magnets; energy products of 320 kJ m^{-3} have been reported (Lee *et al.*, 1985).

Typical intrinsic and flux demagnetization curves for a Magnequench III magnet at various temperatures are shown in Fig. 5.60, showing highly rectangular intrinsic curves and B–H relationships with linear sections, both of which characteristics are of great importance in a wide range of applications.

The magnetic properties of sintered and melt-spun Nd–Fe–B magnets are shown in Table 5.11.

Much higher energy products can be obtained for very carefully prepared laboratory specimens, e.g. Sagawa *et al.* (1987a) obtained an energy product of 403 kJ m^{-3} for a sintered magnet with composition $Nd_{0.13}Fe_{0.81}B_{0.06}$. The intrinsic and B–H demagnetization quadrant for this alloy is shown in Fig. 5.61. Note that the flux coercivity $_BH_c$ is only very slightly less than the intrinsic coercivity $_iH_c$.

Table 5.11 Magnetic properties of sintered and melt-spun Nd–Fe–B alloys.

Nd–Fe–B alloy	B_r (T)	$_BH_c$(k A m^{-1})	$_iH_c$ (kA m^{-1})	$(BH)_{max}$ (kJ m^{-3})
Sintered	1.22	840	1280	280
Melt spun MQI	0.61	424	1200	64
Melt spun MQII	0.80	520	1280	104
Melt spun MQIII	1.17	840	1040	256

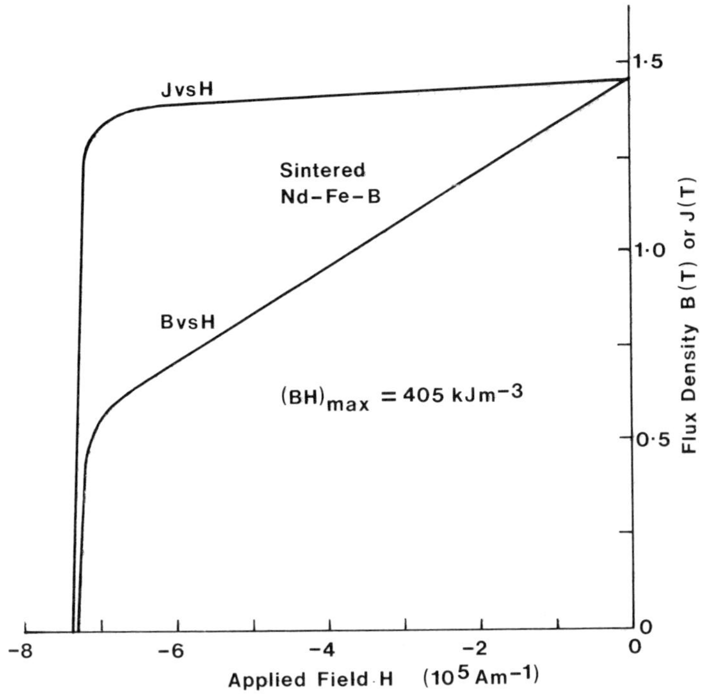

Fig. 5.61 Intrinsic and flux demagnetization curves for a sintered $Nd_{0.13}Fe_{0.81}B_{0.06}$ magnet with an exceptionally high energy product of 403 kJ m^{-3} (after Sagawa *et al.*, 1987a).

5.17.2 Temperature Dependence of Magnetic Properties

Although the magnetic properties of Nd–Fe–B magnets are superior to those of the Sm–Co-based alloys, their temperature stability is certainly inferior. The Curie temperatures of Nd–Fe–B magnets are about 350°C, compared with 700–800°C for Sm–Co-based alloys, but, more seriously, the coercivity decreases very rapidly with increasing temperature as shown in Figs 5.62 and 5.63. This decrease in $_iH_c$ is associated with the decrease in K_1 and J_s, but the correlation is not exact and it may be assumed that the thermal activation energy at the higher temperatures assists the field demagnetization process. The rapid decrease in coercivity with increasing temperature limits the maximum temperature of use to about 100°C. However, additions of Dy, Tb and Al increase the coercivity, while Co additions increase the Curie temperature. Unfortunately, the addition of Co decreases the remanence and increases the cost of the magnet. By using some or all of these additions, the maximum temperature of use can be extended to about 150°C. The highest room-temperature coercivity of 4 MA m^{-1} for a sintered Nd–Fe–B-based alloy yet recorded was obtained for the composition $(Nd_{0.53}Dy_{0.47})_{15}Fe_{77}B_8$ (Sagawa *et al.*, 1987a, b). However, Pinkerton (1986a, b) has observed exceptionally high intrinsic coercivities of more than 6.4 MA m^{-1} in melt-spun Tb–Fe–B- and Dy–Fe–B-based alloys respectively. Unfortunately, the energy products of these alloys are a factor of about 4 lower than those for melt-spun Nd–Fe–B-based alloys.

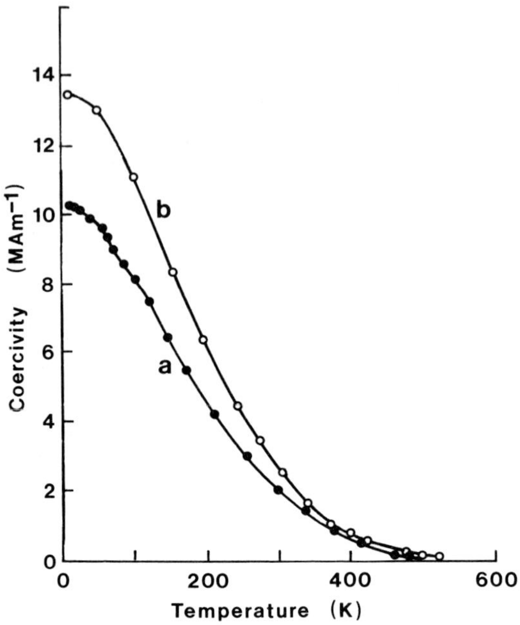

Fig. 5.62 Temperature dependence of intrinsic coercivity of (a) sintered $Nd_{15}Fe_{77}B_8$ and (b) sintered $Pr_{15}Fe_{77}B_8$ magnets (after Kronmüller *et al.*, 1988).

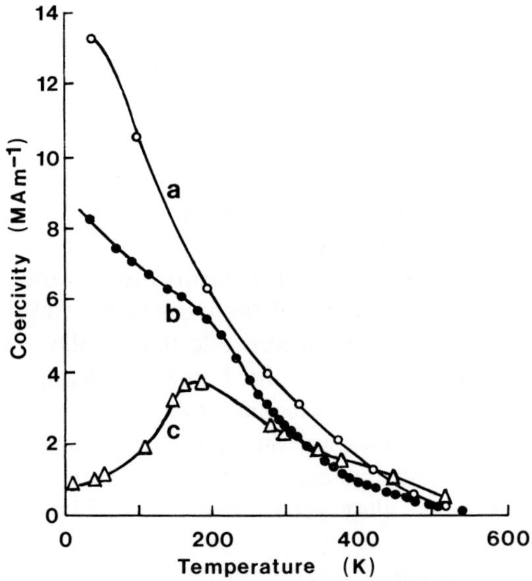

Fig. 5.63 Temperature dependence of intrinsic coercivity of three Nd–Fe–B-based two-phase magnets: (a) sintered $Nd_{16.7}Dy_{1.8}Fe_{75}B_{6.5}$; (b) sintered $Nd_{20}Fe_{71}Al_2B_7$; (c) melt spun $Nd_{15}Fe_{76}B_9$ (after Kronmüller *et al.*, 1988).

5.17.3 Magnetic Viscosity

Another factor that may limit the application of permanent magnets is magnetic viscosity (Gaunt, 1977b; 1986; Gaunt and Roy, 1976). If a permament magnet is subjected to a constant demagnetizing field for a significant period of time, the net magnetization will decrease very slowly owing to the phenomenon of magnetic viscosity. The rate of decrease of magnetization is strongly dependent on the magnitude of the demagnetizing field (compared with the coercivity) and of course on the ambient temperature. Higher temperatures assist the demagnetization process. The effect was first observed in the Alnicos (see the review by McCurrie, 1982). Because of the low Curie temperatures and strong temperature dependence of the coercivities for Nd–Fe–B-based alloys, the effect is of considerable significance and must be taken into account at the design stage, especially if the magnet is likely to be subject to demagnetizing fields at temperatures significantly above room temperature, e.g. Nd–Fe–B-based permanent magnets are widely used in automobile permanent magnet motors. Measurements of the magnetic viscosity in sintered Nd–Fe–B magnets have been made by Jubb and McCurrie (1987), Givord *et al.* (1988a, b, c) and Street *et al.* (1990). The change in magnetization ΔM with time at a fixed temperature in a static reverse field H (less than the coercivity H_c) is given by

$$\Delta M = \text{const} - S \ln t,$$

where S is the coefficient of magnetic viscosity.

Magnetic viscosity measurements, apart from their practical importance, can yield useful information on the mechanisms of magnetization reversal (Herbst, 1991).

5.17.4 Coercivity Mechanisms in Nd–Fe–B Magnets

It has been suggested by several researchers (e.g. Livingston, 1985, 1986; Buschow, 1986; Hadjipanayis and Kim, 1988; Kronmüller *et al.*, 1988; Herbst, 1991; Kronmüller, 1991) that the coercivity in sintered Nd–Fe–B magnets is controlled principally by nucleation of domain walls. Magnetization and demagnetization curves obtained by Sagawa *et al.* (1984a) after the application of increasing applied fields (Fig. 5.64) suggests that the observed increase in coercivity is due to the removal of more and more residual reverse domains. The rapid rise of the initial magnetization curve is also characteristic of a magnet in which the coercivity is controlled principally by domain wall nucleation.

According to Ormerod (1985), there are at least four phases present in a typical Nd–Fe–B sintered magnet, so there is the possibility of domain wall pinning within the grains in addition to the expected domain wall pinning at the grain boundaries, as is believed to occur in sintered $SmCo_5$ magnets. The latter are generally believed to be nucleation-controlled, because the nucleation occurs from the domain walls, which are pinned within the grain boundaries. In sintered Nd–Fe–B magnets both types of nucleation and pinning are likely to occur, but the observed behaviour depends on the preparation, heat treatment and microstructure.

By means of the Kerr magneto-optical effect, Livingston (1985) has shown that the coercivity in sintered Nd–Fe–B magnets is nucleation-controlled, but, unlike $SmCo_5$, the domain walls are nucleated within the grains, so that a multidomain structure similar to that in a thermally demagnetized magnet is observed. The single-domain particle size

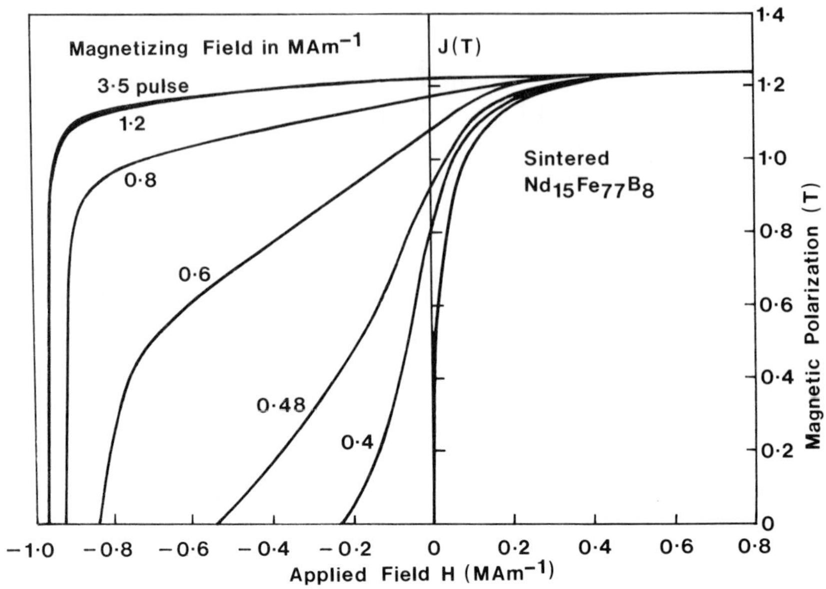

Fig. 5.64 Effects of increasing magnetizing fields on the demagnetization curves of a sintered $Nd_{15}Fe_{77}B_8$ magnet. The magnet was thermally demagnetized by heating above the Curie temperature before each set of measurements (after Sagawa *et al.*, 1984a).

for Nd–Fe–B, as derived from the formula $d_c = 18\gamma/\mu_0 M_s^2$, is about 0.3 μm and is very much smaller than the average grain size of 5–10 μm. It is therefore to be expected that domain wall nucleation will occur within the grains. Furthermore, there seems to be general agreement that the sinter grain boundaries are coated with an Nd-rich phase that broadens into larger regions at three-grain junctions (Livingston, 1986); this is a remnant of the Nd-rich liquid that assisted the densification during the liquid-phase sintering process.

5.18 COERCIVITY MECHANISMS

5.18.1 Domain Wall Nucleation-Controlled Magnets

By means of a rigorous calculation, Brown (1963) has shown that in the absence of shape anisotropy the intrinsic coercivity $_iH_c$ of a material with uniaxial magnetocrystalline anisotropy energy K_1 is equal to the anisotropy field H_A, i.e.

$$_iH_c = H_A = \frac{2K_1}{J_s},$$

where J_s is the saturation magnetic polarization. However, the observed maximum intrinsic coercivities for materials such as $BaFe_{12}O_{19}$, $SmCo_5$ or Nd–Fe–B are only about 20% of the theoretical value (see Fig. 5.65). This can be attributed to the presence of crystallographic defects of various kinds, second phases and surface imperfections.

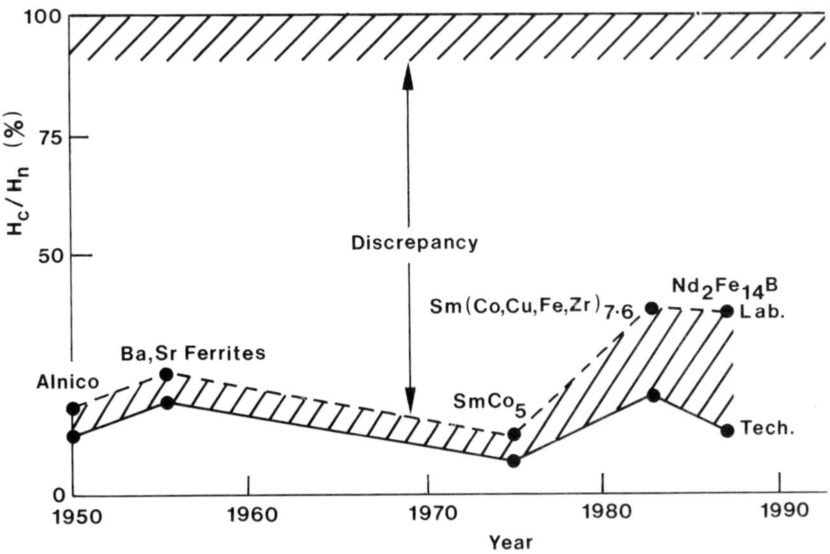

Fig. 5.65 Ratio of coercivity H_c to theoretical nucleation field $H_n = 2K_1/J_s$ for permanent magnets developed during the last four decades: --●--, laboratory magnets; —●—, commercial magnets (after Kronmüller *et al.*, 1988).

The importance of the latter has been demonstrated by the very significant increases in intrinsic coercivity that are observed after etching fine particles of $SmCo_5$ in dilute nitric acid (McCurrie and Carswell, 1973; McCurrie and Willmore, 1979).

The theoretical intrinsic coercivity $_iH_c = 2K_1/J_s$ is also decreased owing to the demagnetizing fields resulting from the shape anisotropy of the specimen – a maximum $H_D = DM_s = M_s$, where D is the demagnetization factor and M_s is the saturation magnetization – and to the demagnetizing fields associated with internal microcracks and grain boundaries. In the latter cases the local demagnetizing field may be as high as $2M_s$ owing to the presence of reversed domains and interaction effects. Kronmüller *et al.* (1988) have suggested that in nucleation-controlled magnets there is no significant domain wall pinning within grains, and that once the domain walls are nucleated they move relatively easily under the influence of applied fields. According to them, the intrinsic coercivity is given by

$$_iH_c = \frac{2K_1}{J_s}\alpha(T) - D_{\text{eff}}M_s,$$

where $\alpha(T)$ (<1) is a temperature-dependent factor that then depends on the nature of the various crystallographic, metallurgical and microstructural parameters of the permanent magnet, and D_{eff} is the average effective load demagnetization factor. The values of $\alpha(T)$ can be determined only by experiment. According to Kronmüller *et al.* (1988), the coercivity of Nd–Fe–B magnets is determined principally by nucleation. The reduction in intrinsic coercivity as compared with the theoretical value is due principally to stray demagnetizing fields and misaligned grains, each individually reducing the intrinsic coercivity by about 2 MA m^{-1}, i.e. a total of about 4 MA m^{-1}. A reduction of only 10% of the theoretical maximum intrinsic coercivity could be attributed to reduced nucleation fields at surface imperfections.

It has also been established by many researchers (e.g. Livingston, 1973, 1975, 1978, 1981b, 1986, 1987; Searle and Garrett, 1974; McCurrie and Mitchell, 1975; den Broeder and Zijlstra, 1976; Kronmüller, 1978; Buschow, 1986, 1988, 1991; Strnat, 1988 and Kumar, 1988) that the coercivity in sintered $SmCo_5$ magnets is determined principally by nucleation, although domain wall pinning does occur at the sinter grain boundaries and is necessary to prevent the rapid spread of reverse domains into contiguous grains. The nucleation may be regarded as taking place from within the sinter grain boundaries, but there is very little impedance to domain wall displacement within the grains themselves. This has been confirmed directly by Fidler (1982a), who has shown by electron microscopy that the grains contain stacking faults and phases such as Sm_2Co_7 and Sm_2Co_{17}, which may act as easy domain wall nucleation sites and offer little or no impedance to domain wall displacement. Thus the precipitation of the Sm_2Co_7 and Sm_2Co_{17} phases must be avoided if good-quality $SmCo_5$ magnets are to be obtained.

The simplest way to determine whether the coercivity of a magnet is controlled principally by nucleation or pinning of domain walls is to measure the dependence of the coercivity on the maximum applied field. This is illustrated in Fig. 5.66. The rapid rise in the intrinsic coercivity for relatively low applied fields is typical of magnets in which it is controlled principally by domain wall nucleation. The effects on the demagnetization curves and the coercivities as the maximum magnetizing field is gradually increased is illustrated in Fig. 5.67. Typical minor hysteresis loops for various field excursions are shown in Fig. 5.68.

It is now generally believed that the coercivities of barium and strontium ferrites, $SmCo_5$, $Sm_2(Co_{0.8}Fe_{0.14}Mn_{0.04}Cr_{0.02})_{17}$ (Menth *et al.*, 1978) and Nd–Fe–B magnets are principally controlled by domain wall nucleation, although pinning *within* the sinter grain boundaries is of vital importance. In some permanent magnets the coercivity may be controlled by both nucleation and pinning effects.

5.18.2 Domain Wall Pinning-Controlled Magnets

Domain wall pinning arises due to variations in the domain wall energy $\gamma = 4(AK_1)^{1/2}$ as a function of position x within the material. The variations in γ are due principally

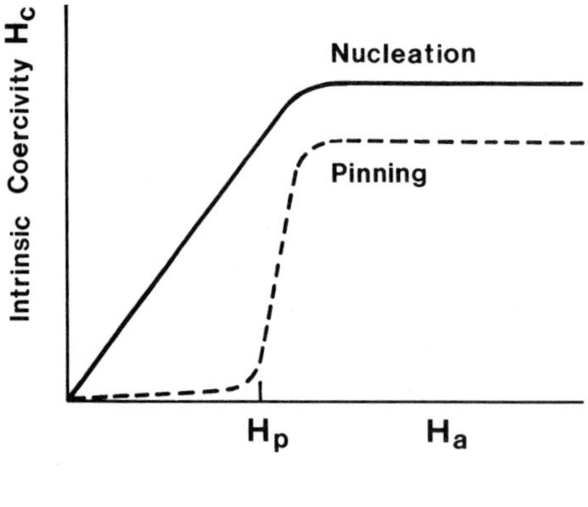

Fig. 5.66 Dependence of intrinsic coercivity on the maximum applied field for magnets in which the coercivity is controlled principally by domain wall nucleation or domain wall pinning. The magnets must be thermally demagnetized between measurements. Note the threshold field H_p for pinning-controlled magnets.

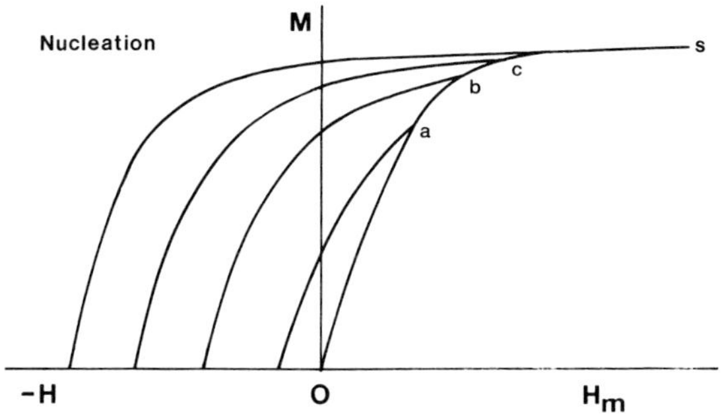

Fig. 5.67 Effects of demagnetization curves and coercivities as the maximum magnetizing field is gradually increased through a, b and c to 'saturation' at s, for magnets in which the coercivities are controlled principally by domain wall nucleation. The magnets must be thermally demagnetized between measurements.

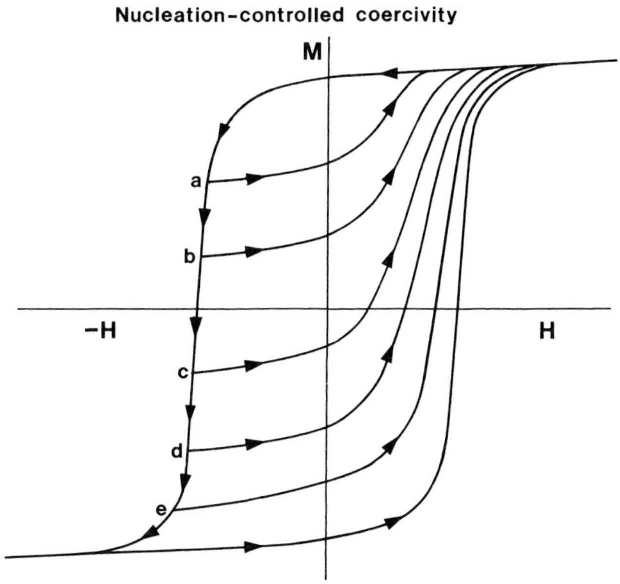

Fig. 5.68 Typical minor hysteresis loops for magnets in which the coercivities are controlled principally by domain wall nucleation. The magnets must be thermally demagnetized between measurements for loops a, b, c, d and e, etc.

to variations in the magnetocrystalline anisotropy energy associated with structural imperfections or compositional fluctuations. Changes in the exchange energy A also make a contribution to changes in γ, but these are expected to be smaller than those due to variations in K_1. The variations in γ act as domain wall traps (γ lower than nearby values) or domain wall barriers (γ higher than nearby values). Both of these impede domain wall motion. The strength of interaction between the domain wall and the traps or barriers depends on the width δ of the domain walls compared with the widths of the barriers. Hilzinger and Kronmüller (1975) and Friedberg and Paul (1975) have shown that the coercivity due to a planar defect of width D is given by

$$H_c = \frac{1}{3\sqrt{3}} \frac{2K_1}{J_s} \frac{\pi D}{\delta} \left(\frac{A}{A'} - \frac{K_1'}{K_1} \right),$$

where D is the width of the planar defect with exchange energy A' and magnetocrystalline anisotropy energy K_1'. The surrounding material has exchange and magnetocrystalline anisotropy energies A and K_1 respectively. The values of A' and K_1' are not known for any alloys – and would in any case be extremely difficult to measure – so that only estimates of H_c can be made by making plausible assumptions concerning the values of A' and K_1'. By assuming that A' and K' differ from A and K_1 by 20–30%, Friedberg and Paul (1975) have obtained intrinsic coercivities for $SmCo_5$ of the correct order of magnitude, although they assumed that the walls were pinned within the sinter grain boundaries prior to the 'nucleation' event, which is now believed to control the coercivity in $SmCo_5$, i.e. the walls were nucleated by unpinning from the sinter grain boundaries.

Consider a domain wall pinning-controlled magnet that has just been cooled from the heat-treatment temperature, i.e. one that has been thermally demagnetized. In this state the domain walls are pinned in the lowest possible energy states. When a small field is applied, only small reversible domain wall displacement or bowing of the wall occur. When the applied field is comparable to the lowest field pinning strength, the domain walls can break away from their pinning sites, and this results in significant irreversible changes in the net magnetization. If there were a single pinning strength throughout the material, i.e. we had an ideal pinning-controlled magnet, there would be a step function change in the coercivity versus applied field graph. However, because there is, in general, a distribution of pinning strengths, the intrinsic coercivity versus applied field curve shows a much less abrupt change, as shown in Fig. 5.66. For magnets in which there is generalized pinning of the domain walls at precipitates within the grains and at the boundaries, there is little increase in the coercivity with the magnetizing field until the latter reaches a critical value H_p known as the pinning field. After this, the coercivity rises very rapidly to its maximum value as shown in Fig. 5.66. The effects on the demagnetization curves and the coercivities as the maximum magnetizing field

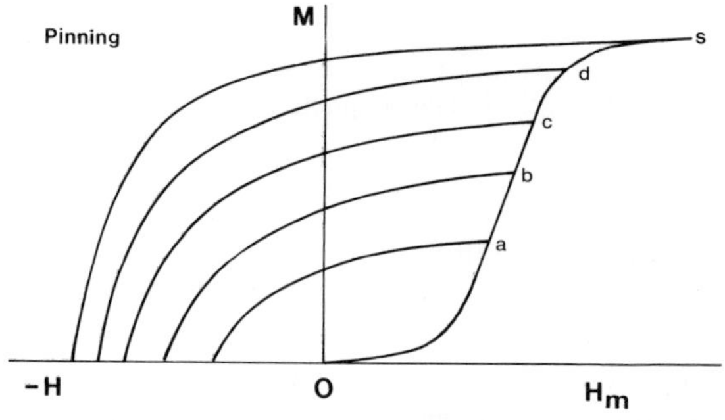

Fig. 5.69 Effects on demagnetization curves and coercivities as the maximum magnetizing field is gradually increased through a, b, c and d to 'saturation' at s, for magnets in which the coercivities are controlled principally by domain wall pinning. The magnets must be thermally demagnetized between measurements.

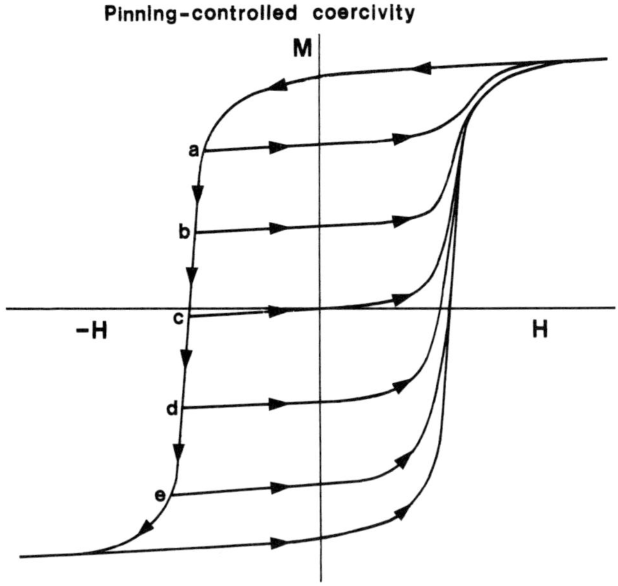

Fig. 5.70 Typical minor hysteresis loops for magnets in which the coercivities are controlled principally by domain wall pinning. The magnets must be thermally demagnetized between measurements for loops a, b, c, d, e, etc.

is gradually increased are illustrated in Fig. 5.69. Typical minor hysteresis loops for various field excursions are shown in Fig. 5.70.

It is now generally believed that the coercivities in $Sm(Co_{0.87}Cu_{0.13})_{7.8}$ (Menth *et al.*, 1978) and magnets based on $Sm(Co, Fe, Cu)_{5-7}$ are controlled principally by domain wall pinning at various crystallographic defects and second phases throughout the grains.

For some high-coercivity magnets based on $Sm_2(Co, Fe, Cu, Zr)_{17}$ and rapidly quenched Nd–Fe–B the domain wall pinning strength may be very non-uniform, in which case a typical initial magnetization curve (C) is as shown in Fig. 5.71. Typical initial magnetization curves for nucleation- and conventional pinning-controlled magnets (A and B respectively) are shown for comparison.

A review of the micromagnetic coercivity theory of domain wall nucleation and pinning in permanent magnets has been published by Kronmüller (1991).

5.18.3 Summary of Intrinsic Coercivity Mechanisms in Lanthanide–Transition Metal Permanent Magnets

As we have already seen, lanthanide–transition metal magnets have compositions based on three main types of intermetallic compounds, namely hexagonal $SmCo_3$, rhombohedral $Sm_2(Co, Fe)_{17}$ and tetragonal $Nd_2Fe_{14}B$. All of these compounds are characterized by high uniaxial magnetocrystalline anisotropy and high saturation magnetic polarization, which are ultimately responsible for their high coercivities. However, the observed coercivities depend critically on the composition, processing technique, heat treatment and the resulting microstructure.

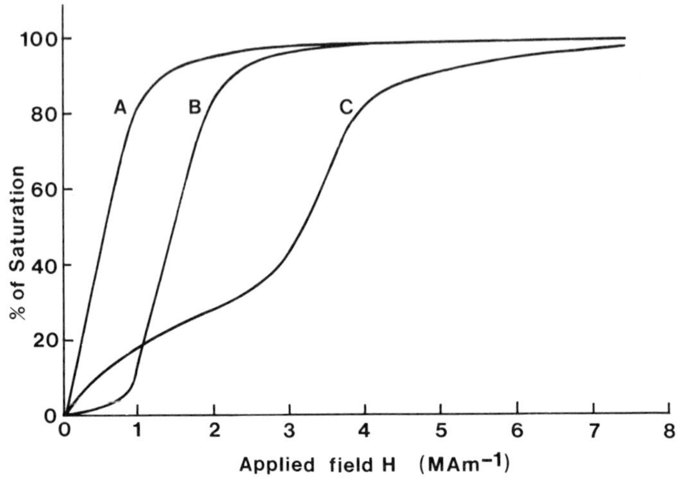

Fig. 5.71 Typical initial magnetization curves for different lanthanide–transition metal permanent magnets: (A) nucleation-controlled magnets with domain walls pinned at grain boundaries such as $SmCo_5$, Nd–Fe–B and their derivatives; (B) pinning-controlled magnets with domain walls pinned by homogeneous precipitates, such as $Sm(Co, Cu)_5$ or the low-coercivity magnets based on $Sm_2(Co, Fe, Cu)_{17}$; (C) pinning-controlled magnets with very non-uniform pinning strengths such as high coercivity magnets based on $Sm_2(Co, Fe, Cu, Zr)_{17}$ and rapidly quenched Nd–Fe–B (after Ervens, 1982).

The distinction between permanent magnets in which the coercivities are controlled by the nucleation or pinning of domain walls is by no means precise. However, lanthanide-based permanent magnet alloys can be very broadly divided into two groups according to their minor hysteresis loops and the dependence of their coercivities on the strength of the magnetizing field. For magnets such as $SmCo_5$ in which the coercivity is principally controlled by the nucleation and rapid displacement of domain walls (even though these are 'nucleated' from within the grain boundaries where they have remained pinned after magnetization) the coercivity increases very rapidly with increasing magnetizing field as shown in Fig. 5.66. For nucleation-controlled magnets, such as sintered Nd–Fe–B and $SmCo_5$, it is essential to ensure that there are no low-anisotropy second phases present, otherwise these will act as low-field wall nucleation sites, e.g. the presence of the phases Sm_2Co_7 and Sm_2Co_{17} in sintered $SmCo_5$ lowers the coercivity very significantly. As mentioned already, it is generally accepted that in sintered Nd–Fe–B magnets the sinter grain boundaries are coated with an Nd-rich phase that broadens into larger regions at three-grain junctions (Livingston, 1986) and is a remnant of the Nd-rich liquid that assisted the densification during the liquid-phase sintering process.

Since the grain size of melt-spun Nd–Fe–B magnets is very much smaller than the critical diameter for single-domain particles, it is generally agreed that the coercivity is due to domain wall pinning at the grain boundaries (Livingston, 1986). The latter, however, are expected to be very different in composition and structure from those in the sintered magnets. Furthermore, it is to be expected that domain wall surface-tension effects will play an important part in determining the coercivity, since wall bowing will occur when the wall is pinned between opposite sides of a grain.

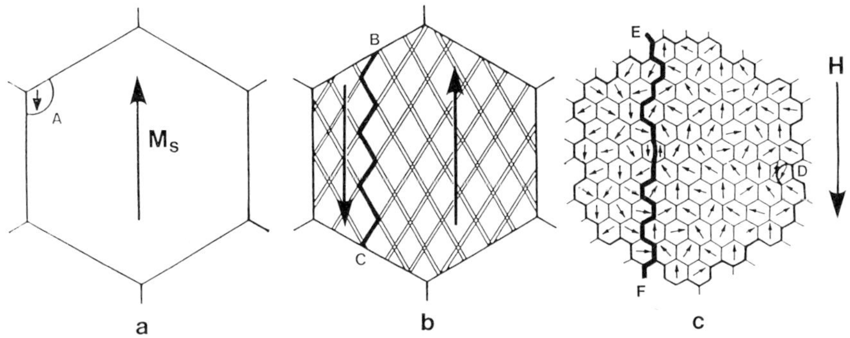

Fig. 5.72 Schematic representation of the three lanthanide–transition metal permanent magnet types, showing magnetization reversal in an applied field H: (a) aligned nucleation-controlled type with reverse domain nucleation at A; (b) aligned pinning-controlled type, with serrated domain wall BC pinned by fine cellular precipitates; (c) isotropic microcrystalline type with a domain wall traversing one grain at D and a domain wall traversing the boundary between downward-directed and upward-directed 'interaction domains' (after Livingston, 1986).

The compositions of pinning-controlled magnets are generally much more complex. The microstructures often consist of cellular structures containing rhombohedral Sm_2TM_{17}-, hexagonal Sm_2TM_{17}- and hexagonal $Sm(Co, Cu)_5$-based phases, which because of their different magnetic properties, act as pinning sites for the domain walls.

The microstructures and coercivity mechanisms for the three principal types of lanthanide–transition metal permanent magnets are summarized in Fig. 5.72.

5.19 PRODUCTION OF LANTHANIDE–TRANSITION METAL PERMANENT MAGNETS

There are a variety of techniques for the production of commercial lanthanide–transition metal permanent magnets. The chosen method will of course depend on whether the magnet is to be cast or sintered and on the magnetic properties required. The very best quality magnets are usually produced by liquid-phase sintering. The various production routes are summarized in Fig. 5.73. Route B uses the fact that most lanthanide–transition metal components readily absorb hydrogen in large quantities at room temperature, which enables the efficient pulverization of the compounds by the hydrogen decrepitation process (McGuiness *et al.*, 1986). Hydrogen desorption during sintering has been investigated by Harris *et al.* (1985, 1987), and it is now well established that the hydrogen desorption process is very effective in producing fine-grained high-coercivity Nd–Fe–B magnets (Buschow, 1991).

5.19.1 Control of Properties of Lanthanide–Transition Metal Permanent Magnets

Since the magnetic properties of lanthanide–transition metal permanent magnets depend strongly on composition and the precise production process, it is possible to engineer magnets with a wide range of properties. The principal effects of various alloying elements are summarized in Table 5.12.

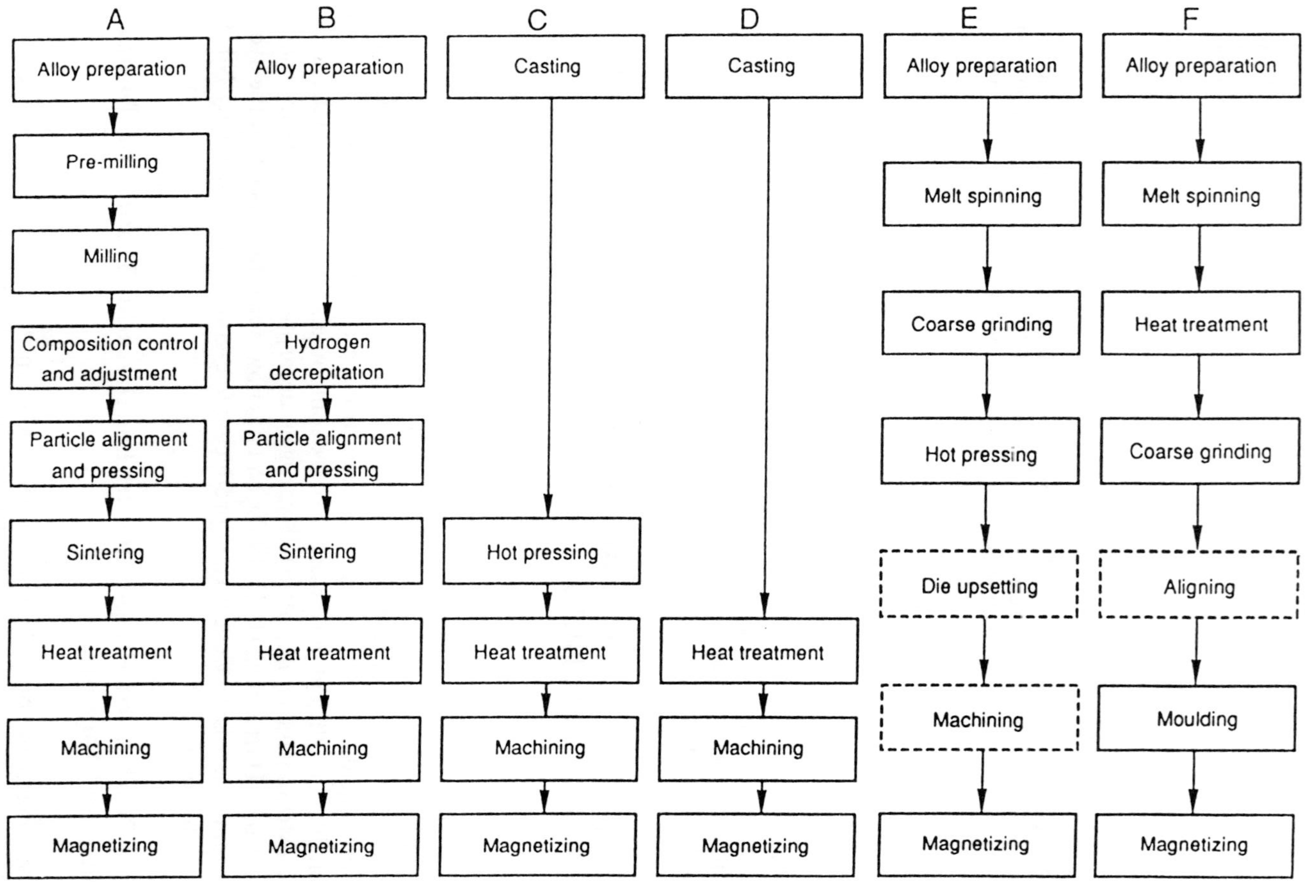

Fig. 5.73 Summary of manufacturing routes in the production of various lanthanide–transition metal permanent magnets (Buschow, 1991).

Table 5.12 Principal effects of alloying additions to lanthanide–transition metal permanent magnets (after Livingston, 1986).

Base compound	Increase T_c	Increase B_r and $(BH)_{max}$	Increase $H_a = 2K_1/J_s$	Compensate dJ_s/dT	Induce coherent precipitation	Decrease cost
$SmCo_5$	—	Pr, Nd	—	Gd, Ho, Dy, Er	Cu	Ce, MM[a]
Sm_2Co_{17}	—	Fe, Pr, Nd	Mn, Cr, Fe, Zr, Ti, Hf	Gd, Ho, Dy, Er	Cu	Ce, MM[a]
$Nd_2Fe_{14}B$	Co	—	Dy, Tb, Al	Ho, Dy	—	Ce, MM,[a] DD[b]

[a] MM, mischmetal (mainly cerium and lanthanum).
[b] DD, didymium (mainly neodymium and praseodymium).

5.20 GENERAL APPLICATIONS OF PERMANENT MAGNETS

It was mentioned at the beginning of this chapter that one of the advantages of using permanent magnets is that the magnetic flux and the associated forces can be supplied in a small space without the necessity for an electrical power supply. This gives a very high degree of reliability to the system, provided that the temperature does not rise significantly. Table 5.13 is a list of applications classified according to the five basic

Table 5.13 Applications of permanent magnets (after Gould, 1971).

Applications dependent on interaction between magnet flux and an electrical conductor; conversion of electrical into mechanical energy, or vice versa
Moving-coil ammeters, voltmeters, fluxmeters, etc. Moving-coil loudspeakers, microphones, vibration generators. Moving-coil transducers. Ribbon microphones, speakers. Electromagnetic flowmeters. Low-speed eddy current brakes for meters, speedometers, oscillation damping. Synchronous AC motors for clocks, timers. DC tachometer generators. DC motors, including stepping motors, servo-motors, watch motors, disk drives, printers, typewriters, cameras, audio and video recorders, actuators for robots and flight control. DC generators. AC generators, including magnetos and other ignition devices. High-speed eddy-current brakes or drives. Mass spectrometers. Electrical stringed instruments. Magnetic detectors. Pulse generators. Gyroscopes.

Applications dependent on magnetic polarization induced by magnet flux, exertion of a force on a soft ferromagnetic material
Magnetizers for small magnets. Reed relays, train-control relay. Separators and filters (extraction of iron impurities and concentration of ores). Snap switches in thermostats. Float switches. Latching relays. Telephone receivers. Polarized bell ringers. Polarized contacts. Control of saturable reactors. Hysteresis drives or brakes (directly or through a partition). Holding magnets, such as chart markers, door fasteners, clamps, magnetic chucks for machining, conveyor systems, 'fishing' magnets (oil wells, etc.), drain plugs, screwdrivers, refrigerator door seals, toys, sheet floaters.

Applications of permanent magnets without pre-magnetization
Rotors of hysteresis motors or hysteresis brakes and drives. Saturistors in AC motors. Recording disks and tapes. Pulse-switched relays (fereed, etc.).

Applications dependent on interaction of magnet flux and electron or ion beams; exertion of a force on moving charged particles
Magnetrons. Klystrons, Omegatrons. Ion pumps. Penning vacuum gauges. TV corrector magnets. Focusing magnets in travelling-wave tubes, ion accelerators, electron microscopes, etc. Magnetic blow-outs.

Applications dependent on interaction of magnet flux and a subsidiary material or device
Nuclear magnetic resonance. Gyromagnetic resonance in microwave devices (e.g. circulators, isolators, etc.). Oxygen analyzers. Transducers based on Hall effect, magneto-resistance or piezo-magnetic effect.

Applications dependent on interaction of magnetic flux with an external flux; alignment with respect to a field
Magnetic compasses. Moving-magnet instruments. Magnetic repulsion bearings. Positioning mechanisms e.g. stepping motors. Magnet-to-magnet drives.

Medical applications
Heart pacemaker devices. Hearing aids. Sternum, neck and head supports. Muscle supports, e.g. eyelids, facial muscles. Clamps for surgery. Wound sutures, etc. Possible use in control of double incontinence. Denture supports. Extraction or retrieval of ferromagnetic objects.

functions that permanent magnets can perform, namely

(1) conversion of electrical energy into mechanical energy and vice versa;
(2) exertion of a force on a ferromagnetic material;
(3) alignment with respect to a magnetic field;
(4) exertion of a force on moving charge carriers.
(5) attraction and repulsion of other permanent magnets

Typical demagnetization curves for the five most important groups of permanent magnet alloys are shown in Fig. 5.74. The relative merits of various permanent magnets classified according to their magnetic properties and cost are shown in Table 5.14. Typical demagnetization curves for various polymer-bonded magnets are shown in Fig. 5.75.

The properties of a wide range of permanent magnets are given in Table 5.15. Applications for which lanthanide–transition metal magnets are particularly suitable are listed in Table 5.16.

In spite of the high cost of lanthanide–transition metal-based alloys, their superior magnetic properties are enabling them to perform an ever-increasing number of tasks. They are replacing many of the older magnets such as the Alnicos, but they are unable to compete with the low-cost barium and strontium ferrites in low-technology applications. The comparative costs of various permanent magnets should be calculated on the basis of the cost per unit of stored magnetic energy.

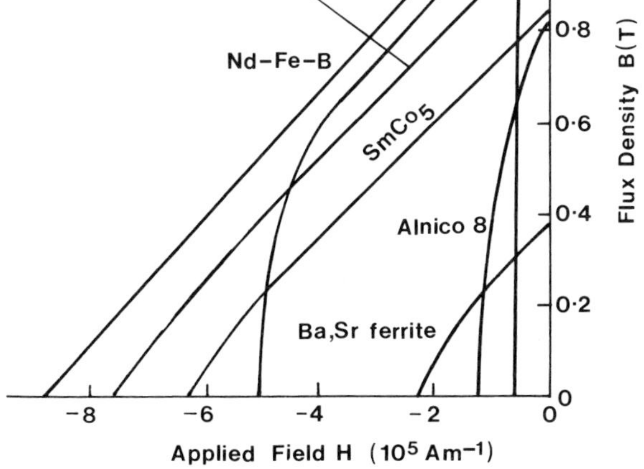

Fig. 5.74 Typical demagnetization curves for the five most important groups of permanent magnets (after Strnat, 1990).

Table 5.14 Comparison of the principal properties for the five most important groups of permanent magnet materials. The material with the best property is shown at the top of the appropriate column, e.g. Nd–Fe–B has the highest $(BH)_{max}$ value (after Gould, 1971).

$(BH)_{max}$	B_r	${}_BH_c$	Temperature-stability	Maximum operating temperature	World consumption (product value)	Low cost (relative cost per unit of stored magnetic energy)
Nd–Fe–B	Alnico 5	$SmCo_5$	Alnico 5	Alnico 5	Ba, Sr ferrites	Ba, Sr ferrites
$Sm_2(TM)_{17}$	Nd–Fe–B	Nd–Fe–B	$Sm_2(TM)_{17}$	$Sm_2(TM)_{17}$	$SmCo_5$	Alnico 5
$SmCo_5$	$Sm_2(TM)_{17}$	$Sm_2(TM)_{17}$	$SmCo_5$	$SmCo_5$	$Sm_2(TM)_{17}$	Nd–Fe–B
Alnico 5	$SmCo_5$	Ba, Sr ferrites	Ba, Sr ferrites	Ba, Sr ferrites	Alnico 5	$Sm_2(TM)_{17}$
Br, Sr ferrites	Ba, Sr ferrites	Alnico 5	Nd–Fe–B	Nd–Fe–B	Nd–Fe–B	$SmCo_5$

TM, transition metals: Fe, Co, Cu, etc.

Table 5.15 Comparison of various permanent magnet materials.

Material	B_r(T)	$_BH_c$ (10^5 A m^{-1})	$_iH_c$ (10^5 A m^{-1})	$(BH)_{max}$ (kJ m^{-3})	T_c (°C)	T_{max} (°C)	Origin of coercivity
Alnico 5	1.25	0.53	0.54	40	850	520	Shape anisotropy, spinodal decomposition
Alnico 9	1.10	1.4	1.45	75	850	520	
$BaFe_{12}O_{19}$	0.4	1.6	1.65	29	450	400	Nucleation-controlled, uniaxial magnetocrystalline anisotropy
$SrFe_{12}O_{19}$	0.4	2.95	3.3	30	450	400	
$SmCo_5$	1.00	7.9	13	196	700	250	
$Sm(Co_{0.76}Fe_{0.10}Cu_{0.14})_{6.8}$	1.04	4.8	5	212	800	300	Pinning-controlled uniaxial magnetocrystalline anisotropy
$Sm(Co_{0.65}Fe_{0.28}Cu_{0.05}Zr_{0.02})_{7.7}$	1.2	10	16	264	800	300	
$Nd_2Fe_{14}B$ (sintered)	1.22	8.4	12.8	280	300	100[a]	Nucleation-controlled, uniaxial magnetocrystalline anisotropy
Nd–Fe–B (melt-spun), MQI	0.61	4.2	12.0	64	300	100[a]	
Nd–Fe–B (melt-spun), MQII	0.80	5.2	12.8	104	300	100[a]	
Nd–Fe–B (melt-spun), MQIII	1.17	8.4	10.4	256	300	100[a]	
Fe–52 wt% Co–14% V (cold-worked, anisotropic), Vicalloy II	1.0	0.42	—	28	~700	~500	Shape anisotropy, precipitation
Fe–24 wt% Cr–15 % Co–3% Mo (anisotropic)	1.54	0.67	—	76	~630	~500	Shape anisotropy, spinodal decomposition
Fe–28 wt% Cr–10.5% Co, Chromindur II	0.98	0.3	—	16	—	—	
Fe–23 wt% Cr–15% Co–3% V–2% Ti	1.35	0.4	—	44	~630	~500	
Cu–20 wt% Ni–20% Fe	0.55	0.4	—	12	410	350	
Pt–23 wt% Co	0.64	4	—	76	480	350	Pinning-controlled uniaxial magnetocrystalline anisotropy
Mn–29.5 wt% Al–0.5% C, anisotropic hot-extruded	0.61	2.16	2.4	56	300	120	

[a] Nd–Fe–B magnets with additions of Dy and Co can be used up to about 150°C.

Table 5.16 Applications for which lanthamide–transition metal magnets are particularly suitable (Melville *et al.*, 1986).

Automotive	Domestic	Electronic and instrumentation	Astro- and aerospace	Biosurgical
DC motor drives	Analogue watches	Sensors	Frictionless bearings	Dentures
Cranking motors	Model aircraft	Contactless switches	Stepping motors	Orthopaedics
Window winders	Head-sets	Spherical motors (robotics)	Couplings	Wound closures
Wipers	Pick-ups	Magnetic resonance spectrometers	Travelling-wave magnetrons	Magnetic-sphincters
Washers	Video recorders	Energy meter disc-repulsion bearings	Klystrons	Stoma seals
Fans	Electric clocks	Automation devices	Autocompasses	Repulsion collars
Seat actuators	Jewellery/ornaments	Electromechanical transducers	Instruments	Magnetic sutures
Speedometers	Hearing aids	Computer print-heads		Ferromagnetic probes
Alternators	Moving-magnet loudspeakers	Crossed-field tubes		Cancer cell separators
	Moving-magnet microphones	Flux-transfer trip devices		Optical implants
		Magnetic latches and switches		Pulsatile magnetomotive artificial hearts
		Dampers		Magnetic resonance body scanners

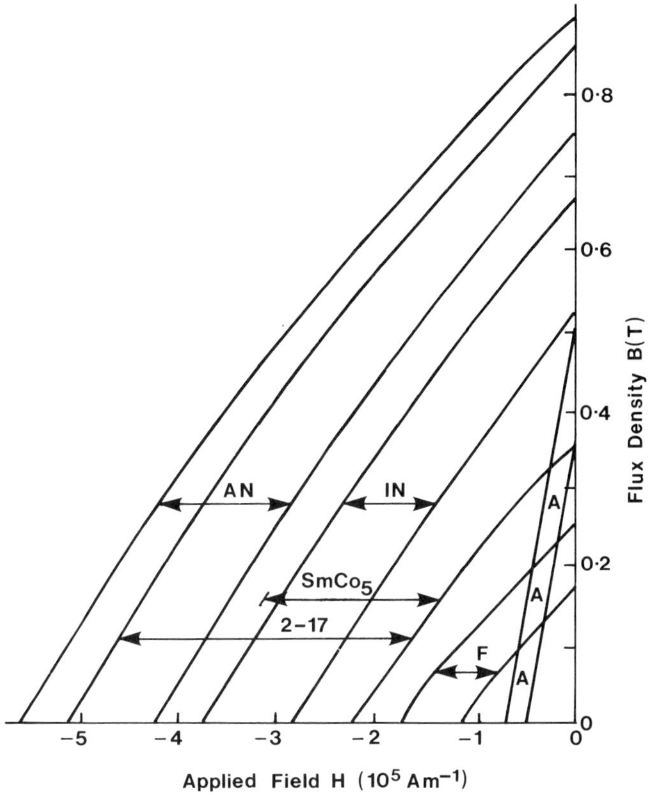

Fig. 5.75 Demagnetization curves for various polymer-bonded permanent magnets, showing ranges of magnetic properties available: AN, anisotropic Nd–Fe–B magnets; IN, isotropic Nd–Fe–B magnets; 2–17, Sm_2TM_{17}-based magnets; A, Alnicos; F, barium and strontium ferrite magnets. The magnets are in or near commercial production (after Strnat, 1990).

5.21 TEMPERATURE DEPENDENCE OF MAGNETIC PROPERTIES OF PERMANENT MAGNETS

The temperature dependence of the magnetic properties of permanent magnets is of great importance in assessing their suitability for particular applications. With the exceptions of $BaFe_{12}O_{19}$ and $SrFe_{12}O_{19}$ (see Figs 5.44 and 5.45), the coercivities of all currently available permanent magnets decrease with increasing temperature. The decrease in the coercivities of Nd–Fe–B-based magnets is particularly serious, and limits the maximum temperature of use to 100°C. Samarium–cobalt-based magnets can be used for short periods without significant deterioration up to about 250–300°C.

Although the Alnicos have been superseded by many magnets with superior coercivities and energy products, their temperature stability is still unmatched. The Alnicos are among the very few permanent magnets that can be used without serious deterioration in their properties up to about 500°C. This important property results from their relatively high Curie temperatures, about 850°C, and the fact that the coercivity is due principally to the shape anisotropy of Fe–Co-rich single-domain particles. The

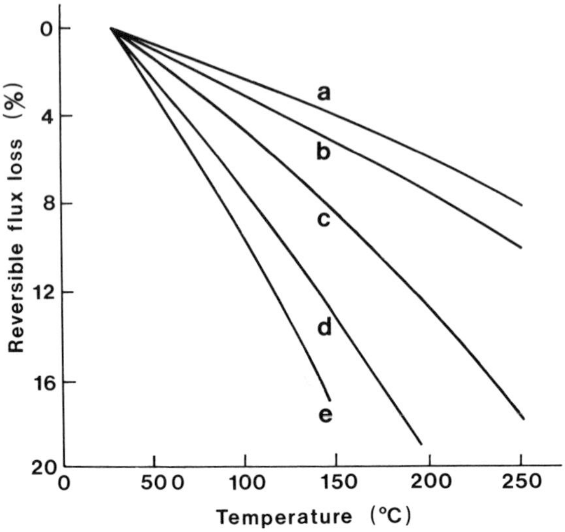

Fig. 5.76 Reversible flux loss as a function of temperature for various lanthanide–transition metal permanent magnets: (a) high- and low-coercivity Sm_2TM_{17}-based magnets; (b) sintered $SmCo_5$; (c) $Sm_{0.5}Ce_{0.5}(Co, Cu, Mn)_7$; (d) $Ce(Co, Cu, Fe)_5$; (e) sintered $Nd_2Fe_{14}B$ (after Strnat, 1990).

Curie temperatures and maximum temperatures for safe operation for various permanent magnets are given in Table 5.15. The reversible flux loss as a function of temperature for the principal types of lanthamide–transition metal magnets is shown in Fig. 5.76.

5.22 SUMMARY OF RESEARCH ON NEW PERMANENT MAGNET ALLOYS

5.22.1 Permanent Magnets such as Sm–Fe–Ti based on the Intermetallic Compound Nd_5Fe_{17}

Although the intermetallic compound Nd_5Fe_{17} is not suitable for direct application as a permament magnet, many of the related ternary compounds are potentially suitable. Schultz *et al.* (1990a) have shown that ternary alloys in the Sm–Fe–Ti system have potentially useful permanent magnet properties, in particlar mechanically alloyed $Sm_{20}Fe_{70}Ti_{10}$, which is metastable. They have obtained a room-temperature coercivity of 4 MA m^{-1} with a remanence of 0.6 T. Katter *et al.* (1990) have shown that compounds in the system $Sm_5(Fe, Ti)_{17}$ may have exceptionally high coercivities – up to 5 MA m^{-1}.

Alloys in the Sm–Fe–Ti system can be prepared by melt spinning, mechanical alloying, plasma spraying and in thin-film form by sputtering.

5.22.2 Permanent Magnets based on $Ln_2Fe_{14}C$ Compounds

The magnetic properties of the carbides $Ln_2Fe_{14}C$ (where Ln is a lanthanide) are very similar to the corresponding borides. The interest in these compounds lies in the possibility of producing very fine particle microstructures by the controlled transformation of the high-temperature phase to the low-temperature $Ln_2Fe_{14}C$ phase, which appears when the alloys are cast. According to Eisses *et al.* (1991), the microstructures can be controlled by annealing, so that microstructures resembling those of sintered magnets can be produced more easily. Thus it is hoped that high-quality magnets can be produced by the relatively cheap route D (see Fig. 5.73), which does not involve powder metallurgy or melt spinning. de Mooij *et al.* (1991) produced a cast ingot of primarily $Nd_2Fe_{14}C$ with a room-temperature intrinsic coercivity of 0.8 MA m^{-1} and a remanence of 0.65 T.

5.22.3 Permanent Magnets based on Fe_3B

The compound Fe_3B is metastable and can be obtained only by rapid quenching, after which it exhibits only a very low coercivity. However, the addition of relatively small quantities (a few per cent) of lanthanide metals such as Tb, La, Nd and Dy enables the development of hard magnetic properties after melt spinning followed by a short anneal. The permanent magnet properties of crystallized amorphous lanthanide–iron–boron alloys were first reported by Koon and Das (1981, 1984), who obtained a coercivity of 0.72 MA m^{-1} and a remanence of 0.5 T for the alloy $Tb_{0.05}La_{0.05}(Fe_{0.82}B_{0.18})_{0.9}$.

Shen *et al.* (1991) obtained a coercivity of 0.29 MA m^{-1} for the alloy $Nd_2DyFe_{77.5}B_{18.5}$. The coercivity is due to the recrystallization of the amorphous material during annealing, in particular the formation of very small crystals of the $Ln_2Fe_{14}B$ phase embedded in an Fe_3B matrix.

Croat *et al.* (1984) investigated the magnetic properties of a new class of permanent magnet alloys based on nanocrystalline Pr–Fe–B and Nd–Fe–B. There is considerable interest in the possibility of producing high energy permanent magnets from recrystallized alloys based on Fe–B with various lanthanide metal additives and in recrystallized Nd–Fe–B and Pr–Fe–B alloys. Alloy systems containing no lanthanide metals are also of interest, e.g. Co–Zr–B.

5.22.4 Permanent Magnets Based on the $ThMn_{12}$ Structure

Iron-rich ternary compounds based on the tetragonal $ThMn_{12}$ structure have been proposed as inexpensive alternatives to other permanent magnets, since they have a relatively low lanthanide content. Compounds with potentially useful permanent magnet properties have the general formula $LnFe_{12-x}M_x$ where $x \approx 2$ when M is Ti, V, Cr, Mo or Si. Only alloys with Ln = Sm have sufficiently high uniaxial magnetocrystalline anisotropy. Although acceptable coercivities have been obtained, the remanences obtained to date are usually too low to be useful. Schultz *et al.* (1990a, b) obtained a coercivity of 0.8 MA m^{-1} and a remanence of 0.5 T for the alloy $Sm_{12}Fe_{73}V_{15}$, which was produced by mechanical alloying. In this process the elemental powders are milled with steel balls and then heat-treated at temperatures up to about 900°C, so that the required alloy composition is produced by diffusion.

5.22.5 Permanent Magnets based on Lanthanide–Iron–Nitrogen

Lanthanide–transition metal intermetallic compounds based on Ln_2Fe_{17} exist throughout the lanthanide series, except for lanthanum itself. They either crystallize with the rhombohedral Th_2Zn_{17} structure (Fig. 5.52b) or with the hexagonal Th_2Ni_{17} structure

(Fig. 5.52b), and may coexist for Ln = Gd, Tb and Dy. The differences between the two structures are small, and depend on the stacking sequence of hexagonal planes. The structures and dimensions for both Th_2Zn_{17} and Th_2Ni_{17} are usually specified in terms of their hexagonal unit cells.

Unfortunately, the Curie temperatures of the Ln_2Fe_{17} compounds are relatively low, ranging from 240 K for Ce to 480 K for Gd. However, the Curie temperatures can be raised quite significantly by partial substitution of iron with various other metals or non-metals (e.g. Al, Si, Ni, Co, H, C or N). A new series of interstitial compounds ($Ln_2Fe_{17}N_{3-x}$) with promising permament magnet properties have been discovered by Coey (1990), Sun *et al.* (1990) and Coey and Sun (1990). The diffusion of nitrogen into interstitial sites in the rhombohedral Sm_2Fe_{17} structure to produce $Sm_2Fe_{17}N_3$ (Fig. 5.77) results in an increase in the saturation magnetic polarization to 1.54 T, comparable to that of $Nd_2Fe_{14}B$, and produces a uniaxial (c-axis) magnetocrystalline anisotropy field of about 15 T. Furthermore, the Curie temperature is increased to 470°C, i.e. 170°C higher than that of $Nd_2Fe_{14}B$. Schultz *et al.* (1991) have succeeded in producing an isotropic magnet based on $Sm_2Fe_{17}N_3$ with an intrinsic coercivity of 2.4 MA m^{-1} and a $(BH)_{max}$ of 87 kJ m^{-3}.

Coey and Sun (1990) have also shown that permanent magnets based on $Sm_2Fe_{17}C_xN_y$ with $0<x<1.5$ and $1<y<3$ show much potential.

Examples of permanent magnet materials currently under investigation are given in Table 5.17, with $SmCo_5$ and $Nd_2Fe_{14}B$ included for comparison. The maximum energy product is of course limited by the saturation magnetization M_s, provided of course that the intrinsic coercivity ${}_iH_c \geqslant \frac{1}{2}M_s$.

The alloy 30% Co–Fe has the highest room-temperature saturation magnetization $M_s = 1.95 \times 10^6$ A m^{-1} ($J_s = 2.45$ T), giving a maximum possible $(BH)_{max}$ of 1200 kJ m^{-3}. To date, the best permanent magnet at room temperature, namely Nd–Fe–B, has a

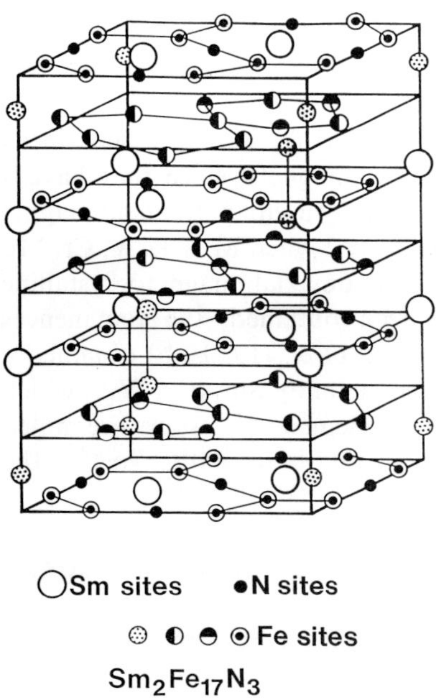

Fig. 5.77 Crystal structure of $Sm_2Fe_{17}N_3$ (Harris, 1991).

Table 5.17 Magnetic characteristics of several lanthanide–transition metal-based magnet alloys. The energy products $(BH)_{max}$ represent the theoretical upper limits (after Buschow, 1991).

Base material	T_c (°C)	B_s (T)	H_A (MA m^{-1})	$(BH)_{max}$ (kJ m^{-1})
$Nd_2Fe_{14}B$	312	1.60	5.4	512
$Nd_2Fe_{14}C$	262	1.50	7.6	450
Fe_3B:Nd	512	1.60	—	512
$SmFe_{11}Ti$	312	1.16	7.4	268
$SmFe_{10}V_2$	337	1.10	4.8	240
$SmFe_{10}Mo_2$	187	0.97	>4.0	188
$Sm_2Fe_{17}C$	267	1.42	4.2	403
$Sm_2Fe_{17}N_{2.7}$	477	1.50	11.2	450
$SmCo_5$	747	1.14	24	258
$Sm(Co, Fe, Cu)_7$	827	1.3	4.0	336
$Nd_2Co_{14}B$	722	1.15	8.0	266

$(BH)_{max}$ of 403 kJ m^{-3}, so there is still considerable scope for research and development.

At present, it is quite impossible to predict which of the above-mentioned alloys will eventually prove to possess all the required magnetic, mechanical and chemical properties to be commercially viable. There is clearly the possibility of whole new generations of permanent magnets based on these and similar compounds, and a great deal of research is currently in progress.

Thorough and excellent reviews of new developments in permanent magnet materials have been published by Buschow (1991), Strnat (1988, 1990), Herbst (1991), Stadelmaier (1990) and Stadelmaier and Henig (1991).

REFERENCES

Aharoni, A. (1966). *Phys. Stat. Sol.* **16**, 1.

Biedermann, E. von and Kneller, E. (1956a). *Z. Metallk.* **47**, 289.

Biedermann, E. von and Kneller, E. (1956b). *Z. Metallk.* **47**, 760.

Bretthauer, K. (1968). *Tech. Mitt. Krupp* **26**, 1.

Brissonneau, P. and Schlenker, M. (1967). *C. R. Acad. Sci. Paris* **264**, 282.

Brissonneau, P., Blanchard, A. and Bartholin, H. (1966). *IEEE Trans. Magn.* **MAG-2**, 479.

Brissonneau, P., Blanchard, A. and Shlenker, M. (1968). *J. Appl. Phys.* **39**, 1266.

Brown, W. F. (1963). *Micromagnetics* (Wiley–Interscience, New York), pp. 68–84.

Brown, W. F. (1969). *Ann. NY Acad. Sci.* **147**, 463.

Buschow, K. H. J. (1986). *Mater. Sci. Rep.* **1**, 1.

Buschow, K. H. J. (1988). *Ferromagnetic Materials: A Handbook on the Properties of Magnetically Ordered Substances*, Vol. 4, ed. E. P. Wohlfarth and K. H. J. Buschow (Elsevier, Amsterdam), p. 1.

Buschow, K. H. J. (1991). *Rep. Prog. Phys.* **54**, 1123.

Buschow, K. H. J. and Velge, W. J. J. (1969). *Z. Angew. Phys.* **26**, 157.

Butler, E. P. and Thomas, G. (1970). *Acta Met.* **18**, 347.

Cech, R. E. (1974). *J. Metals* **26**, 32.

Chin, G. Y., Nesbitt, E. A., Wernick, J. H. and Mendorf, D. R. (1969). *J. Appl. Phys.* **40**, 760.

Chin, G. Y., Plewes, J. T. and Wonsiewicz, B. C. (1978). *J. Appl. Phys.* **49**, 2046.

Cochardt, A. (1963). *J. Appl. Phys.* **34**, 1273.
Cochardt, A. (1966). *J. Appl. Phys.* **37**, 1112.
Coey, J. M. D. (1990). *J. Phys. Cond. Matt.* **2**, 6465.
Coey, J. M. D. and Sun, H. (1990). *J. Magn. Magn. Mater.* **87**, L251.
Croat, J. J. (1988). *J. Mater. Engng* **10**, 7.
Croat, J. J., Herbst, J. F., Lee, R. W. and Pinkerton, F. E. (1984). *J. Appl. Phys.* **55**, 2078.
Dannöhl, W. and Neumann, H. (1938). *Z. Metallk.* **30**, 217.
de Vos, K. J. (1969). *Magnetism and Metallurgy*, Vol. 1, ed. A. E. Berkowitz and E. Kneller (Academic Press, London), p. 473.
de Mooij, D. B., Eisses, J., Buschow, K. H. J. and Martinek, G. (1991). *J. Less-Common Met.* (in press).
den Broeder, F. J. A and Zijlstra, H. (1976). *J. Appl. Phys.* **47**, 2688.
Domazer, H. G. (1974). *Proc. 3rd Eur. Conf. Hard Magnetic Materials*, ed. H. Zijlstra (Bond voor Materialenkennis, Den Haag), p. 140.
Drozzhina, V. and Janus, R. (1935). *Nature* **13**, 3.
Durst, K. D. and Kronmüller, H. (1987). *J. Magn. Magn. Mater.* **68**, 63.
Ebeling, R. and Krause, H. (1978). VDI-Berichte No. 3098, Düsseldorf.
Edwards, A. (1962). *Permanent Magnets and Magnetism*, ed. D. Hadfield (Iliffe, London), p. 191.
Eisses, J., de Mooij, D. B., Buschow, K. H. J. and Martinek, G. (1991). *J. Less-Common Met.* **171**, 17.
Eurin, P. and Pauleve, J. (1969) *IEEE Trans. Magn.* **MAG-5**, 216.
Ervens, W. (1982). *Tech. Mitt. Krupp Forschungsberichte* **40**, 99.
Fahlenbrach, H. (1960). *Techn. Mitt. Krupp. Forsch. Ber.* **18**, 91.
Fahlenbrach, H. (1962). *Techn. Mitt. Krupp. Forsch. Ber.* **20**, 141.
Fahlenbrach, H. (1964). *Cobalt* **25**, 187.
Fahlenbrach, H. and Baran, W. (1964). *Z. Angew. Phys.* **17**, 178.
Fayling, R. E. (1979). *IEEE Trans. Magn.* **MAG-15**, 1567.
Fidler, J. (1982). *Phil. Mag.* **B46**, 565.
Fidler, J. and Skalicky, P. (1982). *J. Magn. Magn. Mater.* **27**, 127.
Frei, E. H., Shtrikman, S. and Treves, D. (1957). *Phys. Rev.* **106**, 446.
Friedberg, R. and Paul, D. I. (1975). *Phys. Rev. Lett.* **34**, 1234.
Gaunt, P. (1966). *Phil. Mag.* **13**, 579.
Gaunt, P. (1986). *J. Appl. Phys.* **59**, 4129.
Gaunt, P. (1977a). Private communication.
Gaunt, P. (1977b) *J. Appl. Phys.* **48**, 3470.
Gaunt, P. and Mylvaganum, C. K. (1977). *J. Appl. Phys.* **48**, 2587.
Gaunt, P. and Roy, G. J. (1976) *Phil. Mag.* **34**, 781.
Givord, D., Tenaud, P. and Viadieu, T. (1988a). *J. Magn. Magn. Mater.* **72**, 247.
Givord, D., Tenaud, P. and Viadieu, T. (1988b). *IEEE Trans. Magn.* **MAG-24**, 1921.
Givord, D., Heiden, C., Hoehler, A., Tenaud, P., Viadieu, T. and Zeibig, K. (1988c). *IEEE Trans. Magn.* **MAG-24**, 1918.
Givord, D., Lu, O., Rossignol, M. F., Tenaud, P. and Viadieu, T. (1990). *J. Magn. Magn. Mater.* **83**, 183.
Gould, J. E. (1971). *Cobalt Alloy Permanent Magnets* (Centre d'Information du Cobalt, Brussels).
Hadjipanayis, G. C. (1988). *J. Appl. Phys.* **63**, 3310.
Hadjipanayis, G. C. and Gaunt, P. (1979). **50**, 2358.
Hadjipanayis, G. C. and Kim, A. (1988). *J. Appl. Phys.* **63**, 3310.
Hadjipanayis, G. C., Singleton, E. W. and Tang, Z. X. (1980). *J. Magn. Magn. Mater.* **81**, 318.
Hadjipanayis, G. C. (1990). *J. Magn. Magn. Mater.* **92**, L43.
Haneda, K. and Kojima, H. (1973). *J. Appl. Phys.* **44**, 3760.
Haneda, K., Miyakawa, C. and Kojima, H. (1974). *J. Amer. Ceram. Soc.* **57**, 354.

Harris, I. R. (1991). *Physics World* **4**, 27.
Harris, I. R., Noble, C. and Bailey, T. (1985). *J. Less-Common Met.* **106**, L1.
Harris, I. R., McGuinness, P. J., Jone, D. G. R. and Abell, J. S. (1987). *Physica Scripta* **T19**, 439.
Heck, C. (1974). *Magnetic Materials and their Applications* (Butterworths, London).
Heimke, G. (1976). *Keramische Magnete* (Springer-Verlag, Vienna).
Henkel, O. (1966a). *Z. Angew. Phys.* **21**, 52.
Henkel, O. (1966b). *Z. Angew. Phys.* **21**, 112.
Herbst, J. F. (1991). *Rev. Mod. Phys.* **63**, 819.
Herbst, J. F., Croat, J. J., Pinkerton, F. E. and Yelon, W. B. (1984). *Phys. Rev.* **B29**, 4176.
Hilzinger, H. R. (1977). *Appl. Phys.* **12**, 253.
Hilzinger, H.-R. and Kronmüller, H. (1975). *Phys. Lett.* **51A**, 59.
Hoffer, G. I. and Strnat, K. J. (1966). *IEEE Trans. Magn.* **MAG-2**, 487.
Homma, M., Horikoshi, E., Minowa, T. and Okada, M. (1980). *Appl. Phys. Lett.* **37**, 92.
Hubbard, W. M., Adams, E. and Gilfrich, J. (1960). *J. Appl. Phys.* **31** (Suppl.), 368S.
Jahn, L. and Müller, H. G. (1969). *Phys. Stat. Sol.* **35**, 723.
Jakubovics, J. P. and Jolly, T. W. (1977). *Physica* **86–88B,** 1357.
Jellinghaus, W. (1936). *Z. Tech. Phys.* **17**, 33.
Jin, S. and Tiefel, T. H. (1980). *IEEE Trans. Magn.* **MAG-16**, 1062.
Joffe, I. and Gaunt, P. (1971). *J. Phys. D: Appl. Phys.* **4**, 586.
Jonas, B. and Meerkamp van Embden, H. J. (1941). *Philips Tech. Rev.* **6**, 8.
Jubb, G. A. and McCurrie, R. A. (1987). *IEEE Trans. Magn.* **MAG-23**, 1801.
Kamino, K., Kawaguchi, T. and Nagakura, M. (1966). *IEEE Trans. Magn.* **MAG-2**, 506.
Kaneko, H., Homma, M. and Suzuki, S. (1968). *Trans. Japan. Inst. Met.* **9**, 124.
Kaneko, H., Homma, M. and Nakamura, K. (1971). *AIP Conf. Proc.* **5**, 1088.
Kaneko, H., Homma, M., Okada, M., Nakamura, S. and Ikuta, N. (1975). *AIP Conf. Proc.* **29**, 620.
Katter, M., Wecker, J., Schultz, L. and Grossinger, R. (1990). *Appl. Phys. Lett.* **56**, 1377.
Kawaguchi, T. and Ogawa, K. (1971). *IEEE Trans. Magn.* **MAG-7**, 668.
Koch, A. J. J., Hokkeling, P., van der Steeg, M. G. and de Vos, K. J. (1960). *J. Appl. Phys.* **315**, 75.
Kojima, H. (1982). *Ferromagnetic Materials: A Handbook on the Properties of Magnetically Ordered Substances*, Vol. 3, ed. E. P. Wohlfarth (North-Holland, Amsterdam), p. 305.
Kondorsky, E. (1952). *Dokl. Akad. Nauk SSR* **80**, 197; **82**, 365.
Kono, H. (1958). *J. Phys. Soc. Japan*, **13**, 1444.
Kools, F. (1992). *Concise Encyclopedia of Magnetic and Superconducting Materials*, ed. J. Evetts (Pergamon, Oxford).
Koon, N. C. and Das, B. N. (1981). *Appl. Phys. Lett.* **39**, 840.
Koon, N. C. and Das, B. N. (1984). *J. Appl. Phys.* **55**, 2063.
Köster, W. (1932). *Arch. Eisenhüttenw.* **6**, 17.
Köster, W. and Wachtel, E. (1960). *Z. Metallk* **51**, 271.
Krijtenberg, G. S. (1965). *Proc. 1st Europeam Conf. Vortrag Nr. 213 uber Magnetismus in Wien.*
Kronmüller, H. (1991). *Supermagnets, Hard Magnetic Materials*, ed. G. J. Long and F. Grandjean (Kluwer, Dordrecht).
Kronmüller, H. (1978). *J. Magn. Magn. Mater.* **7**, 341.
Kronmüller, H., Durst, K.-D. and Sagawa, M. (1988). *J. Magn. Magn. Mater.* **74**, 291.
Kronmüller, H. and Hilzinger, H. R. (1976). *J. Magn. Magn. Mater.* **2**, 3.
Kumar, K. (1988). *J. Appl. Phys.* **63**, R13.
Kumar, K., Das, D. and Wettstein, E. (1978). *IEEE Trans. Magn.* **MAG-14**, 788.
Kütterer, R., Hilzinger, H. R. and Kronmüller, H. (1977). *J. Magn. Magn. Mater.* **4**, 1.
Lee, R. W., Brewer, E. G. and Schaffel, N. A. (1985). *IEEE Trans. Magn.* **MAG-21**, 1958.
Li, D., Mildrum, H. F. and Strnat, K. J. (1988). *J. Appl. Phys.* **63**, 3984.
Livingston, J. D. (1973a). *AIP Conf. Proc.* **10**, 643.
Livingston, J. D. (1973b). *Phys. Stat. Sol.* (a) **18**, 579.

Livingston, J. D. (1975). *J. Appl. Phys.* **46**, 5259.
Livingston, J. D. (1978). *IEEE Trans. Magn.* **MAG-14**, 668.
Livingston, J. D. (1981a). *Prog. Mater. Sci.* **7**, 243.
Livingston, J. D. (1981b). *J. Appl. Phys.* **52**, 2544.
Livingston, J. D. (1986). *Soft and Hard Magnetic Materials with Applications: American Society for Metals Symposium Proceedings* (American Society for Metals, Metals Park, Ohio), p. 71.
Livingston, J. D. (1987). *IEEE Trans. Magn.* **MAG-23**, 2109.
Livingston, J. D. and Martin, D. L. (1977). *J. Appl. Phys.* **48**, 1350.
McCaig, M. and Clegg, A. G. (1987). *Permanent Magnets in Theory and Practice*, 2nd edn (Pentech, London).
McCurrie, R. A. (1982). *Ferromagnetic Materials: A Handbook on the Properties of Magnetically Ordered Substances*, Vol. 3, ed. E. P. Wohlfarth (North-Holland, Amsterdam), p. 107.
McCurrie, R. A. (1992). *Encyclopedia of Magnetic and Superconducting Materials*, ed. J. E. Evetts (Pergamon Press, Oxford).
McCurrie, R. A. and Carswell, G. P. (1970). *J. Mater. Sci. Lett.* **5**, 825.
McCurrie, R. A. and Carswell, G. P. (1973). *Phil. Mag.* **28**, 611.
McCurrie, R. A. and Gaunt, P. (1965). *Proc. Int. Conf. Magnetism, Nottingham, Sept. 1964* (Institute of Physics and the Physical Society, London), p. 780.
McCurrie, R. A. and Gaunt, P. (1966). *Phil. Mag.* **13**, 567.
McCurrie, R. A. and Gaunt, P. (1969). *Phil. Mag.* **19**, 339.
McCurrie, R. A., Rickman, J., Dunk, P. and Hawkridge, D. G. (1978). *IEEE Trans. Magn.* **MAG-14**, 682.
McCurrie, R. A. and Jackson, S. (1980). *IEEE Trans. Magn.* **MAG-16**, 1310.
McCurrie, R. A. and Jackson, S. (1987). *J. Appl. Phys.* **62**, 627.
McCurrie, R. A. and Mitchell, R. K. (1975). *IEEE Trans. Magn.* **MAG-11**, 1408.
McCurrie, R. A. and Willmore, L. E. (1979) *J. Appl. Phys.*, **50**, 3560.
McGuiness, P. J., Harris, I. R., Rosendaal, E., Ormerod, J. and Ward, M. (1986). *J. Mater. Sci.* **21**, 4107.
Martin, D. L. (1958) *Trans. Metall. Soc. AIME*, **212**, 4780.
Mee, C. D. and Jeschke, J. C. (1963). *J. Appl. Phys.* **34**, 1271.
Melville, D., Ormerod, J., Walkden, A. and Hadfield, D. (1986). BNF Metals Technology Centre Report, BNF 6.101.0/2, Wantage, Oxfordshire.
Menth, A., Nagel, H. and Perkins, R. S. (1978). *Ann. Rev. Mater. Sci.* **8**, 21.
Mildrum, H. F., Graves, G. A. and Abdelnour, Z. A. (1981). *Proc. 5th Int. Workshop on Rare Earth–Cobalt Permanent Magnets and Their Applications, Roanoke, Virginia*, ed. K. J. Strnat (University of Dayton, Ohio), p. 313.
Mintern, R. A. (1961). *Platinum Metals Rev.* **5**, 82.
Mishima, T. (1932). *Ohm* **19**, 353.
Mishin, D. D. (1967). *Phys. Stat. Sol.* **20**, K9.
Mishin, D. D., Shmelkov, A. P. and Ivanov, O. A. (1965). *Uchen. Zap. Ural. Univ.* **1**, 47.
Mishin, D. D. (1967). *Phys. Stat. Sol.* **20**, K9.
Moskowitz, L. (1976). *Permanent Magnet Design and Application Handbook* (Cahners, Boston).
Naastepad, P. A. (1966). *Z. Angew. Phys.* **21**, 104.
Narasimhan, K. S. V. L. (1985). *J. Appl. Phys.* **57**, 4081.
Nesbitt, E. A. (1946). *Trans. Met. Soc. AIME* **166**, 415.
Nesbit, E. A. and Kelsall, G. A. (1940). *Phys. Rev.* **58**, 202.
Nesbit, E. A., Chin, G. Y. and Jaffe, D. (1968). *J. Appl. Phys.* **39**, 1268.
Nesbitt, E. A. and Wernick, J. H. (1973). *Rare-Earth Permanent Magnets* (Academic Press, New York).
Neumann, H., Buchner, A. and Reinboth, H. (1937). *Z. Metallk.* **29**, 173.
Newkirk, J. B., Geisler, A. H., Martin, D. L. and Smoluchowski, R. (1950). *Trans. AIME, J.*

Metals **188**, 1249.
Newkirk, J. B., Smoluchowski, R., Geisler, A. H. and Martin, D. L. (1951a). *J. Appl. Phys.* **22**, 290.
Newkirk, J. B., Smoluchowski, R., Geisler, A. H. and Martin, D. L. (1951b). *Acta Cryst.* **4**, 507.
Ohtani, T., Kato, N., Kojima, S., Kojima, K., Sakamoto, Y, Konno, I., Tsukahara, M. and Kubo, T. (1977). *IEEE Trans. Magn.* **MAG-13**, 1328.
Ojima, T., Tomizawa, S., Yoniyama, T. and Hori, T. (1977). *IEEE Trans. Magn.* **MAG-13**, 1317.
Okada, M., Kassai, M., Suzuki, Y., Sasaki, T. and Henmi, Z. (1972). *Trans. Japan. Inst. Met.* **13**, 391.
Okada, M., Kassai, M., Sasaki, T. and Henmi, Z. (1973). *Conf. Magnetism and Magnetic Materials* (A.I.P., Conf. Proc. No. 18).
Okada, M., Thomas, G., Homma, M. and Kaneko, H. (1978). *IEEE Trans. Magn. MAG*-14, 245.
Oliver, D. A. and Shedden, H. J. (1938). *Nature* **142**, 209.
Olsen, K. M. and Stoffers, R. C. (1971). *J. Appl. Phys.* **42**, 1792.
Ormerod, J. (1985). *Less-Common Met.* **111**, 49.
Parker, R. J. (1990). *Advances in Permanent Magnetism* (Wiley, New York).
Parker, R. J. and Studders, R. J. (1962). *Permanent Magnets and Their Applications* (Wiley, New York).
Pinkerton, F. E. (1986a). *J. Magn. Magn. Mater.* **54–57**, 579.
Pinkerton, F. E. (1986b) *IEEE Trans Magn.* **MAG-22**, 922.
Pinnel, M. R. (1992). *Concise Encylopedia of Magnetic and Superconducting Materials*, ed. J. E. Evetts (Pergamon Press, Oxford), p. 475.
Pinnell, M. R. and Bennett, J. E. (1973). *Bell Syst. Tech. J.* **52**, 1325.
Pinnel, M. R. (1976). *IEEE Trans. Magn.* **MAG-12**, 789.
Pinnel, M. R. (1992). *Magnetic and Superconducting Materials*, ed. J. E. Evetts (Pergamon Press, Oxford).
Prokop, G. S. and Soffa, W. A. (1978). *IEEE Trans. Magn.* **MAG-14**, 794.
Radeloff, C. and Pfeiffer, I. (1975). *IEEE Trans. Magn.* **MAG-11**, 1417.
Ralph, B. and Brandon, G. (1965). *Proc. Journées des Applications du Cobalt, 9–11 June 1964* (Brussels, 1965), p. 76.
Rathenau, G. W. (1953). *Rev. Mod. Phys.* **25**, 297.
Rauscher, G. and Radeloff, C. (1989). *Sensors: A Comprehensive Survey*, ed. W. Göpel, J. Hesse and J. N. Zemel, Vol. 5. *Magnetic Sensors*, ed. R. Boll and K. J. Overshott (VCH, Weinheim).
Ray, A. E. (1984). *IEEE Trans. Magn.* **MAG-20**, 1614.
Richter, H. G. and Dietrich, H. E. (1968). *IEEE Trans. Magn.* **MAG-4**, 263.
Roos, W. and Hempel, K. A. (1980). *J. Magn. Magn. Mater.* **15**–18, 1457.
Roos, W., Voigt, C., Dederichs, H. and Hempel, K. A. (1980). *J. Magn. Magn. Mater.* **15**–18, 1455.
Rossiter, P. L. and Houghton, M. E. (1984). *Metals Forum* **7**, 187.
Sagawa, M., Fujimura, S., Togawa, N., Yamomoto, H. and Matsuura, Y. (1984a). *J. Appl. Phys.* **55**, 2083.
Sagawa, M., Fujimura, S., Yamamoto, H., Matsuura, Y. and Hiraga, K. (1984b). *IEEE Trans. Magn.* **MAG-20**, 1584.
Sagawa, M., Hirosawa, S., Yamamoto, H., Matsuura, Y., Fujimura, S., Tokuhara, H. and Hiraga, K. (1986). *IEEE Trans. Magn.* **MAG-22**, 910.
Sagawa, M., Hirosawa, S., Yamamoto, H., Fujimura, S. and Matsuura, Y. (1987a). *Japan J. Appl. Phys.* **26**, 785.
Sagawa, M., Hirosawa, S., Tokuhara, K., Yamamoto, H., Fujimura, S., Tsubokawa, Y. and Shimizu, R. (1987). *J. Appl. Phys.* **61**, 3559.
Sakamoto, Y., Kojima, S., Kojima, K., Ohtani, T. and Kubo, T. (1979). *J. Appl. Phys.* **50**, 2355.
Sakamoto, Y., Ibata, A., Kojima, S. and Ohtani, T. (1980). *IEEE Trans. Magn.* **MAG-16**, 1056.
Sasaki, T., Okada, M., Kassai, M. and Henmi, Z. (1970). *J. Appl. Phys.* **41**, 1105.
Schippan, R. A. and Hempel, K. A. (1983). *J. Magn. Magn. Mater.* **38**, 319.

Schnitzke, K., Schultz, L., Wecker, J. and Katter, M. (1991). *Appl. Phys. Lett.* **57**, 2853.
Schüler, K. and Brinkmann, K. (1970). *Dauermagnete-Werkstoffe und Anwendungen* (Springer-Verlag, Berlin).
Schultz, L. and Katter, M. (1991). *Supermagnets, Hard Magnetic Materials*, ed. G. J. Long and F. Grandjean (Kluwer, Dordrecht).
Schultz, L., Schnitzke, K. and Wecker, J. (1990a). *J. Magn. Magn. Mater.* **83**, 254.
Schultz, L., Schnitzke, K. and Wecker, J. (1990b). *Appl. Phys. Lett.* **56**, 868.
Searle, C. W. and Garrett, H. J. (1974). *J. Appl. Phys.* **45**, 5037.
Seljesater, K. S. and Rogers, B. A. (1932). *Trans. Am. Soc. Steel Treating* **91**, 553.
Shen, B. G., Zhang, J. X., Yang, L. Y., Wo, F., Ning, T. S., Zhao, J. G., Guo, H. Q. and Zhan, W. S. (1991). *J. Less-Common Met.* **167**, 339.
Shilling, J. W. and Soffa, W. A. (1973). *IEEE Trans. Magn.* **MAG-9**, 362.
Shimizu, S. and Hashimoto, E. (1968). *J. Appl. Phys.* **39**, 2369.
Shimizu, S. and Horiuchi, S. (1970). *Metall. Trans.* **1**, 330.
Shur, Y. S., Luzhinskaya, M. G. and Shubina, L. A. (1957a). *Izv. Akad. Nauk SSSR, Ser. Fiz.* **21**, 1275.
Shur, Y. S., Luzhinskaya, M. G. and Shubina, L. A. (1947b). *Trudy Institute Fiz. Met. Akad. Nauk SSSR* **4**, 54, 60.
Shur, Y. S., Magat, L. M., Ivanova, G. V., Mitsek, A. I., Yermolinko, A. S. and Ivanov, O. A. (1968). *Fiz. Met. Metalloved.* **26**, 56.
Sixtus, K. J. and Tonks, L. (1931). *Phys. Rev.* **37**, 930.
Sixtus, K. J., Kronenberg, K. J. and Tenzer, R. K. (1956) *J. Appl. Phys.* **27**, 1051.
Smit, J. and Wijn, H. P. J. (1959). *Ferrites* (Philips Technical Library, Eindhoven).
Southworth, H. N. and Ralph, B. (1966). *Phil. Mag.* **14**, 383.
Stäblein, H. (1982) *Ferromagnetic Materials: A Handbook on the Properties of Magnetically Ordered Substances*, Vol. 3, ed. E. P. Wohlfarth (North-Holland, Amsterdam), p. 441.
Stadelmaier, H. H. (1990). *J. Mater. Eng.*, **12**, 185.
Stadelmaier, H. H. and Henig, E. T. (1991). *J. Metals* **43**, 32.
Stoner, E. C. and Wohlfarth, E. P. (1948). *Phil. Trans. R. Soc. Lond.* **A240**, 599.
Street, R., Folks, L. and Hirosawa, S. (1990). *Proc. 6th Int. Symp. Magnetic Anisotropy and Coercivity in Rare-Earth–Transition-Metal Alloys*, ed. S. G. Sankar (Carnegie–Mellon University, Pittsburgh), p. 1.
Strnat, K. J. (1988). *Ferromagnetic Materials: A Handbook on the Properties of Magnetically Ordered Substances*, Vol. 4, ed. E. P. Wohlfarth and K. H. J. Buschow (Elsevier, Amsterdam), p. 131.
Strnat, K. J. (1990). *Proc. IEEE* **78**, 923.
Struijts, A. L. (1968). *Ceramic Microstructures*, ed. R. M. Fuerath and J. A. Pask (Wiley, New York), p. 443.
Struijts, A. L. Rathenau, G. W. and Weber, G. H. (1954). *Philips Tech. Rev.* **16**, 141.
Sun, H., Coey, J. M. D., Otani, Y. and Hurley, D. P. F. (1990). *J. Phys.: Cond. Mater.* **2**, 6465.
Suzuki, Y., Sagawa, M., Okada, M., Henmi, Z. (1978). *IEEE Trans. Magn.* **MAG-14**.
Tokuyoshi, M. (1971). *IEEE Trans. Magn.* **MAG-8**, 664.
Tsong, T. T. and Müller, E. W. (1967). *J. Appl. Phys.* **38**, 545.
van den Broek, C. A. M. and Struijts, A. L. (1977). *Philips Tech. Rev.* **37**, 157.
van Landuyt, J., van Tendeloo, G., van den Broek, J. J., Donkersloot, H. and Zijlstra, H. (1978). *IEEE Trans. Magn.* **MAG-14**, 679.
Ward, M. and Taylor, J. S. (1989) *New Permanent Magnets and their Applications*, Colloquium, 9 Jan 1989, Digest No. 1989/2, IEE, London.
Went, J. J., Rathenau, G. W., Gorter, E. W. and Van Oosterhout, G. W. (1951/52). *Philips Tech. Rev.* **13**, 194.
Wernick, J. H. and Chin, G. Y. (1992). *Concise Encyclopedia of Magnetic and Superconducting Materials*, ed. J. Evetts (Pergamon, Oxford), pp. 197 and 211.

Wiegand, J. R. (1981). *Switchable Magnetic Device*, US Patent 4 247 601.
Wonsiewicz, B. C., Plewes, J. T. and Chin, G. Y. (1979). *IEEE Trans. Magn.* **MAG-15**, 950.
Yamaguchi, A., Tanaka, Y., Yanagimoto, K., Sakaguchi, J. and Kato, N. (1989). *J. Japan Metals Soc.* **28**, 422.
Yamamoto, H. (1972). US Patent 3 661 567.
Zijlstra, H. (1982). *Ferromagnetic Materials: A Handbook on the Properties of Magnetically Ordered Substances*, Vol. 3, ed. E. P. Wohlfarth (North-Holland, Amsterdam).
Zijlstra, H. and Haanstra, H. B. (1966). *J. Appl. Phys.* **37**, 2853.

Index